W9-BNH-349

UNDERSTANDING CONSTRUCTION DRAWINGS
Single and Multifamily Dwellings

MARK W. HUTH

OUACHITA TECHNICAL COLLEGE

DELMAR PUBLISHERS INC.
2 COMPUTER DRIVE, WEST — BOX 15-015
ALBANY, NEW YORK 12212

NOTICE TO THE READER

Publisher does not warrant or guarantee,any of the products described herein or perform any independent analysis in connection with any of the product information contained herein. Publisher does not assume, and expressly disclaims, any obligation to obtain and include information other than that provided to it by the manufacturer.

The reader is expressly warned to consider and adopt all safety precautions that might be indicated by the activities described herein and to avoid all potential hazards. By following the instructions contained herein, the reader willingly assumes all risks in connection with such instructions.

The publisher makes no representations or warranties of any kind, including but not limited to, the warranties of fitness for particular purpose or merchantability, nor are any such representations implied with respect to the material set forth herein, and the publisher takes no responsibility with respect to such material. The publisher shall not be liable for any special, consequential or exemplary damages resulting, in whole or in part, from the readers' use of, or reliance upon, this material.

COPYRIGHT © 1983
BY DELMAR PUBLISHERS INC.

All rights reserved. No part of this work covered by the copyright hereon may be reproduced or used in any form or by any means — graphic, electronic, or mechanical, including photocopying, recording, taping, or information storage and retrieval systems — without written permission of the publisher.

10 9 8 7 6 5

LIBRARY OF CONGRESS CATALOG CARD NUMBER: 82-46008
ISBN: 0-8273-1584-8

Printed in the United States of America
Published simultaneously in Canada
by Nelson Canada,
A division of International Thomson Limited

T
355
H87
1983

Preface

UNDERSTANDING CONSTRUCTION DRAWINGS will help you develop print reading skills in all aspects of residential construction. Each of the units in this textbook has four parts. The objectives listed at the beginning of the unit tell you what you should be able to do after you study the unit. These are the topics you should look for as you read the second part — the instructional part of the unit. The instructional part of the unit is highlighted with subheadings and many illustrations. The third part is a checklist of tasks you should be able to perform by the end of the unit. These "Check Your Progress Lists" are a restatement of the beginning objectives. The final part of each unit is the assignment. Because the units are short, the assignments follow each important topic. Most of the more than 500 questions in the assignments are designed to give you actual print reading practice.

The units are arranged in the same order as the information is needed when using construction drawings. Each unit explains how to find and use the information needed for the construction stage represented by that unit.

Many of the units refer you to the large construction prints. These are actual prints for buildings that were constructed. At first the maze of lines on each print may seem overwhelming. By studying certain parts of the drawings in each unit, you will soon feel comfortable with these prints as well as the prints for other projects.

At the back of the text you will find several helpful aides for studying construction drawings. The glossary defines all of the new technical terms introduced throughout the textbook. Each of these terms is defined when it is used for the first time. However, if you need to refresh your memory, turn to the glossary. There is also a complete list of the construction abbreviations used on the prints along with their meanings. Other pages in the appendix include most of the symbols used to show equipment and materials on drawings. The last part of the appendix is a series of math reviews. You should refer to these reviews whenever you need help with the math required for the assignments. These reviews explain the math needed to complete the assignments in this textbook.

Many companies and individuals contributed ideas, materials, and illustrations for use in this book. The author is grateful to each of these contributors and expresses special thanks to Charles Talcott, President, National Home Planners, for many construction drawings; Clark Forrest Butts, Architect at Berkus-Group Architects, for the Hidden Valley drawings; Robert Kurzon, A.I.A., for the duplex and lake house drawings, and for much technical advice; and Marjorie Huth for the countless hours spent in editing and typing.

Contents

Part I

DRAWINGS— THE LANGUAGE OF INDUSTRY

Part I helps you develop a foundation upon which to build skills and knowledge in reading the drawings used in the construction industry. The topics of the various units in this section are the basic concepts upon which all construction drawings are read and interpreted. The details of construction will be explored in Parts II and III.

Many of the assignment questions in this section refer to the drawings of the Duplex included in the drawing packet that accompanies this textbook. The Duplex was designed as income property for a small investor. It was built on a corner lot in a small city in upstate New York. The Duplex is an easy-to-understand building. Its one-story, rectangular design requires only a minimum of views; you can quickly become familiar with the Duplex drawings.

The Design-Construction Process
Section 1

| | UNIT 1   *The Design-construction Sequence and the Design Professions* |

OBJECTIVES

After completing this unit, you will be able to perform the following tasks:

- Name the professions included in the design and planning of a house or light commercial building.

- List the major functions of each of these professions in the design and planning process.

- Identify the profession or agency that should be contacted for specific information about a building under construction.

The construction industry employs about 15 percent of the working people in the United States and Canada. More than 60 percent of these workers are involved in new construction. The rest are involved in repairing, remodeling, and maintenance. As the needs of our society change, the demand for different kinds of construction increases. Homeowners and businesses demand more energy-efficient buildings. The shift toward automation and the use of computers in business and industry mean that more offices are needed. Our national centers of commerce and industry are shifting. These are only a few of the reasons that new housing starts are considered important indicators of our economic health.

There are four main classifications of construction: light, heavy, industrial, and civil. *Light construction* includes single-family homes, small apartment buildings, condominiums, and small commercial buildings, Figure 1-1. *Heavy construction* includes high-rise office and apartment buildings, hotels, large stores and shopping centers, and other large buildings. *Industrial construction* includes structures other than buildings, such as refineries and paper mills, that are built for industry.

Figure 1-1 Light construction
Courtesy of Weyerhaeuser Co.

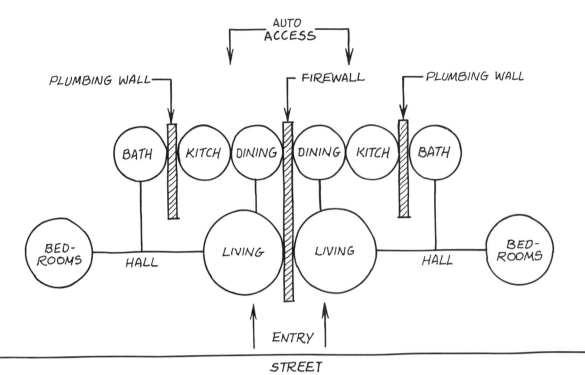

Figure 1-2 Balloon sketch of Duplex

Civil construction is more closely linked with the land and refers to highways, bridges, airports, and dams, for example. This book refers specifically to light construction. The principles apply to drawings for all types of construction, however.

THE DESIGN PROCESS

The design process starts with the owner. The owner has definite ideas about what is needed, but may not be expert at describing that need or desire in terms the builder can understand. The owner contacts an architect to help describe the building.

The architect serves as the owner's agent throughout the design and construction process. Architects combine their knowledge of construction — of both the mechanics and the business — with artistic or aesthetic knowledge and ability. They design buildings for appearance and use.

The architect helps the owner determine how much space is needed, how many rooms are needed for now and in the future, what type of building best suits the owner's life-style or business needs, and what the costs will be. As the owner's needs take shape, the architect makes rough sketches to describe the planned building. At first these may be balloon diagrams, Figure 1-2, to show traffic flow and the number of rooms. Eventually, the design of the building begins to take shape, Figure 1-3.

Before all of the details of the design can be finalized, other construction professionals become involved. Most communities have building codes which specify requirements to insure that buildings are safe from fire hazards, earthquakes, termites, surface water, and other concerns of the community. There are several organizations that publish model building codes, Figure 1-4. These are called *model codes* because they are simply models to be followed by building departments. The codes have no authority until they are adopted by a government agency. A community may adopt the total model code or may choose specific parts of the code. This then becomes the *local building code*.

The local building code is administered by a building department of the local government. Building inspectors, working for the building department, review the architect's plans before construction begins and inspect the construction throughout its progress to insure that the code is followed.

Most communities also have zoning laws. A *zoning law* divides the community into zones where only certain types of buildings are permitted. Zoning laws prevent such problems as factories and shopping centers being built in the same neighborhood as homes.

Building departments usually require that very specific procedures are followed for each construction project. A building permit is required before construction begins. The building permit notifies the building department about planned construction. Then, the building department can make sure that the building complies with all the local zoning laws and building codes. When the building

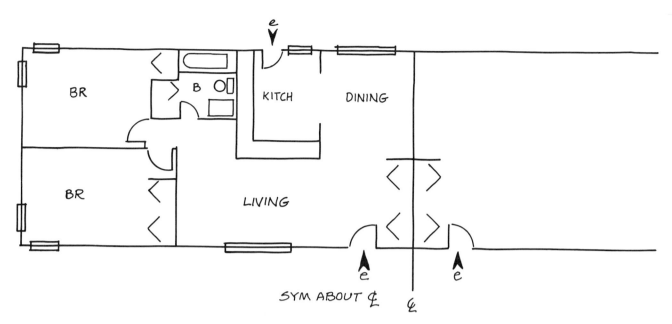

Figure 1-3 Straight line sketch of Duplex

department approves the construction, it issues a *certificate of occupancy*. This certificate is not issued until the building department is satisfied that the construction has been completed according to the local code. The owner is not permitted to move into the new building until the certificate of occupancy has been issued.

If the building is more complex than a home or simple frame building, engineers may be hired to help design the structural, mechanical, electrical, or other aspects of the building. Consulting engineers specialize in certain aspects of construction and are employed by architects to provide specific services. Finally, architects and their consultants prepare construction drawings that show all aspects of the building. These drawings tell the contractor specifically what to build.

Many homes are built from stock plans available from catalogs of house designs, building materials dealers, or magazines, Figure 1-5. However, many states require a registered architect to approve the design and supervise the construction.

STARTING CONSTRUCTION

After the architect and the owner decide on a final design, the owner obtains financing. The most common way of financing a home is through a mortgage. A *mortgage* is a guarantee that the loan will be paid in installments. If the loan is not paid, the lender has the right to sell the building in order to recover the money owed. In return for the use of the lender's money, the borrower pays interest — a percentage of the outstanding balance of the loan.

When financing has been arranged (sometimes before it is finalized), a contractor is hired. Usually a general contractor is hired with overall responsibility for completing the project. The general contractor in turn hires subcontractors to complete certain parts of the project. All stages of construction may be subcontracted. The parts of home construction most often subcontracted are plumbing and heating, electrical, drywall, painting and decorating, and landscaping. The relationships of all of the members of the design and construction team are shown in Figure 1-6.

Figure 1-4 Building codes cover all aspects of construction.

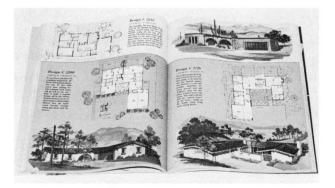

Figure 1-5 Stock plans can be ordered from catalogs.

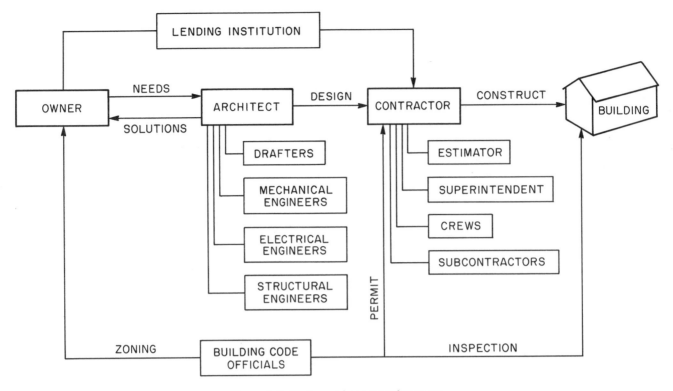

Figure 1-6 Design and construction team

✓ CHECK YOUR PROGRESS

Can you perform these tasks?

☐ List construction design professions.

☐ Describe what work is done by each of these professions.

☐ Name the profession responsible for each major part of the design-construction progress.

ASSIGNMENT

1. Who acts as the owner's agent while the building is being constructed?

2. Who designs the structural aspects of a commercial building?

3. Who would normally hire an electrical engineer for the design of a store?

4. Who is generally responsible for obtaining financing for a small building?

5. To whom would the general contractor go if there were a problem with the foundation design for a home?

6. If local building codes require specific features for earthquake protection, who is responsible for seeing that they are included in a home design?

7. Whom would the owner inform about last-minute changes in the interior trim when the building is under construction?

8. What regulations specify what parts of the community are to be reserved for single-family homes only?

9. Who issues the building permit?

10. What regulations are intended to insure that all new construction is safe?

UNIT 2 Drawing Reproduction and Storage

After completing this unit, you will be able to perform the following tasks:

- Explain the importance of drawings in construction.
- Describe the methods of reproducing drawings.
- Handle prints properly to avoid damaging them.

The construction of a building involves many people working at different locations. Architects, engineers, and owners design the project; building code officials insure that it is safe and suits the community; bankers and owners finance it; and building trades workers construct it. Each of these groups needs to communicate with the others.

THE IMPORTANCE OF DRAWINGS

Construction drawings are used for communication among those working on a project. For a small, single-family home, there may be as few as two or three sheets of drawings. On a large project, there may be several hundred sheets of drawings. These working drawings give information about the size, shape, and location of each piece of the building, Figure 2-1.

In order to insure that these drawings are interpreted the same by everyone who reads them, standard *conventions* (rules) of drafting are followed. Such things as the thickness (weight) of lines, the location of dimensions, and the position of views can affect the meaning of a drawing. So that everyone involved can read and understand the construction drawings correctly, there should be accurate copies for everyone. The copies should be kept in good condition throughout the construction processes.

Not all information can be conveyed best by drawings. Some information is prepared in written form, called *specifications* or simply *specs*, Figure 2-2. Just as the drawings show where each piece of a building is to be placed, the specs tell what types of materials are to be used and give directions for their use. For a house, the specs are normally only a few pages long.

DRAWING REPRODUCTION

When the drafter in the architect's or engineer's office completes a set of working drawings, copies or prints must be made to give to all involved personnel. The original working drawings are stored in the architect's office for future reference. Making these copies is called reproduction. Drawing reproduction can be done in several ways and may

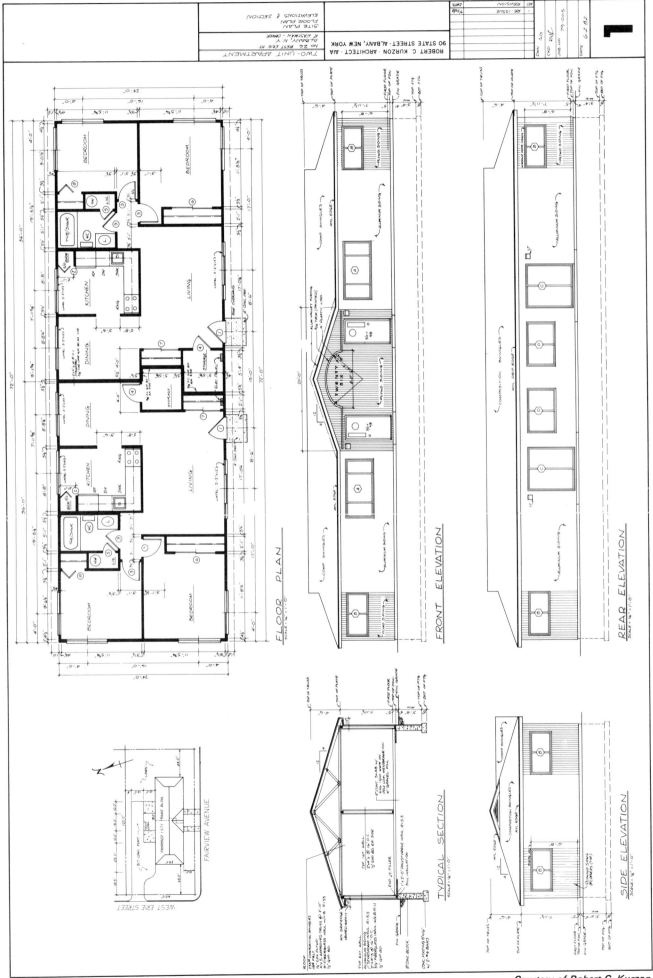

Courtesy of Robert C. Kurzon

Figure 2-1 Typical working drawing

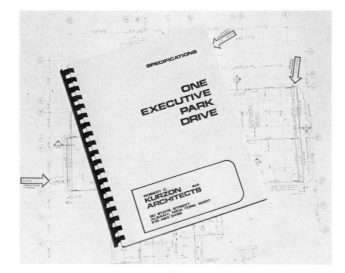

**Figure 2-2 Construction specifications
are written instructions.**

include enlarging or reducing the size of the drawings. Only the most commonly used methods are discussed here.

Blueprinting

The original drawing is done on translucent (nearly transparent) paper (*vellum*) or plastic film. The original is held against a sheet of light-sensitive paper in a glass frame. The two sheets are exposed to a bright light for a specified period of time. The light strikes the light-sensitive sheet everywhere except where the lines are drawn on the original. Wherever there are marks on the original, the light does not strike the sensitized surface.

After exposure, the sensitized paper is washed with water. Where the sensitized coating was exposed to light, it hardens and turns blue in washing. Where the coating was shielded from the light, it remains soft and is washed away by the water. This leaves a blue background with white lines, Figure 2-3. The blueprint is then washed with a chemical fixing agent to prevent the print from further developing. As a final step, the fixing agent is washed off and the blueprint is hung to dry.

For many years, the blueprint process was used for almost all drawing reproduction. Although other processes have largely replaced blueprinting, all reproductions of drawings are commonly referred to as blueprints.

Diazo Process

A more widely used method of reproduction uses ammonia vapor as a developing agent. This is the *diazo process*. In this process, the sensitized paper and the original drawing are exposed to a strong light, as in the blueprint process. After exposure, the sensitized paper is exposed to ammonia vapor, Figures 2-4 and 2-5. No fixing or washing is necessary. The finished print is free from the distortion usually caused by the washing and drying operations.

A print made by the diazo process is the reverse of a blueprint. The background is white and the lines are blue or black.

Photo-reproduction Printing

Recently, to fill the need for more efficient ways to produce large quantities of quality prints, a photographic method is sometimes used. The original is photographed on a large copy camera. Then the negative is developed as in regular photography. Instead of the negative being printed on photographic paper, it is printed on an offset printing plate. This plate is attached to an offset printing press. Offset printing is widely used for books, magazines, and newspapers. A single offset plate can be used to print several hundred copies in less than one hour, if necessary. Another advantage of the photographic process is that it allows easy reductions in size; thereby, it reduces mailing and storage costs. The prints that accompany this textbook were reproduced by photo-offset printing.

Both the photo-offset and diazo processes result in *white prints* (white background with colored lines). Because blueprints were once so widely used, white prints are often called blueprints. It is not correct, however, to refer to all drawings as plans. A plan is a specific type of drawing which will be discussed later.

HANDLING PRINTS

Although construction drawings are printed on durable paper, they should be handled with care to ensure their long life. Do not fold or crease prints. Do not use prints for notetaking or as scrap paper for calculations. It is a good idea to have notepaper handy at all times while studying this textbook and while working in the field. Keep prints clean and dry. Store them where they will be out of the way, but easy to find when needed. If flat storage is not available, prints can be rolled for storage and handling. (The prints accompanying this text were folded for publication and mass shipping.) Never work from the original working drawings — they should be reserved for making authorized changes and additional prints.

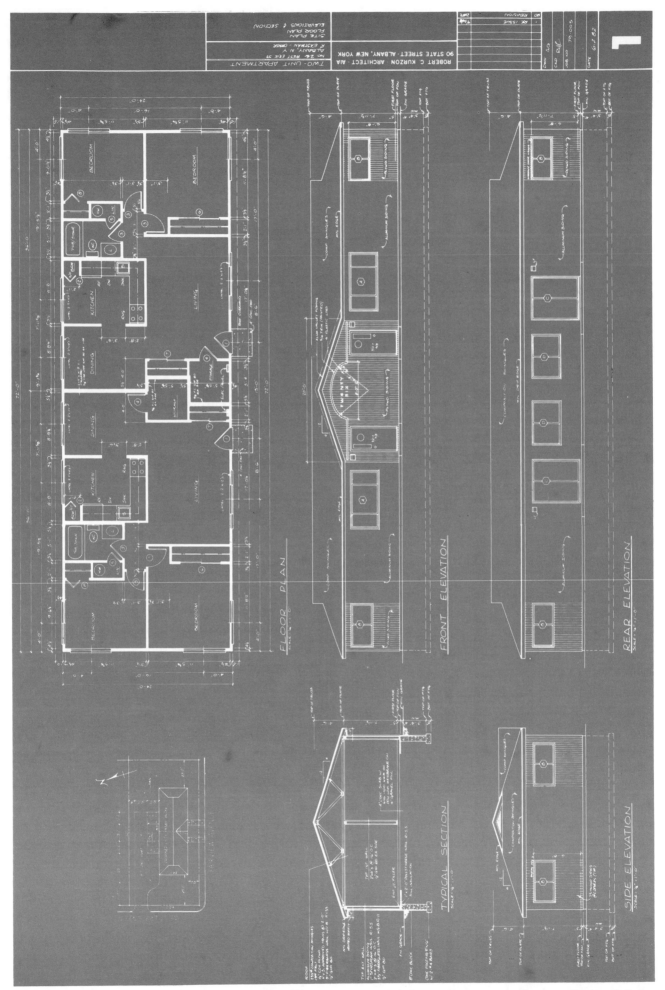

Courtesy of Robert C. Kurzon

Figure 2-3 Blueprint

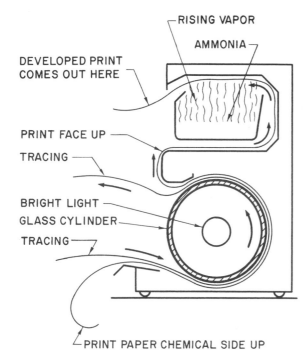

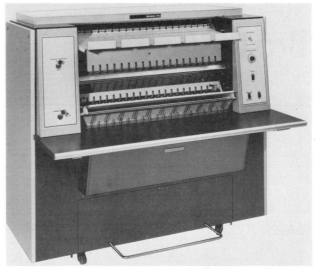

Courtesy of AM Bruning

Figure 2-5 Diazo copier

ROLLERS MOVE THE TRACING AND PRINT
AROUND THE LIGHT AND MOVE THE PRINT
PAST THE RISING AMMONIA VAPOR

Figure 2-4 The diazo process

√ **CHECK YOUR PROGRESS** ─────────────────────────────

Can you perform these tasks?

☐ Explain why drawings are necessary in construction.

☐ Describe the blueprint process.

☐ Describe the diazo process.

☐ Describe photo-offset printing.

☐ Handle prints properly.

 ASSIGNMENT ────────────────────────────────

1. What is the main purpose of construction drawings?

2. What three basic characteristics of a building are shown on drawings?

3. Why is it important for construction drawings to be made according to established rules?

4. What kind of sheet must be used for the original working drawing if blueprints or diazo prints are to be made?

5. What is a blueprint?

6. How is a diazo print different from a blueprint?

7. Briefly describe the diazo process.

8. What process is best suited for reducing the size of prints?

9. What is done with the original drawings after all prints are made?

10. What is a white print?

Basic Views

Section 2

UNIT 3   Theory of Projection

OBJECTIVES

After completing this unit, you will be able to perform the following tasks:

- Recognize perspective, oblique, isometric, and orthographic drawings.

- Distinguish between lines drawn true length and foreshortened lines.

- Identify plan views, elevations, and sections.

PICTORIAL DRAWINGS

The easiest kinds of drawings to visualize are those that look like the objects they depict, *pictorial drawings*. Among the types of pictorial drawings used to describe construction, *perspective drawings* are the most realistic. Most people know that parallel lines appear to get closer together as they go farther from the viewer, Figure 3-1. If these *converging* (getting closer together) lines go far enough, they appear to meet. This is called their *vanishing point* — the point at which the object vanishes, Figure 3-2. To make the appearance

Figure 3-1 Parallel lines appear to get closer together as they get farther away

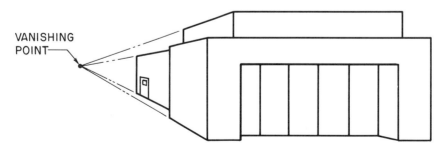

VANISHING POINT

Figure 3-2 One-point perspective

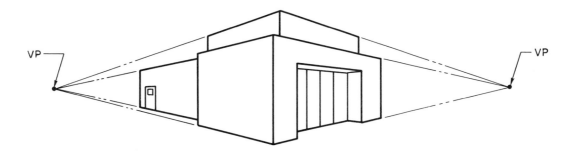

Figure 3-3 Two-point perspective

more realistic, some perspective drawings have two vanishing points, Figure 3-3.

Obviously, not all of the lines in a perspective drawing are drawn to scale. For example, the building shown in Figures 3-2 and 3-3 is the same height at both ends. However, the lines on the page are not the same length. Lines that are drawn shorter than their actual proportion are called *foreshortened lines*. Those that are in true proportion are called *true-length lines*. Because of the foreshortened lines of perspective drawings, this kind of drawing is not practical to use in a working drawing from which to build a building. However, the realism of perspective drawings makes them ideal for showing potential owners what the building will look like. Perspectives used for this purpose are often painted and are called *renderings*, Figure 3-4.

A more useful type of pictorial drawing for construction purposes is the *isometric drawing*. In an isometric drawing, vertical lines are drawn vertically and horizontal lines are drawn at an angle of 30 degrees from horizontal, Figure 3-5. All lines on one of these isometric axes are drawn in proportion to their actual length. Notice that both ends of the building in Figure 3-5 are drawn the same size. Isometric drawings tend to look out of proportion because we are used to seeing the converging lines of a perspective drawing.

Isometric drawings are often used to show construction details, Figure 3-6. The ability to draw simple isometric sketches is a useful skill for communicating on the job site. Try sketching a brick in isometric as shown in Figure 3-7.

Step 1. Sketch a Y with the top lines about 30° from horizontal.

Step 2. Sketch the bottom edges parallel to the top edges.

Step 3. Mark off the width on the left top and bottom edges. This will be about twice the height.

Figure 3-4 Rendering of a building

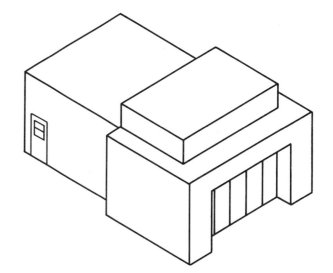

Figure 3-5 Isometric of building in Figure 3-1

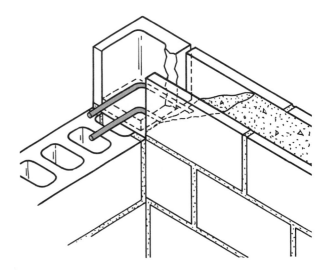

Figure 3-6 Isometric construction detail
Courtesy of Stearns Manufacturing Company, Inc.

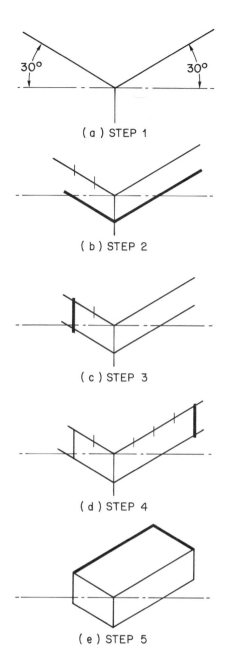

(a) STEP 1

(b) STEP 2

(c) STEP 3

(d) STEP 4

(e) STEP 5

Figure 3-7 Sketching an isometric brick

(a) GABLE ROOF BUILDING

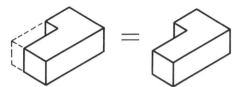

(b) ELL SHAPED BUILDING

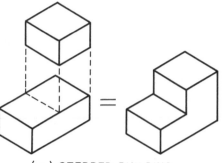

(c) STEPPED BUILDING

Figure 3-8 Variations on the isometric brick

(a)   FRONT VIEW OF CROWN MOLDING

(b)   OBLIQUE VIEW OF CROWN MOLDING

Figure 3-9 Oblique drawings

Step 4. Mark off the length on the right top and bottom edges. The length will be about twice the width.

Step 5. Sketch the two remaining vertical lines and the back edges.

Other isometric shapes can be sketched by adding to or subtracting from this basic isometric brick, Figure 3-8. Angled surfaces are sketched by locating their edges; then, connecting them.

When an irregular shape is to be shown in a pictorial drawing, an *oblique drawing* may be best. In oblique drawings, the most irregular surface is drawn in proportion as though it were flat against the drawing surface. Parallel lines are added to show the depth of the drawing, Figure 3-9.

ORTHOGRAPHIC PROJECTION

To show all information accurately and to keep all lines and angles in proportion, most construction drawings are drawn by *orthographic projection*. Orthographic projection is most often explained by imagining the object to be drawn inside a glass box. The corners and the lines representing the edges of the object are then projected onto the

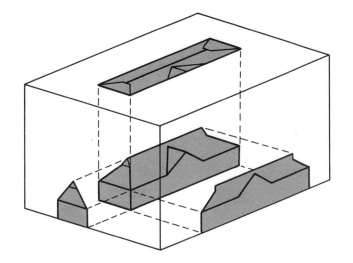

**Figure 3-10 Duplex inside a glass box: method of
orthographic projection of roof, front side**

sides of the box, Figure 3-10. If the box is unfolded, the images projected onto its sides will be on a single plane, as on a sheet of paper. In other words, in orthographic projection each view of an object shows only one side (or top or bottom) of the object.

All surfaces that are parallel to the plane of projection (the surface of the box) are shown in proportion to their actual size and shape. However, surfaces that are not parallel to the plane of projection are not shown in proportion. For example, both of the roofs in the top views of Figure 3-11 appear to be the same size and shape, but they are quite different. To find the actual shape of the roof you must look at the end view.

The names of the various views depend on what type of drawing is used. In machine drawings, views are named for their relative positions — front,

side, top. In construction drawings, they are called plans and elevations. A *plan* view is comparable to a top view in machine drawings, Figure 3-12. A set of drawings for a building usually includes plan views of the site (lot), the floor layout, and the foundation. *Elevations* are drawings that show height. For example, a drawing that shows what would be seen standing in front of a house is a building elevation, Figure 3-13. Elevations are also used to show cabinets and interior features.

Because not all features of construction can be seen in plan views and elevations from the outside of a building, many construction drawings are section views. A section view, usually referred to simply as a *section*, shows what would be exposed if a cut were made through the object, Figure 3-14. Actually, a floor plan is a type of section view, Figure 3-15.

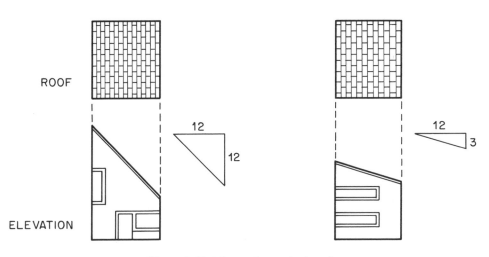

Figure 3-11 Views of two shed roofs

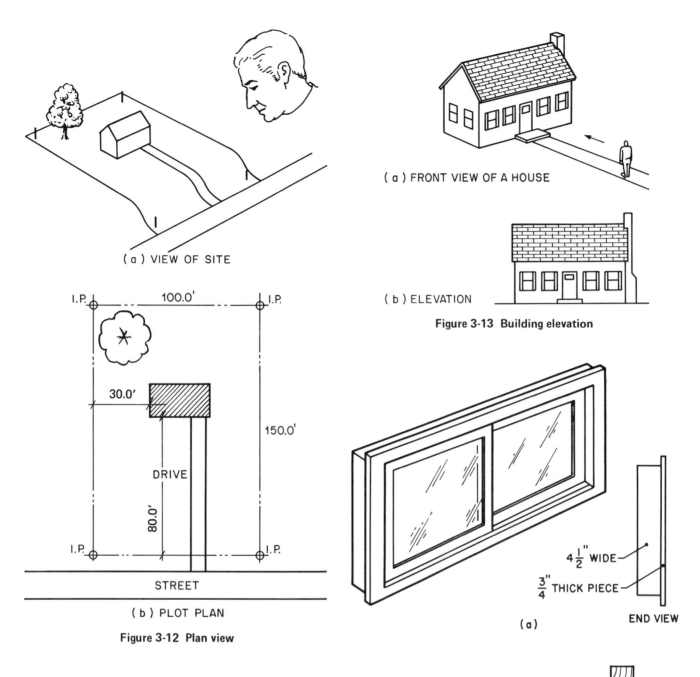

(a) VIEW OF SITE

(a) FRONT VIEW OF A HOUSE

(b) ELEVATION

Figure 3-13 Building elevation

(b) PLOT PLAN

Figure 3-12 Plan view

(a)

END VIEW

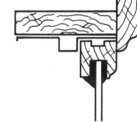

(b) SECTION

Figure 3-14 Section of a window sash

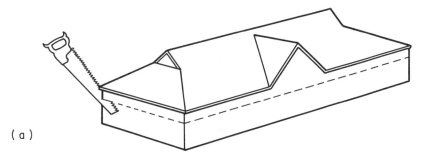

(a)

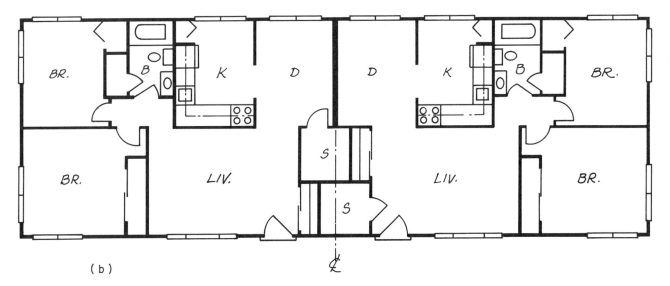

(b)

Figure 3-15 A floor plan is actually a section view of the building.

✓ CHECK YOUR PROGRESS

Can you perform these tasks?

- ☐ Identify perspective drawings.
- ☐ Identify oblique drawings.
- ☐ Identify isometric drawings.
- ☐ Identify orthographic drawings.
- ☐ Identify lines on drawings as either true length or foreshortened.
- ☐ Identify plan views.
- ☐ Identify elevation views.
- ☐ Identify section views.

ASSIGNMENT

1. Identify each of the drawings in Figure 3-16 as perspective, oblique, isometric, or orthographic.

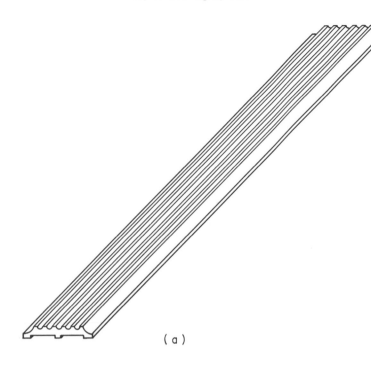

(a)

(b)

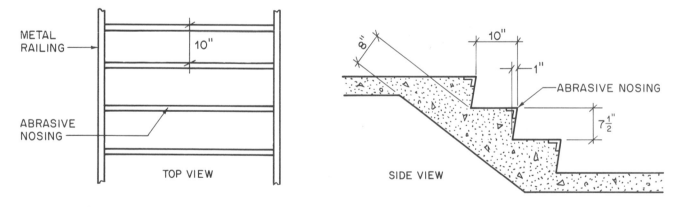

METAL RAILING

10"

ABRASIVE NOSING

TOP VIEW

8"

10"

1"

ABRASIVE NOSING

$7\frac{1}{2}$"

SIDE VIEW

(c)

Figure 3-16

2. Identify each of the drawings in Figure 3-17 as elevation, plan, or section.

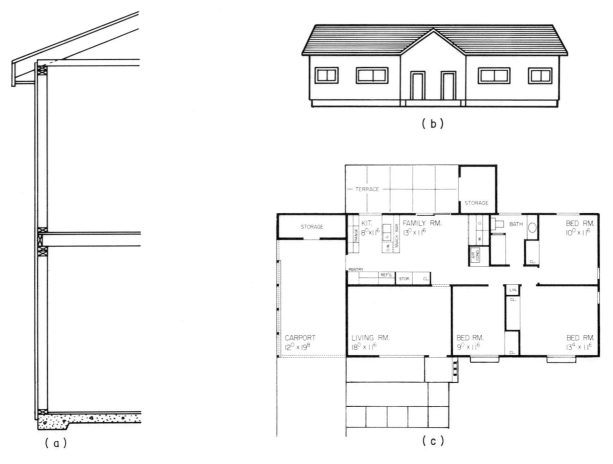

Figure 3-17

3. In the view of the house shown in Figure 3-18, which lines are true length?

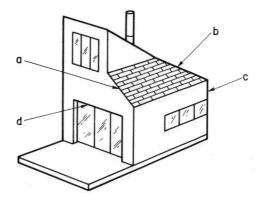

Figure 3-18

4. What type of drawings are usually used for presentation drawings?

5. What type of pictorial drawings are easiest to draw on the job site?

6. What type of drawings are used for working drawings?

UNIT 4 Scales

OBJECTIVES

After completing this unit, you will be able to perform the following tasks:

- Identify the scale used on a construction drawing.

- Read an architect's scale.

SCALE DRAWINGS

Because construction projects are too large to be drawn full size on a sheet of paper, everything must be drawn proportionately smaller than it really is. For example, floor plans for a house are frequently drawn 1/48 of the actual size. This is called *drawing to scale*. At a scale of 1/4'' = 1'-0'', 1/4 inch on the drawing represents 1 foot on the actual building. When it is necessary to fit a large object on a drawing, a small scale is used. Smaller objects and drawings that must show more detail are drawn to a larger scale. The floor plan in Figure 4-1 was drawn to a scale of 1/4'' = 1'-0''. The detail drawing in Figure 4-2 was drawn to a scale of

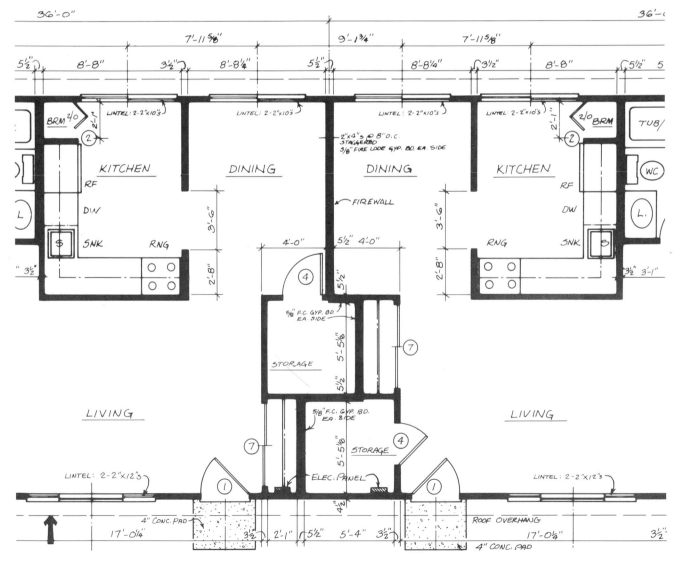

Courtesy of Robert C. Kurzon

Figure 4-1 Portion of a floor plan with a firewall

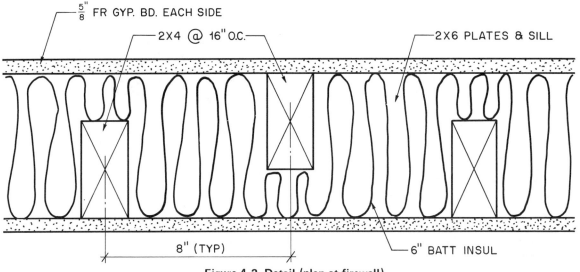

Figure 4-2 Detail (plan at firewall)

$3'' = 1'-0''$ to show the construction of one of the walls on the floor plan.

The scale to which a drawing is made is noted on the drawing. The scale is usually indicated alongside or beneath the title of the view. On some drawings, the scale is shown by including a drawing that looks something like a ruler. This graphic scale has graduations representing feet and inches drawn to the scale of the view, Figure 4-3. If the drawing is enlarged or reduced, the graphic scale is also enlarged or reduced. The graduations on the scale indicator can be marked on the edge of a sheet of paper, then stepped off on the drawing, Figure 4-4. They may also be transferred to the drawing with dividers, Figure 4-5.

READING AN ARCHITECT'S SCALE

All necessary dimensions should be shown on the drawings. The instrument used to measure scale drawings is called an *architect's scale*, Figure 4-6. The triangular scale includes eleven scales frequently used on drawings.

Full Scale			
$3/32''$	$= 1'-0''$	$3/16''$	$= 1'-0''$
$1/8''$	$= 1'-0''$	$1/4''$	$= 1'-0''$
$3/8''$	$= 1'-0''$	$3/4''$	$= 1'-0''$
$1/2''$	$= 1'-0''$	$1''$	$= 1'-0''$
$1\ 1/2''$	$= 1'-0''$	$3''$	$= 1'-0''$

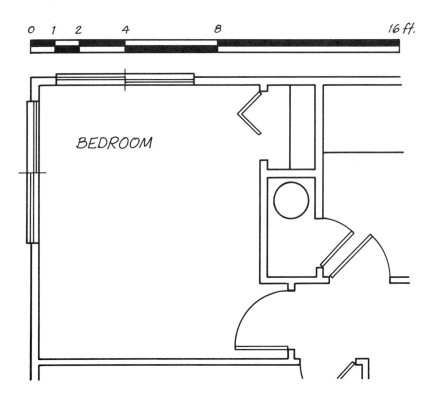

Figure 4-3 Graphic scale

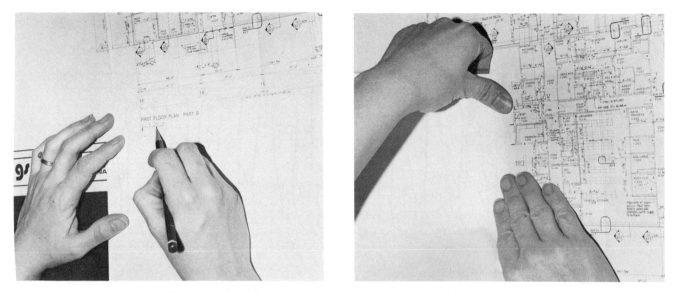

Figure 4-4 Marking the graduations on the edge of a piece of paper

Two scales are combined on each face, except for the full-size scale which is fully divided into sixteenths. The combined scales work together because one is twice as large as the other, and their zero points and extra divided units are on opposite ends of the scale.

The fraction, or number, near the zero at each end of the scale indicates the unit length in inches that is used on the drawing to represent one foot of the actual building. The extra unit near the zero end of the scale is subdivided into twelfths of a foot (inches) as well as fractions of inches on the larger scales.

To read the architect's triangular scale, turn it to the 1/4-inch scale. The scale is divided on the left from the zero toward the 1/4 mark so that each line represents one inch. Counting the marks from the zero toward the 1/4 mark, there are twelve lines marked on the scale. Each one of these lines is one inch on the 1/4" = 1'-0" scale.

The fraction 1/8 is on the opposite end of the same scale, Figure 4-7. This is the 1/8-inch scale and is read in the opposite direction. Notice that the divided unit is only half as large as the one on the 1/4-inch end of the scale. Counting the lines from zero toward the 1/8 mark, there are only six lines. This means that each line represents two inches at the 1/8-inch scale.

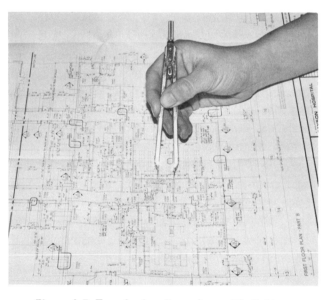

Figure 4-5 Transferring dimensions with dividers

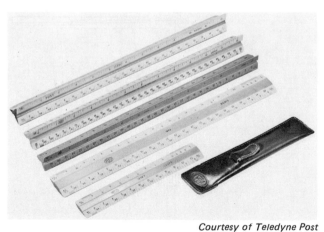

Courtesy of Teledyne Post
Figure 4-6 Architect's scales

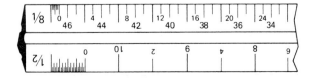

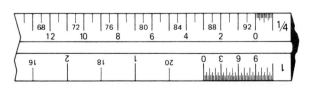

Figure 4-7 Architect's triangular scale

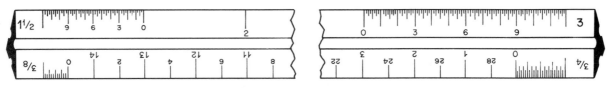

Figure 4-8   Architect's triangular scale showing 1 1/2" and 3" scales

Now look at the 1 1/2-inch scale, Figure 4-8. The divided unit is broken into twelfths of a foot (inches) and also fractional parts of an inch. Reading from the zero toward the number 1 1/2, notice the figures 3, 6, and 9. These figures represent the measurements of 3 inches, 6 inches, and 9 inches at the 1 1/2" = 1'-0" scale. From the zero to the first long mark that represents one inch (which is the same length as the mark shown at 3) are 4 lines. This means that each line on the scale is equal to 1/4 of an inch. Reading the zero to the 3, read each line as follows: 1/4, 1/2, 3/4, 1, 1 1/4, 1 1/2, 1 3/4, 2, 2 1/4, 2 1/2, 2 3/4, and 3 inches.

√ CHECK YOUR PROGRESS

Can you perform these tasks?
- ☐ Locate the scale notations on drawings.
- ☐ Use an architect's scale to measure objects drawn to scale.

 ## ASSIGNMENT

1–10.   What are the dimensions indicated on the scale in Figure 4-9?

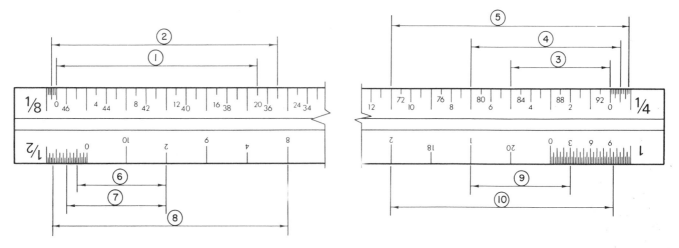

Figure 4-9

11.   What scales are used for the following views of the Duplex? (Refer to the Duplex drawings in the packet.)
  a. Floor plan
  b. Site plan
  c. Front elevation
  d. Typical wall section

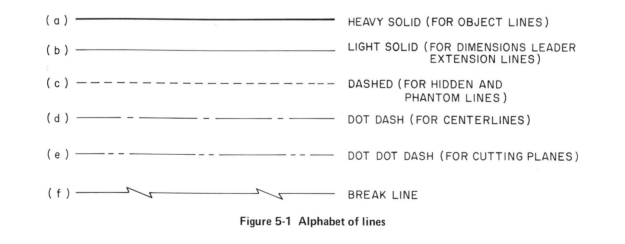

UNIT 5 Alphabet of Lines

OBJECTIVES

After completing this unit, you will be able to identify and understand the meaning of the listed lines:

- Object lines
- Dashed lines (hidden and phantom)
- Extension lines and dimension lines
- Centerlines
- Leaders
- Cutting-plane lines

The fact that drawings are used in construction for the communication of information was discussed earlier. The drawings, then, serve as a language for the construction industry. The basis for any language is its alphabet. The English language uses an alphabet made up of twenty-six letters. The language of construction drawings uses an *alphabet of lines*, Figure 5-1.

The weight or thickness of lines is sometimes varied to show their relative importance. For example, in Figure 5-2 notice that the basic outline of the building is heavier than the windows and doors. This difference in line weight sometimes helps distinguish the basic shape of an object from surface details.

OBJECT LINES

Object lines are used to show the shape of an object. All visible edges are represented by object lines. All of the lines in Figure 5-2 are object lines. Drawings usually include many solid lines that are not object lines, however. Some of these other solid lines are discussed here. Others are discussed later.

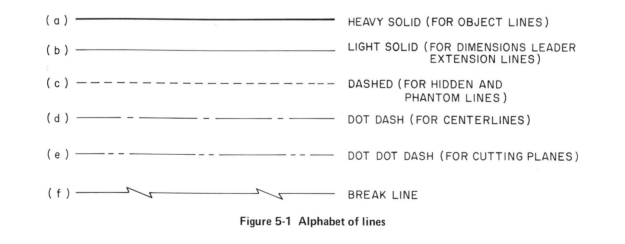

(a) ———————————— HEAVY SOLID (FOR OBJECT LINES)

(b) ———————————— LIGHT SOLID (FOR DIMENSIONS LEADER EXTENSION LINES)

(c) — — — — — — — — DASHED (FOR HIDDEN AND PHANTOM LINES)

(d) ——— - ——— - ——— - ——— DOT DASH (FOR CENTERLINES)

(e) ——— - - ——— - - ——— - - ——— DOT DOT DASH (FOR CUTTING PLANES)

(f) ———⌁———⌁——— BREAK LINE

Figure 5-1 Alphabet of lines

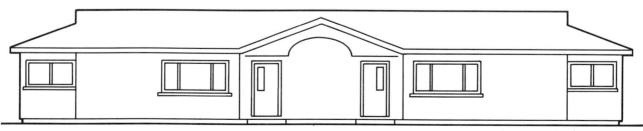

Courtesy of Robert C. Kurzon

Figure 5-2 Elevation outlined

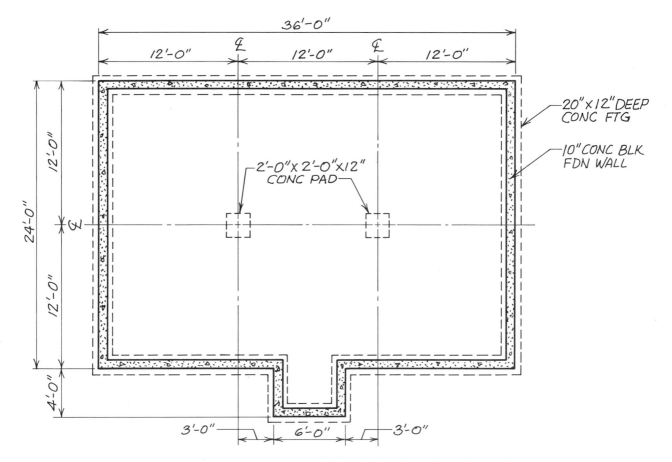

Figure 5-3 The dashed lines on this foundation plan indicate the footing.

DASHED LINES

Dashed lines have more than one purpose in construction drawings. One type of dashed line, the *hidden line*, is used to show the edges of objects that would not otherwise be visible in the view shown. Hidden lines are drawn as a series of evenly sized short dashes, Figure 5-3. If a construction drawing were to include hidden lines for all con-cealed edges, the drawing would be cluttered and hard to read. Therefore, only the most important features are shown by hidden lines.

Another type of dashed line is used to show important overhead construction, Figure 5-4. These dashed lines are called *phantom lines*. The objects they show are not hidden in the view — they are simply not in the view. For example, the most practical way to show exposed beams on a living room ceiling may be to show them on the floor plan with phantom lines. Phantom lines are also used to show alternate positions of objects, Figure 5-5. To avoid confusion, the dashed lines may be

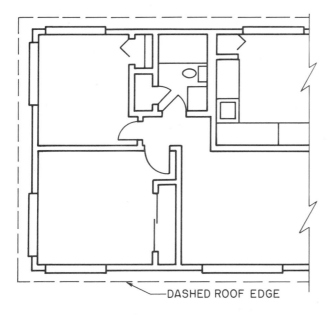

Figure 5-4 The dashed lines on this floor plan indicate the edge of the roof overhang.

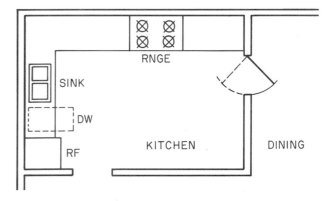

Figure 5-5 The dashed lines here are phantom lines to show alternate positions of the double-acting door and the door of the dishwasher.

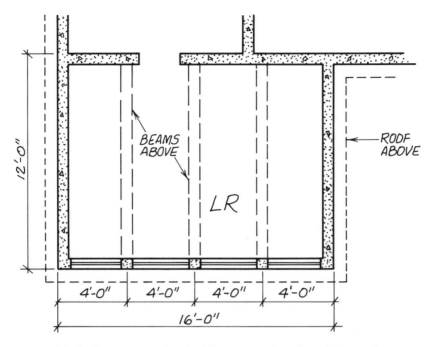

Figure 5-6 Different types of dashed lines are used to show different features.

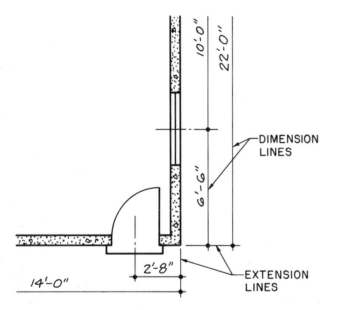

Figure 5-7 Dimension and extension lines

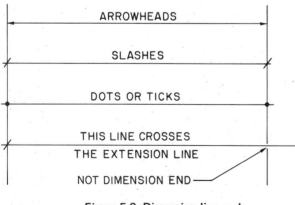

Figure 5-8 Dimension line ends

made up of different weights and different length dashes depending on the purpose, Figure 5-6.

EXTENSION LINES AND DIMENSION LINES

Extension lines are thin, solid lines that project from an object to show the extent or limits of a dimension. Extension lines should not quite touch the object they indicate, Figure 5-7.

Dimension lines are solid lines of the same weight as extension lines. A dimension line is drawn from one extension line to the next. The dimension (distance between the extension lines) is lettered above the dimension line. On construction drawings, dimensions are expressed in feet and inches. The ends of dimension lines are drawn in one of three ways as shown in Figure 5-8.

Dimensions that can be added together to come up with one overall dimension are called *chain dimensions*. The dimension lines for chain dimensions are kept in line as much as possible. This makes it easier to find the dimensions that must be added to find the overall dimension.

CENTERLINES

Centerlines are made up of long and short dashes. They are used to show the centers of round or cylindrical objects. Centerlines are also used to indicate that an object is *symmetrical*, or the same on both sides of the center, Figure 5-9. To show the center of a round object, two centerlines are used so that their short dashes cross in the center, Figure 5-10.

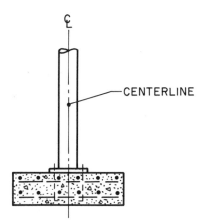

Figure 5-9 This centerline indicates that the column is symmetrical, or the same, on both sides of the centerline.

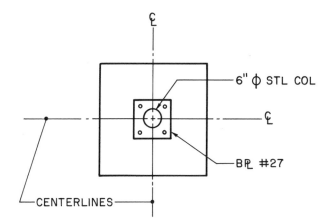

Figure 5-10 When centerlines show the center of a round object, the short dashes of two centerlines cross.

To lay out an *arc* or part of a circle, the radius must be known. The *radius* of an arc is the distance from the center to the edge of the arc. On construction drawings, the center of an arc is shown by crossing centerlines. The radius is dimensioned on a thin line from the center to the edge of the arc, Figure 5-11.

Rather than clutter the drawing with unnecessary lines, only the short, crossing dashes of the centerlines are shown. If the centerlines are needed to dimension the location of the center, only the needed centerlines are extended.

LEADERS

Some construction details are too small to allow enough room for clear dimensioning by the methods described earlier. To overcome this problem, the dimension is shown in a clear area of the drawing. A thin line called a *leader* shows where the dimension belongs, Figure 5-12.

CUTTING-PLANE LINES

It was established earlier that section views are needed to show interior detail. In order to show where the imaginary cut was made, a *cutting-plane line* is drawn on the view through which the cut was made, Figure 5-13. A cutting-plane line is usually a heavy line with long dashes and pairs of short dashes. Some drafters, however, use a solid, heavy line. In either case cutting-plane lines always have some identification at their ends. Cutting-plane-line identification symbols are discussed in the next unit.

Some section views may not be referenced by a cutting-plane line on any other view. These are *typical sections* that would be the same if drawn from an imaginary cut in any part of the building, Figure 5-14.

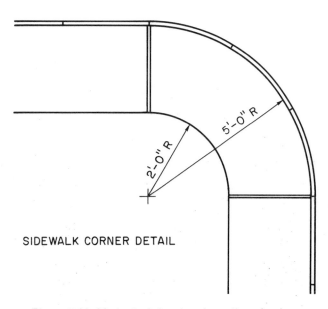

SIDEWALK CORNER DETAIL

Figure 5-11 Method of showing the radius of an arc

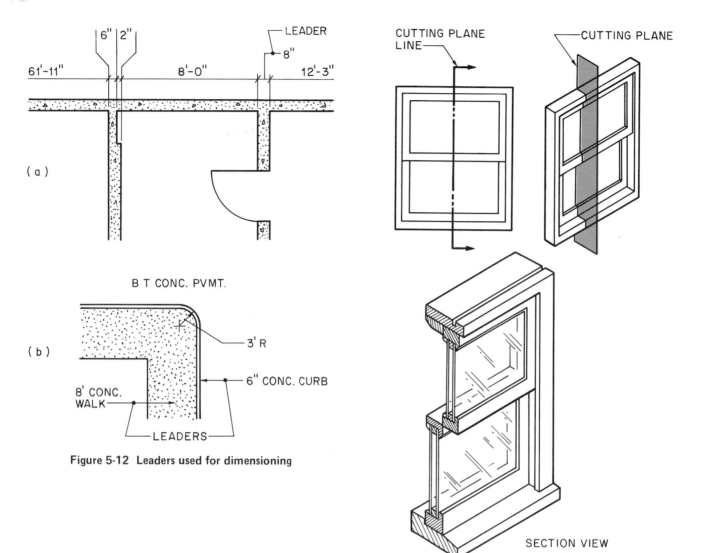

(a)

B T CONC. PVMT.

(b)

3' R

6" CONC. CURB

8' CONC. WALK

LEADERS

LEADER

8"

61'-11"   6" 2"   8'-0"   12'-3"

CUTTING PLANE LINE

CUTTING PLANE

SECTION VIEW

Figure 5-12 Leaders used for dimensioning

Figure 5-13 A cutting-plane line indicates where the imaginary cut is made and how it is viewed.

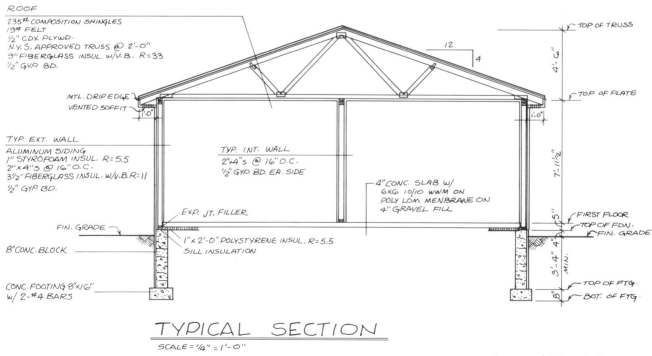

ROOF
235# COMPOSITION SHINGLES
19# FELT
½" CDX PLYWD.
N.Y.S. APPROVED TRUSS @ 2'-0"
9" FIBERGLASS INSUL. w/V.B. R=33
½" GYP. BD.

MTL. DRIP EDGE
VENTED SOFFIT

TYP. EXT. WALL
ALUMINUM SIDING
1" STYROFOAM INSUL. R=5.5
2"x4"S @ 16" O.C.
3½" FIBERGLASS INSUL. w/V.B. R=11
½" GYP. BD.

TYP. INT. WALL
2"x4"S @ 16" O.C.
½" GYP. BD. EA. SIDE

4" CONC. SLAB w/
6x6 10/10 WWM ON
POLY LOM MENBRANE ON
4" GRAVEL FILL

EXP. JT. FILLER

FIN. GRADE

8" CONC. BLOCK

1" x 2'-0" POLYSTYRENE INSUL. R=5.5
SILL INSULATION

CONC. FOOTING 8"X16"
w/ 2-#4 BARS

TOP OF TRUSS
12
4
TOP OF PLATE
4'-6"
7'-11½"
FIRST FLOOR
TOP OF FDN.
FIN. GRADE
5"
3'-4" MIN.
4"
TOP OF FTG.
8"
BOT. OF FTG.

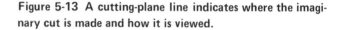

TYPICAL SECTION
SCALE = ¼" = 1'-0"

Courtesy of Robert C. Kurzon

Figure 5-14 Building section

√ CHECK YOUR PROGRESS ─────────────────────────

Can you perform these tasks?

☐ Identify and explain the use of object lines.

☐ Identify and explain the use of hidden lines.

☐ Identify and explain the use of phantom lines.

☐ Identify and explain the use of dimension and extension lines.

☐ Identify and explain the use of centerlines.

☐ Identify and explain the use of leaders.

☐ Identify and explain the use of cutting-plane lines.

 ASSIGNMENT ─────────────────────────────

Refer to the drawings of the Duplex in the packet. For each of the lines numbered A5.1 through A5.10, identify the kind of line and briefly describe its purpose on these drawings. The broad arrows with A5 numbers are for use in this assignment.

Example: A5.E, object line, shows end of building.

UNIT 6 Use of Symbols

OBJECTIVES ─────────────────────

After completing this unit, you will be able to identify and understand the meaning of the listed symbols:

• Door and window symbols

• Materials symbols

• Electrical and mechanical symbols

• Reference marks for coordinating drawings

• Abbreviations

An alphabet of lines allows for clear communication through drawings; the use of standard symbols makes for even better communication. Many features of construction cannot be drawn exactly as they appear on the building. Therefore, standard symbols are used to show various materials, plumbing fixtures and fittings, electrical devices, windows, doors, and other common objects. Notes are added to drawings to give additional explanations.

It is not important to memorize all of the symbols and abbreviations used in construction before you learn to read drawings. You should, however, memorize a few of the most common symbols and abbreviations so that you may learn the principles involved in their use. Additional symbols and abbreviations can be looked up as they are needed. The illustrations shown here represent only a few of the more common symbols and abbreviations. A more complete reference is given in the Appendix.

DOOR AND WINDOW SYMBOLS

Door and window symbols show the type of door or window used and the direction the door or

OUACHITA TECHNICAL COLLEGE

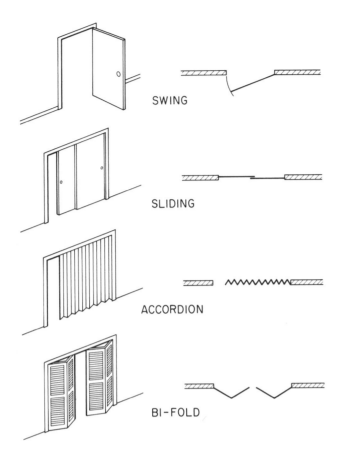

SWING

SLIDING

ACCORDION

BI-FOLD

Figure 6-1 Types of doors and their plan symbols

window opens. There are three basic ways for household doors to open — swing, slide, or fold, Figure 6-1. Within each of these basic types there are variations that can be readily understood from their symbols. The direction a swing-type door opens is shown by an arc representing the path of the door.

There are seven basic types of windows. They are named according to how they open, Figure 6-2. The symbols for hinged windows — awning, casement, and hopper — indicate the direction they open. In elevation, the symbols include dashed lines that come to a point at the hinged side.

The sizes of windows and doors are shown either on the symbol or in a separate schedule. Door and window schedules are explained later in Units 22 and 28. When the size is given with the symbol, the width is given first and the height, second. Figure 6-3 explains a typical door or window dimension callout. These are nominal dimensions. A *nominal dimension* is only an approximate dimension used for comparison of relative sizes. The actual dimensions should be obtained from the manufacturer before construction begins.

MATERIALS SYMBOLS

The drawing of an object shows its shape and location. The outline of the drawing may be filled in with a material symbol to show what the object is made of, Figure 6-4. Many materials are represented by one symbol in elevations and another symbol in sections. Examples of such symbols are concrete block and brick. Other materials look pretty much the same when viewed from any direction, so their symbols are drawn the same in sections and elevations. An example of a material with the same symbol regardless of the view is earth.

When a large area is made up of one material, it is common to only draw the symbol in a part of the area, Figure 6-5. Some drafters simplify this even further by using a note to indicate what material is used and omitting the symbol altogether.

ELECTRICAL AND MECHANICAL SYMBOLS

The electrical and mechanical systems in a building include wiring, electrical devices, piping, pipe fittings, plumbing fixtures, registers, and heating and air-conditioning ducts. It is not practical to draw these items as they would actually appear, so standard symbols have been devised to indicate them. Electrical and mechanical information is covered in depth in Section 12; you need to understand only the most common symbols for now.

The electrical system in a house includes wiring as well as devices such as switches, receptacles, light fixtures, and appliances. Wiring is indicated by lines that show the general path of current. These lines are not shown in their actual position. They simply indicate which switches control which lights, for example. Outlets (receptacles) and switches are usually shown in their approximate positions. Major fixtures and appliances are shown in their actual positions. A few of the most common electrical symbols are shown in Figure 6-6.

Mechanical systems — plumbing and HVAC (heating, ventilating, and air conditioning) — are not usually shown in much detail on drawings for single-family homes. However, some of the most important features may be shown. Piping is shown by lines; different types of lines represent different kinds of piping. Symbols for pipe fittings are the same basic shape as the fittings they represent. A short line, or *hash mark*, represents the joint between the pipe and the fitting. Plumbing fixtures are drawn pretty much as the actual fixture appears. A few plumbing symbols are shown in Figure 6-7.

PLAN   ELEVATION   PICTORAL

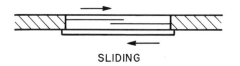

DOUBLE HUNG

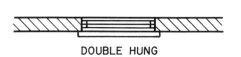

SLIDING

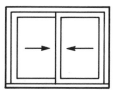

AWNING

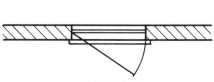

CASEMENT

HOPPER

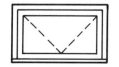

JALOUSIE

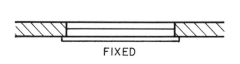

FIXED

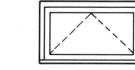

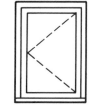

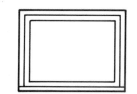

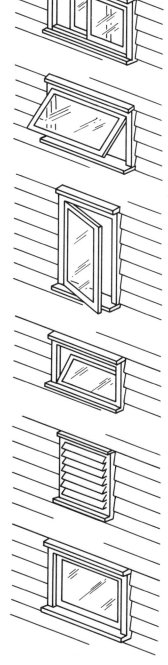

Figure 6-2 Window symbols

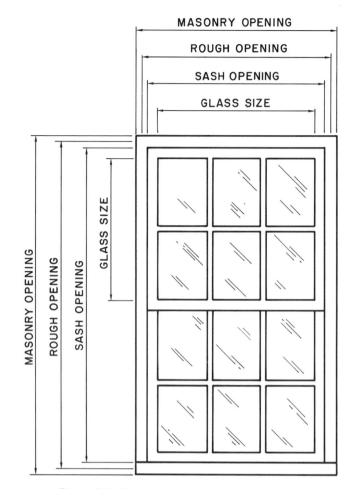

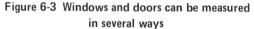

Figure 6-3 Windows and doors can be measured
in several ways

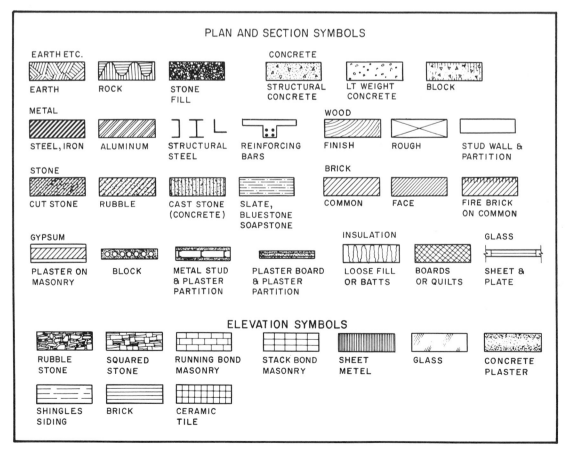

Figure 6-4 Material symbols

Figure 6-5 Only part of the area is covered by the brick symbol, although the entire building will be brick.

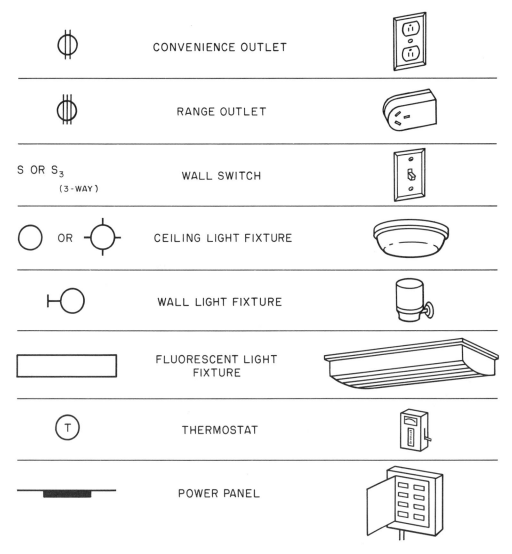

Figure 6-6 Some common electrical symbols

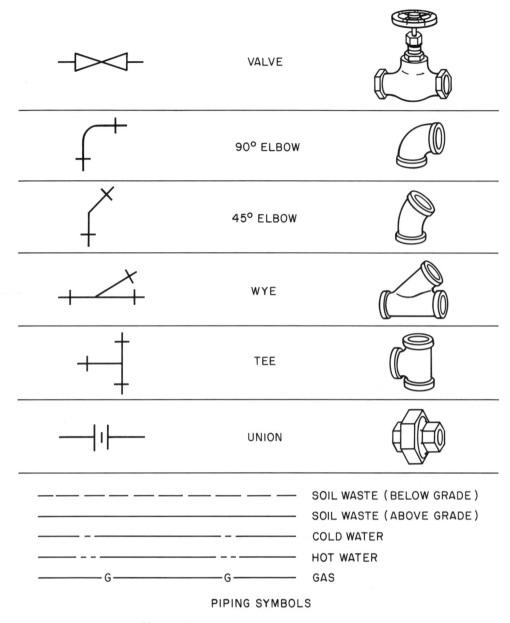

PIPING SYMBOLS

Figure 6-7 Some common plumbing symbols

REFERENCE MARKS

A set of drawings for a complex house may include several sheets of section and detail drawings. These sections and details do not have much meaning without some way of knowing what part of the building they are meant to show. Callouts, called *reference marks*, on plans and elevations indicate where details or sections of important features have been drawn. To be able to use these reference marks for coordinating drawings, you must first understand the numbering system used on the drawings. The simplest numbering system for drawings consists of numbering the drawing sheets and naming each of the views. For example, Sheet 1 might include a site plan and foundation plan; Sheet 2, floor plans; and Sheet 3, elevations.

On large, complex sets of drawings the sheets are numbered according to the kind of drawings shown. Architectural drawing sheets are numbered A-1, A-2, and so on for all the sheets. Electrical drawings are numbered E-1, E-2, and E-3. A view number identifies each separate drawing or view on the sheet. Figure 6-8 shows drawing 5 on sheet A-4.

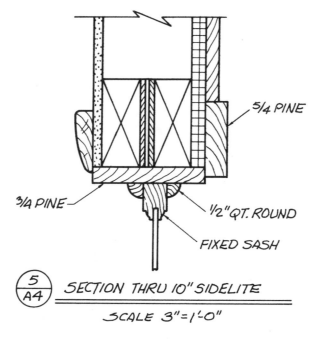

5/4 PINE

3/4 PINE

1/2" QT. ROUND

FIXED SASH

⌀ 5/A4 SECTION THRU 10" SIDELITE

SCALE 3"=1'-0"

Figure 6-8 This is drawing 5 on sheet A-4.

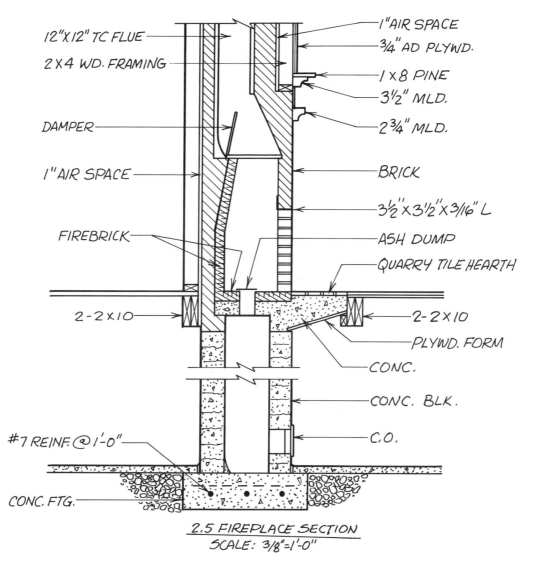

12"X12" TC FLUE

2 X 4 WD. FRAMING

DAMPER

1" AIR SPACE

FIREBRICK

2-2 X 10

#7 REINF. @ 1'-0"

CONC. FTG.

1" AIR SPACE

3/4" AD PLYWD.

1 X 8 PINE

3 1/2" MLD.

2 3/4" MLD.

BRICK

3 1/2" X 3 1/2" X 3/16" L

ASH DUMP

QUARRY TILE HEARTH

2-2 X 10

PLYWD. FORM

CONC.

CONC. BLK.

C.O.

2.5 FIREPLACE SECTION

SCALE: 3/8"=1'-0"

Figure 6-9 This section view is drawing 2 on sheet 5.

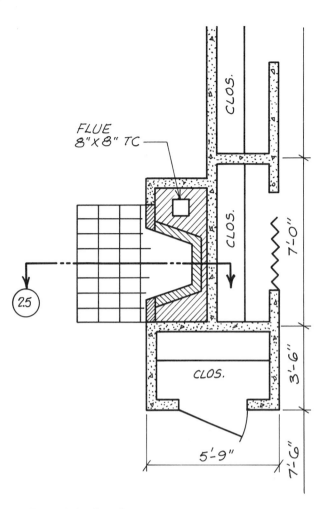

Figure 6-10 Plan for fireplace detailed in Figure 6-9

Because most of the drawings for a house are architectural, and the drawing set is fairly small, letters indicating the type of drawing are not usually included. Instead, the views are numbered and a second number shows on which sheet it appears. For example, the fourth drawing on the third sheet would be 4/3, 4.3, or 4-3.

Numbering each view and the sheet on which it appears makes it easy to reference a section or detail to another drawing. The identification of a section view is given with the cutting-plane line showing where it is taken from. For example, the section view shown in Figure 6-9 shows the fireplace at the cutting-plane line in Figure 6-10. This numbering system is also used for details that can-

not be located by a cutting-plane line. The detail drawing of the cornice (edge of the roof) in Figure 6-11 is drawing 4 on sheet A-4, Figure 6-12.

ABBREVIATIONS

Drawings for construction include many notes and labels of parts. These notes and labels are usually abbreviated as much as possible to avoid crowding the drawing. The abbreviations used on drawings are usually a shortened form of the word and are easily understood. For example, BLDG stands for building. The abbreviations used throughout this textbook and on the related drawings are defined in the Appendix.

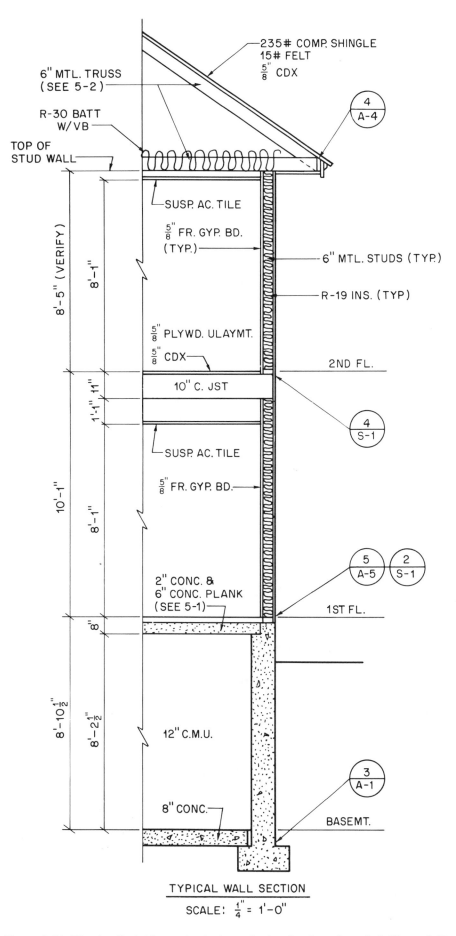

235# COMP. SHINGLE
15# FELT
$\frac{5}{8}''$ CDX

6" MTL. TRUSS
(SEE 5-2)

R-30 BATT
W/VB

TOP OF
STUD WALL

$\frac{4}{A-4}$

SUSP. AC. TILE

$\frac{5}{8}''$ FR. GYP. BD.
(TYP.)

6" MTL. STUDS (TYP.)

R-19 INS. (TYP)

$\frac{5}{8}''$ PLYWD. ULAYMT.
$\frac{5}{8}''$ CDX

2ND FL.

10" C. JST

$\frac{4}{S-1}$

SUSP. AC. TILE

$\frac{5}{8}''$ FR. GYP. BD.

8'-5" (VERIFY)

8'-1"

1'-1" 11"

10'-1"

8'-1"

$\frac{5}{A-5}$ $\frac{2}{S-1}$

2" CONC. &
6" CONC. PLANK
(SEE 5-1)

1ST FL.

8"

8'-10$\frac{1}{2}$"

8'-2$\frac{1}{2}$"

12" C.M.U.

$\frac{3}{A-1}$

8" CONC.

BASEMT.

TYPICAL WALL SECTION
SCALE: $\frac{1}{4}''$ = 1'-0"

Figure 6-11 The detail of this cornice is shown in drawing 4 on sheet A-4, Figure 6-12.

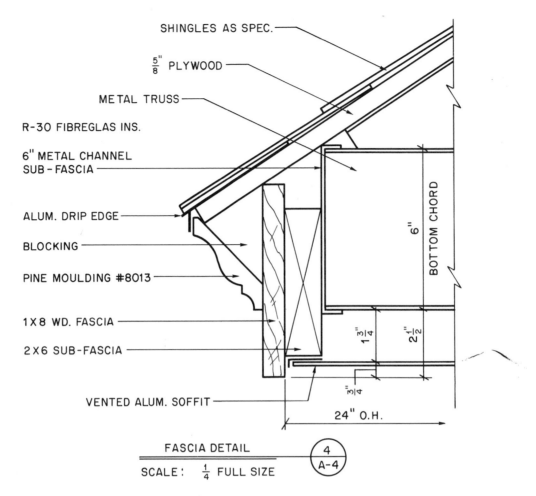

SHINGLES AS SPEC.

$\frac{5}{8}$" PLYWOOD

METAL TRUSS

R-30 FIBREGLAS INS.

6" METAL CHANNEL
SUB-FASCIA

ALUM. DRIP EDGE

BLOCKING

PINE MOULDING #8013

1 X 8 WD. FASCIA

2 X 6 SUB-FASCIA

VENTED ALUM. SOFFIT

BOTTOM CHORD

6"

$1\frac{3}{4}$"

$2\frac{1}{2}$"

$\frac{3}{4}$"

24" O.H.

FASCIA DETAIL

SCALE: $\frac{1}{4}$ FULL SIZE

4
A-4

Figure 6-12 This is the detail of the cornice in Figure 6-11.

✓ CHECK YOUR PROGRESS

Can you perform these tasks?

- ☐ Identify window types by their symbols.
- ☐ Identify materials by their symbols.
- ☐ Identify the most common electrical equipment by its symbols.
- ☐ Identify the most common plumbing equipment by its symbols.
- ☐ Reference details by their symbols.
- ☐ Define several common abbreviations used on construction drawings.

938 _1_ ½

ASSIGNMENT

1. What is represented by each of these symbols?

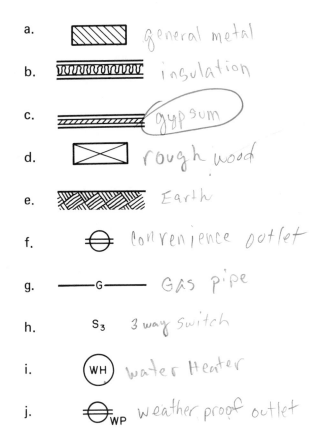

a. general metal

b. insulation

c. gypsum

d. rough wood

e. Earth

f. convenience outlet

g. —G— Gas pipe

h. S₃ 3 way switch

i. (WH) water Heater

j. weather proof outlet

2. What is meant by each of these abbreviations?
   a. GYP. BD. gypsum board
   b. FOUND. Foundation
   c. FIN. FL. finish Floor
   d. O.C. on centers
   e. REINF. reinforced
   f. EXT. exterior
   g. COL. Coloumns
   h. DIA. diameter
   i. ELEV. elevation
   j. CONC. concrete

3. Where in a set of drawings would you find a detail numbered 6.4? What page and drawings

4. Where in a set of drawings would you find a detail numbered $\frac{5}{M\text{-}3}$?

   fifth page, drawing M-3 ½

UNIT 7 Plan Views

OBJECTIVES

After completing this unit, you will be able to explain the general kinds of information shown on the listed plans:

- Site plans
- Foundation plans
- Floor plans

You learned earlier in Unit 3 that plans are drawings that show an object as viewed from above. Many of the detail and section drawings in a set show parts of the building from above. Some of the plan views that show an entire building are discussed here. This brief explanation will help you feel more comfortable with plans, although it does not cover plans in depth. You will use plans frequently throughout your study of the remainder of this textbook. Each of the remaining units helps you understand plan views more thoroughly.

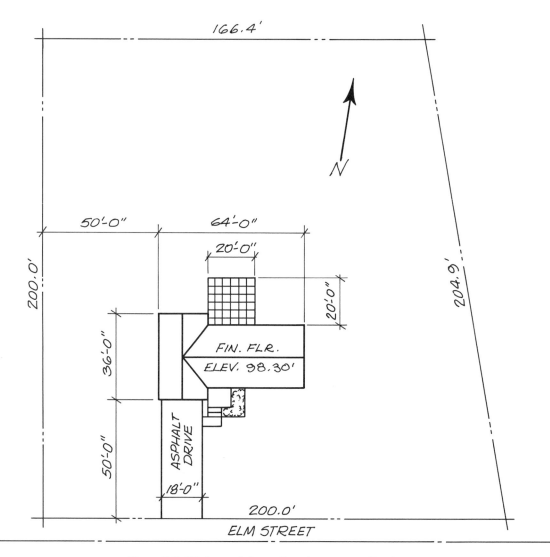

Figure 7-1 Minimum information shown on a site plan

SITE PLANS

A site plan gives information about the site on which the building is to be constructed. The boundaries of the site (property lines) are shown. The property line is usually a heavy line with one or two short dashes between longer line segments. The lengths of the boundaries are noted next to the line symbol. Property descriptions are often the result of a survey by a surveyor or civil engineer. These professionals usually work with decimal parts of feet, rather than feet and inches. Therefore, site dimensions are usually stated in tenths or hundredths of feet, Figure 7-1.

A symbol or arrow of some type indicates what compass direction the site faces. Unless this north arrow includes a correction for the difference between true north and magnetic north, it may be only an approximation. However, it is sufficient to show the general direction the site faces.

The site plan also indicates where the building is positioned on the site. As a minimum, the dimensions to the front and one side of the site are given. The overall dimensions of the building are also included. Anyone reading the site plan will have this basic information without referring to the other drawings. If the finished site is to include walks, drives, or patios, these are also described by their overall dimensions.

FOUNDATION PLANS

A foundation plan is like a floor plan, but of the foundation instead of the living spaces. It shows the foundation walls and any other structural work to be done below the living spaces.

There are two types of foundations that are commonly used in homes and other small buildings. One type has a concrete base, called the *footing*, supporting foundation walls, Figure 7-2. The other is the slab-on-grade type. A *slab-on-grade* foundation consists of a concrete slab placed directly on the soil with little or no other support. Slabs on grade are usually thickened at their edges and wherever they must support a heavy load, Figure 7-3.

When the footing-and-wall type foundation is used, girders are used to provide intermediate support to the structure above, Figure 7-4. The girder is shown on the foundation plan by phantom lines and a note describing it.

The foundation plan includes all of the dimensions necessary to lay out the footings and foundation walls. The footings follow the walls and may be shown on the plan. If they are shown, it is usually by means of hidden lines to show their outline only. In addition to the layout of the foundation walls, dimensions are given for openings for windows, doors, and ventilators. Notes on the plan indicate areas that are not to be excavated, concrete-slab floors, and other important information about the foundation, Figure 7-5. See Section 4 for more detailed information about foundations.

FLOOR PLANS

A floor plan is similar to a foundation plan. It is a section view taken at a height that shows the placement of walls, windows, doors, cabinets, and other important features. A separate floor plan is included for each floor of the building. The floor

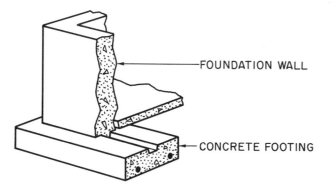

Figure 7-2 Footing and foundation wall

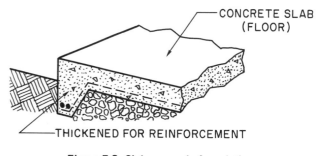

Figure 7-3 Slab-on-grade foundation

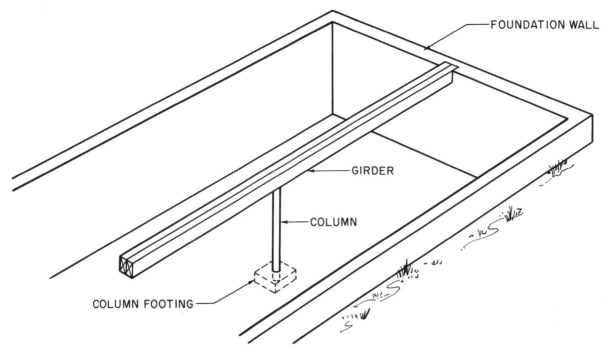

FOUNDATION WALL

GIRDER

COLUMN

COLUMN FOOTING

Figure 7-4 A girder provides intermediate support between the foundation walls.

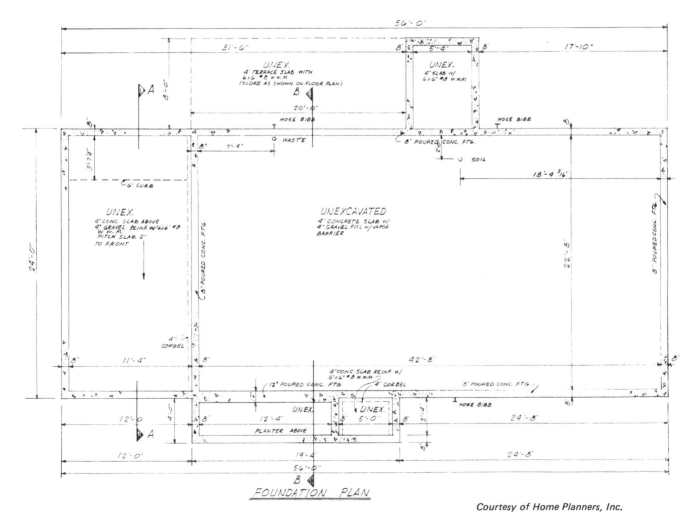

Courtesy of Home Planners, Inc.

Figure 7-5 Foundation plan

plans provide more information about the building than any of the other drawings.

Building Layout

The floor plans show the locations of all of the walls, doors, and windows. Therefore, the floor plans show how the building is divided into rooms, and how to get from one room to another. Before attempting to read any of the specific information on the floor plans, it is wise to familiarize yourself with the general layout of the building.

To quickly familiarize yourself with a floor plan, imagine that you are walking through the house. For example, imagine yourself standing in the front door of the left side of the Duplex — plans for which are included in the drawing packet with this text. You are looking across the living room. There is a closet on your right and a large window on your left. Straight ahead is the dining room with doors into a storage room and the kitchen. Looking in the kitchen doorway (notice there is no door in this doorway), there are cabinets, a sink, and refrigerator on the opposite wall. More cabinets and a range are located on the left. Now, walk out of the kitchen and into the bedroom area. There are three doors; one leads into a large front bedroom with a long closet, another opens into a smaller bedroom, and the third opens into the bathroom. The bathroom includes a linen closet with bifold doors.

Dimensions

Dimensions are given for the sizes and locations of all walls, partitions, doors, windows, and other important features. On frame construction, exterior walls are usually dimensioned to the outside face of the wall framing. If the walls are to be covered with stucco or masonry veneer, this material is outside the dimensioned face of the wall frame. Interior partitions may be dimensioned to their centerlines or to the face of the studs. (*Studs* are the vertical members in a wall frame.) Windows and doors may be dimensioned about their centerlines, Figure 7-6, or to the edges of the openings.

Solid masonry construction is dimensioned entirely to the face of the masonry, Figure 7-7. Masonry openings for doors and windows are dimensioned to the edge of the openings.

Other Features of Floor Plans

The floor plan includes as much information as possible without making it cluttered and hard to read. Doors and windows are shown by their symbols as explained in Unit 6. Cabinets are shown in their proper positions. The cabinets are explained further by cabinet elevations and details, which are discussed in Unit 9. If the building includes stairs, these are shown on the floor plan. Important overhead construction is also indicated on the floor plans. If the ceiling is framed with joists, their size, direction, and spacing is shown on the floor plan. Architectural features such as exposed beams, arches in doorways, or unusual roof lines may be shown by phantom lines.

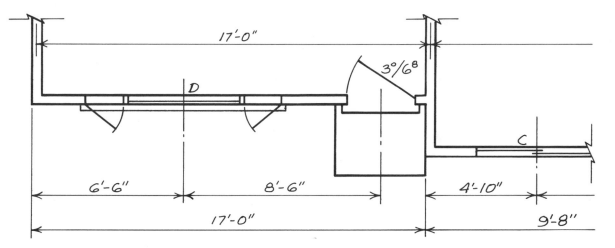

Figure 7-6   Frame construction dimensioning

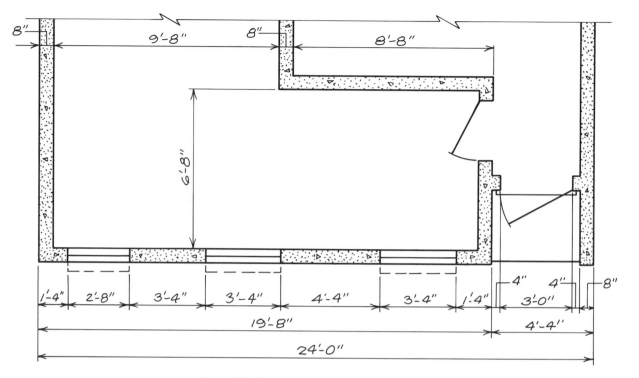

Figure 7-7 Masonry construction dimensioning

✓ CHECK YOUR PROGRESS

Can you perform these tasks?

- ☐ Describe property boundaries from a simple site plan.

- ☐ Tell which direction a site faces from the site plan.

- ☐ Describe the position of a building within the site using the dimensions on a site plan.

- ☐ List the overall dimensions of a foundation from the information on a foundation plan.

- ☐ Identify girders on a foundation plan.

- ☐ Describe the locations of windows shown on plans.

- ☐ Describe the arrangement of rooms shown on a simple, one-story floor plan.

- ☐ Describe other major features shown on plans.

83.4% (handwritten, top of page)
2 (handwritten)

ASSIGNMENT

Refer to the drawings for the Duplex (which are included in the accompanying packet) to complete the assignment.

1. In what direction does the Duplex face? *South*
2. What is the length and width of the Duplex site? *60.0' X 120.0'*
3. How far is the front of the Duplex from the front property line? *10.0'*
4. What is the overall length and width of the Duplex? *24.0' X 72.0'*
5. What are the inside dimensions of the front bedroom? *16'-5¾" X 16'-8½"*
6. What is the thickness of the partitions between the bedrooms? *3½"*
7. What is the thickness of the interior wall between the two dining rooms? *5½"*
8. With two exceptions, the apartments in the Duplex are exactly reversed. What are the two exceptions? *Storage, closet*
9. What is the distance from the west end of the Duplex to the centerline of the west, front entrance? *29'-6"*
10. What is indicated by the small rectangle on the floor plan outside each main entrance? *4" conc. pad*
11. What is the distance from the ends of the Duplex to the centerlines of the 6^0 x 6^8 sliding glass doors? *26'-3½"*
12. What is indicated by the dashed line just outside the front and back walls on the floor plan of the Duplex? *roof overhang*

UNIT 8   *Elevations*

OBJECTIVES

After completing this unit, you will be able to perform the following tasks:

- Orient building elevations to building plans.
- Explain the kinds of information shown on elevations.

Drawings that show the height of objects are called *elevations*. However, when builders and architects refer to building elevations, they mean the exterior elevation drawings of the building, Figure 8-1. A set of working drawings usually includes an elevation of each of the four sides of the building. If the building is very complex, there may be more than four elevations. If the building is simple, there may be only two elevations — the front and one side.

ORIENTING ELEVATIONS

It is important to determine the relationship of one drawing to another. This is called *orienting* the drawings. For example, if you know which elevation is the front, you must be able to picture how it relates to the front of the floor plan.

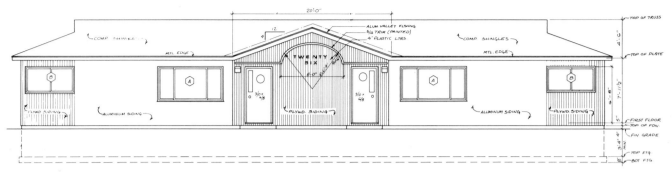

FRONT ELEVATION

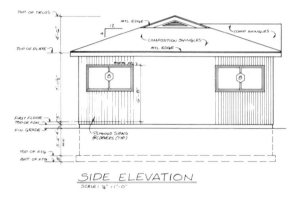

SIDE ELEVATION
SCALE: ⅛" = 1'-0"

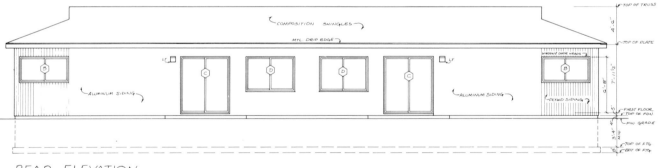

REAR ELEVATION
SCALE: ¾" = 1'-0"

Courtesy of Robert C. Kurzon

Figure 8-1 Building elevations

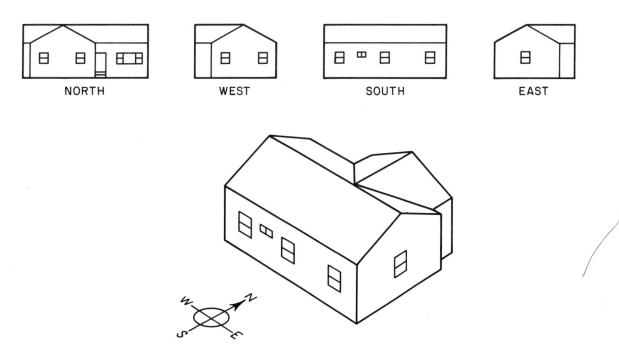

NORTH WEST SOUTH EAST

Figure 8-2 Elevations are usually named according to their compass directions.

Elevations are usually named according to compass directions, Figure 8-2. The side of the house that faces north is the north elevation, and the side that faces south is the south elevation, for example. When the elevations are named according to compass direction, they can be oriented to the floor plan, foundation plan, and site plan by the north arrow on those plans. It might help to label the edges of the plans according to the north arrow, Figure 8-3.

It is not always possible to label elevations according to compass direction, however. When drawings are prepared to be sold through a catalog or when they are for use on several sites, the compass directions cannot be included. In this case, the elevations are named according to their position

Courtesy of Robert C. Kurzon

Figure 8-3 Plan labeled to help orientation to north arrow.

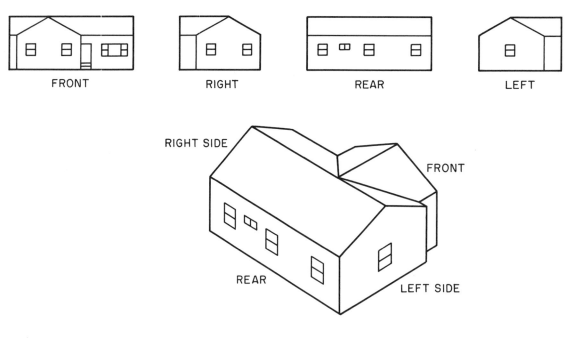

FRONT   RIGHT   REAR   LEFT

Figure 8-4 Elevations can be named according to their relative positions.

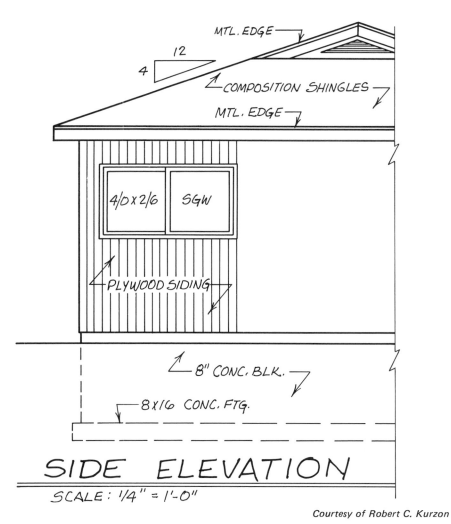

SIDE ELEVATION

SCALE: 1/4" = 1'-0"

Courtesy of Robert C. Kurzon

Figure 8-5 Underground portion of building is shown with dashed lines.

as you face the building, Figure 8-4. To orient these elevations to the plans, find the front on the plans. The front is usually at the bottom of the sheet, but it can be checked by the location of the main entrance.

INFORMATION ON BUILDING ELEVATIONS

Building elevations are normally quite simple. Although the elevations do not include a lot of detailed dimensions and notes, they show the finished appearance of the building better than other views. Therefore, elevations are a great aid in understanding the rest of the drawing set.

The elevations show most of the building, as it will actually appear, with solid lines. However, the underground portion of the foundation is shown as hidden lines, Figure 8-5. The footing is shown as a rectangle of dashed lines at the bottom of the foundation walls.

The surface of the ground is shown by a heavy solid line, called a *grade line*. The grade line usually includes one or more notes to indicate the elevation above sea level or another reference point, Figure 8-6. *Elevation* used in this sense is a level, or

height — not a type of drawing. All references to the height of the ground or the level of key parts of the building are in terms of elevation. Methods for measuring site elevations are discussed later in Unit 11.

Some important dimensions are included on the building elevations. Most of them are given in a string at the end of one or more elevations, Figure 8-7. The dimensions most often included are listed:

- Thickness of footing

- Height of foundation walls

- Top of foundation to finished first floor

- Finished floor to ceiling or top of plate (The *plate* is the uppermost framing member in the wall.)

- Finished floor to bottom of window headers (The *headers* are the framing across the top of a window opening.)

- Roof overhang at eaves

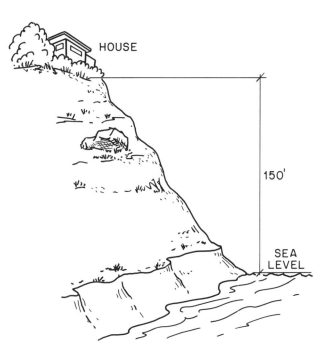

Figure 8-6 The elevation of this site is 150'.

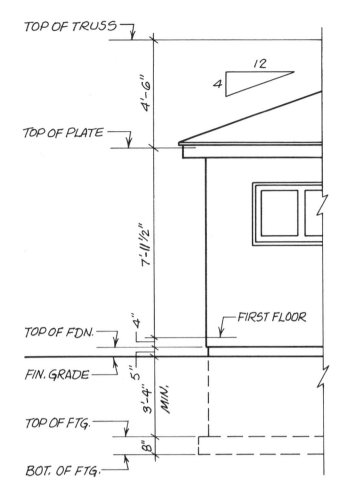

TOP OF TRUSS

12
4

TOP OF PLATE

4'-6"

7'-11½"

FIRST FLOOR

TOP OF FDN.

4"

FIN. GRADE

5"

3'-4"
MIN.

TOP OF FTG.

8"

BOT. OF FTG.

Figure 8-7 Dimensions on an elevation

✓ CHECK YOUR PROGRESS

Can you perform these tasks?

☐ Explain which side of a plan view is represented by a building elevation.

☐ Identify footings and foundations on building elevations.

☐ Find the height of a foundation wall on the building elevations.

☐ Find the dimension from the floor to the top of a wall on building elevations.

☐ Find the amount of roof overhang shown on building elevations.

☐ Describe the appearance of a building from information given on the building elevations.

108⁰

ASSIGNMENT

Refer to the drawings of the Duplex in the packet to complete this assignment.

1. Which elevation is the north elevation?   *rear elevation*
2. In what compass direction does the left end of the Duplex face?   *west*
3. What is the dimension from the surface of the floor to the top of the wall   *7'-11½"*
   framing?
4. What is the thickness of the floor?   *5"*
5. How far does the foundation project above the ground?   *4"*
6. How far below the surface of the ground does the foundation wall extend?   *3'-4"*
7. What is the total height of the foundation walls?   *3'-8"*
8. What is the minimum depth of the bottom of the footings?   *8"*

UNIT 9   *Sections and Details*

OBJECTIVES

After completing this unit, you will be able to perform the following tasks:

- Find and explain information shown on section views.

- Find and explain information shown on large-scale details.

- Orient sections and details to the other plans and elevations.

It is not possible to show all of the details of construction on foundation plans, floor plans, and building elevations. Those drawings are meant to show the relationships of the major building elements to one another. To show how individual pieces fit together, it is necessary to use larger-scale drawings and section views. These drawings are usually grouped together in the drawing set. They are referred to as *sections and details*, Figure 9-1.

SECTIONS

Nearly all sets of drawings include, at least, a typical wall section. The typical section may be a section view of one wall, or it may be a full section of the building. Full sections are named by the direction in which the imaginary cut is made. Figure 9-2 shows a transverse section. A *transverse section* is taken from an imaginary cut across the width of the building. Transverse sections are sometimes called *cross sections*. A full section taken from a lengthwise cut through the building is called a *longitudinal section*, Figure 9-3.

Full sections and wall sections normally have only a few dimensions, but have many notes with leaders to identify the parts of the wall. The following is a list of the kinds of information that are included on typical wall sections with most sets of drawings:

- Footing size and material (This may be specified by building codes.)

- Foundation wall thickness, height, and material

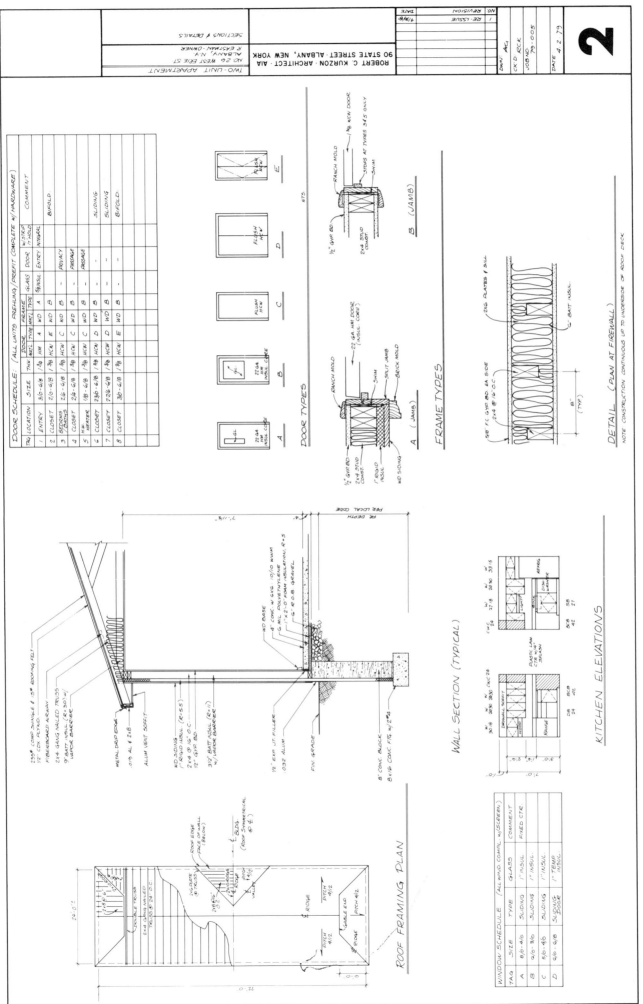

Figure 9-1 Typical sheet of sections and details for a small building

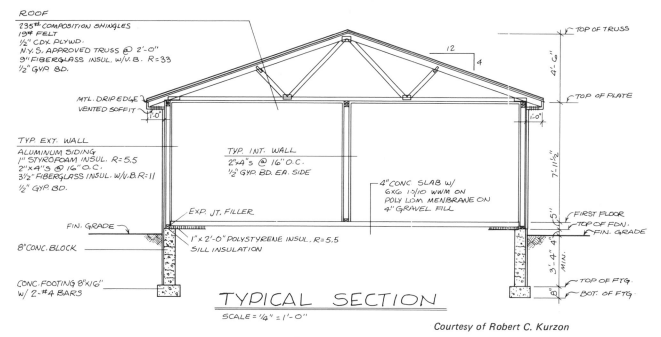

Figure 9-2 Transverse section

- Insulation, waterproofing, and interior finish for foundation walls
- Fill and waterproofing under concrete floors
- Concrete floor thickness, material, and reinforcement
- Sizes of floor framing materials
- Sizes of wall framing materials
- Wall covering (sheathing, siding, stucco, masonry, and interior wall finish) and insulation
- Cornice construction — materials and sizes (The *cornice* is the construction at the roof eaves.)
- Ceiling construction and insulation

Other section drawings are included as necessary to explain special features of construction. Wherever wall construction varies from the typical wall section, another wall section should be in-cluded. Section views are used to show any special construction that cannot be shown on normal plans and elevations. Figure 9-4 is an example of a special section in elevation. This section view is said to be *in elevation* because it shows the height of the ridge construction. Figure 9-5 is *in plan* because it shows the interior of the fireplace as viewed from above.

OTHER LARGE-SCALE DETAILS

Sometimes necessary information can be conveyed without showing the interior construction. A large scale may be all that is needed to show the necessary details. The most common examples of this are on cabinet installation drawings, Figure 9-6. Cabinet elevations show how the cabinets are located, without showing the interior construction.

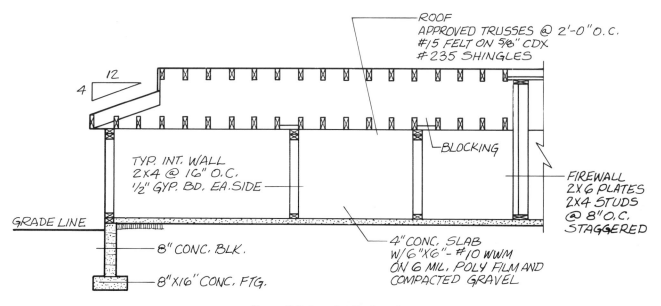

Figure 9-3 Longitudinal section

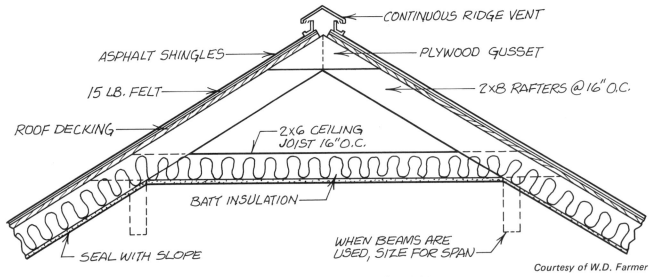

CONTINUOUS RIDGE VENT

ASPHALT SHINGLES

PLYWOOD GUSSET

15 LB. FELT

2×8 RAFTERS @ 16" O.C.

ROOF DECKING

2×6 CEILING JOIST 16" O.C.

BATT INSULATION

SEAL WITH SLOPE

WHEN BEAMS ARE USED, SIZE FOR SPAN

Courtesy of W.D. Farmer

Figure 9-4 Special section of ventilated ridge

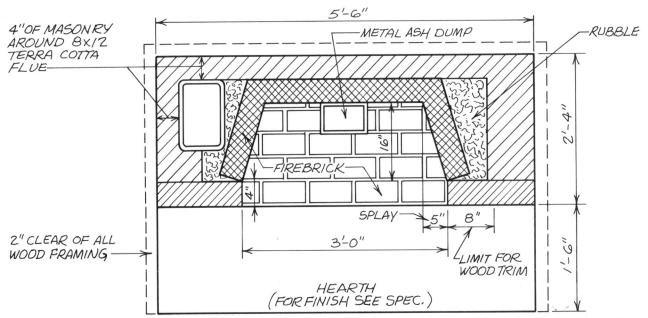

5'-6"

4" OF MASONRY AROUND 8×12 TERRA COTTA FLUE

METAL ASH DUMP

RUBBLE

2'-4"

16"

FIREBRICK

4"

SPLAY

5" 8"

2" CLEAR OF ALL WOOD FRAMING

3'-0"

LIMIT FOR WOOD TRIM

1'-6"

HEARTH (FOR FINISH SEE SPEC.)

Courtesy of W.D. Farmer

Figure 9-5 A section in plan

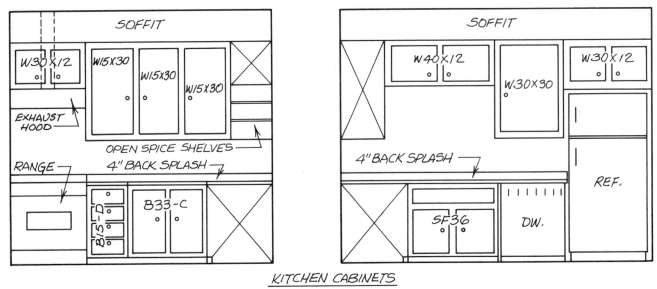

SOFFIT

W30×12

W15×30

W15×30

W15×30

EXHAUST HOOD

OPEN SPICE SHELVES

RANGE

4" BACK SPLASH

B15-D

B33-C

SOFFIT

W40×12

W30×30

W30×12

4" BACK SPLASH

REF.

SF36

DW.

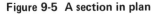

KITCHEN CABINETS

SCALE: 3/8" = 1'-0"

Figure 9-6 Cabinet elevations

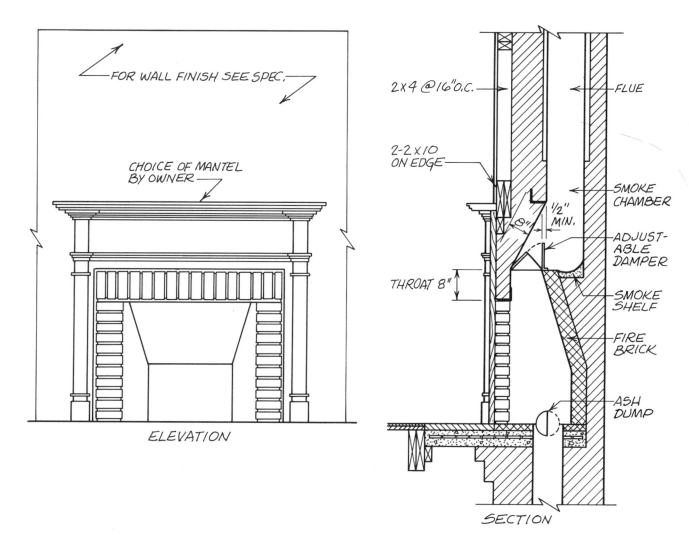

Courtesy of W.D. Farmer

Figure 9-7 Fireplace details

Many details are best shown by combining elevations and sections or by using isometric drawings. Figure 9-7 shows an example of an elevation and a section used together to explain the con- struction of a fireplace. Figure 9-8 shows an isometric detail drawing that includes sections to show interior construction.

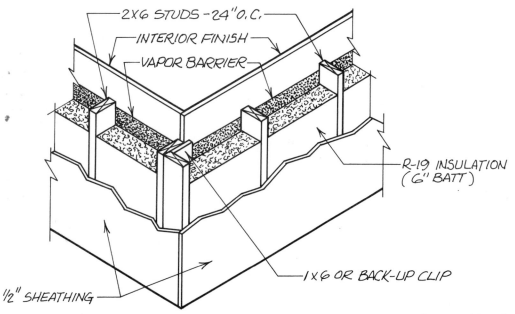

Courtesy of W.D. Farmer

Figure 9-8 Isometric section

ORIENTING SECTIONS AND DETAILS

As explained earlier, some sections and details are labeled as typical. These drawings describe the construction that is used throughout most of the building.

Details and sections that refer to only one place in the building are identified by a reference mark. As was pointed out earlier in Unit 5, sections are usually referenced by a cutting-plane line. This line shows where the section was taken from. Arrows on the ends of the cutting-plane line indicate what direction the imaginary cut is viewed from.

A reference mark near the arrow indicates where the detail drawing is shown. These reference marks used for orienting details may vary from one set of drawings to another. It is important, although not usually difficult, to study the drawings and learn how the architect references details. Usually a system of sheet numbers and view numbers is used. One such numbering system was explained earlier.

Some basic principles of details and sections have been discussed here. You will gain more practice later in reading details and sections.

✓ CHECK YOUR PROGRESS

Can you perform these tasks?

☐ Find the part of a building from which a section view was drawn.

☐ Explain the notes commonly included on typical wall sections.

☐ Explain whether a section view is in elevation or plan.

☐ Find the part of a building from which a large-scale detail was drawn.

ASSIGNMENT

Refer to the drawings of the Duplex in the packet to complete the assignment

1. What kind of section drawing is the Typical Wall Section on Sheet 2? _transverse_
2. What kind and size material is to be used for the foundation walls? _8" conc. block_
3. What is used between the concrete-slab floor and the exterior wall framing? _1/2" exp. Jt. filler_
4. What kind and size of insulation is used around the foundation? Is this used on the inside or outside of the foundation? _polystyrene_ _Typical_
5. What kind and size of material is to be used on the inside of the frame walls? _gypsum board_
6. Sheet 2 includes a firewall detail. Where in the Duplex is this firewall? _in the middle_
7. What is the distance between the centerlines of the studs in the firewall? _8"_
8. What is the total thickness of the firewall? (Remember that a 2x6 is actually 5 1/2" wide.) _27"_
9. Were the cabinet elevations drawn of the kitchen on the east side or the west side of the Duplex? _east_
10. How would the kitchen elevations be different if they were drawn from the other kitchen? _they would be on the west side_
11. What is the distance from the kitchen countertops to the bottom of the wall cabinets?
12. How far does the roof overhang project beyond the exterior walls?

Part II

READING DRAWINGS FOR TRADE INFORMATION

In Part II, you will examine all of the information necessary to build a moderately complex single-family home. The sequence of the units in Part II follows the sequence of actual construction. In some cases, all of the information necessary for a particular phase of construction can be found on one sheet of drawings. Other phases require cross-referencing among several drawings. The relationships among the various drawings are discussed as the need to cross-reference them arises.

The assignments in this section refer to the Lake House drawings provided in the packet. The Lake House was designed as a vacation home on a lake in Virginia. The design is moderately complex, involving several floor levels and some interesting construction techniques.

Site Preparation and Earthwork
Section 3

UNIT 10   Boundary Description

OBJECTIVES

After completing this unit, you will be able to perform the following tasks:

- Read and interpret property descriptions.
- Interpret boundary information found on a site plan.

The first step in using any parcel of land is to determine the boundaries of the parcel according to accepted practice. A grid system of land subdivision is often used. In the grid system, land is divided into townships which measure six miles on a side. *Townships* are numbered according to a township number and a range number, Figure 10-1. Each township is divided into 36 sections of one square mile each. Subdivisions within the section are described by their position within the section.

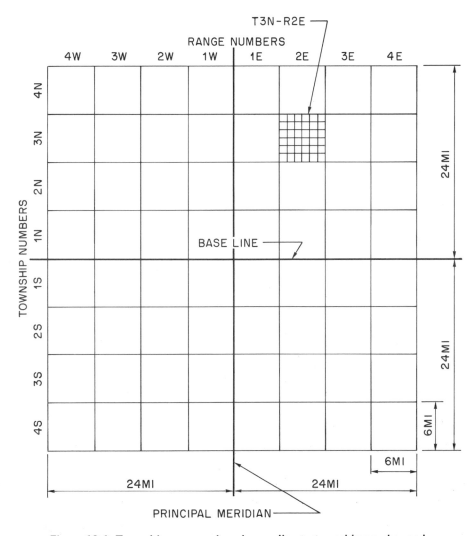

Figure 10-1 Townships are numbered according to township number and range number. Each township contains 36 sections.

The older states along the East coast use a system of metes and bounds devised during the colonization of America. *Metes* refers to measurement from fixed monuments. *Bounds* refers to description relative to adjoining property boundaries. The property description for the Lake House in Virginia uses the metes and bounds system, Figure 10-2.

In order to completely describe a property, both the length and the direction of the boundaries must be determined. The site plan is drawn from measurements first taken by a surveyor. Therefore, the lengths of property lines are usually expressed in feet and decimal parts of a foot instead of in feet and inches.

The direction of a property line is usually expressed as a bearing angle. The *bearing* of a line is the angle between the line and north or south. Bearing angles are measured from north or south de-pending upon which keeps the bearing under 90 degrees, Figure 10-3. Angles are expressed in degrees (°), minutes ('), and seconds ("). There are 360 degrees in a complete circle, 60 minutes in a degree, and 60 seconds in a minute.

The point of beginning (P.O.B.) may or may not be shown on the site plan. If the point of beginning is not shown on the plan, start at a convenient corner. Corners are usually marked with an iron pin (I.P.) or some permanent feature. The approximate direction of the boundaries can be found with a hand-held compass. This approximation should be accurate enough to aid in finding the marker (iron pin, manhole cover, concrete marker, or similar item) at the next corner. Proceed around the perimeter in this manner to find all corners. All construction activity should be kept within the property boundaries unless permission is first obtained from neighboring landowners.

STATE OF VIRGINIA
COUNTY OF CHENANGO

THIS IS TO CERTIFY THAT: we have examined the records of the offices of the County and District Courts of this county and find:

A good and sufficient description of the property is: All that property situated in Chenango County, Virginia described as follows:

A certain tract of land containing 0.6 acres out of the Alan Tomac 21 acre tract; the said 0.6 acre tract being more particularly described as follows:

1. Beginning at an existing iron pipe on the southeast corner of said 21 acre tract and 5.50 feet from the high water mark of Grant Lake;

2. then west along the north shore of Grant Lake, 102.40 feet to an iron pipe 5.53 feet from the high water mark of said lake;

3. then N 03°55' W;261.30 feet to an iron pipe at the north line of an existing right of way through said Tomac tract;

4. then N 86°05'E ;100.00 feet to an iron pipe on the west boundary of the property now or formerly owned by Norbert Coon;

5. then S 03°55'E, along the west line of the property now or formerly owned by Norbert Coon; 245.00 feet to the point of beginning.

We find the title recorded in:
    Harry Wong
who acquired the above described land by deed from Alan Tomac, dated January 21, 1982, recorded in Volume 3610, page 123, of the deed records of said county.

Figure 10-2 Lake House property description

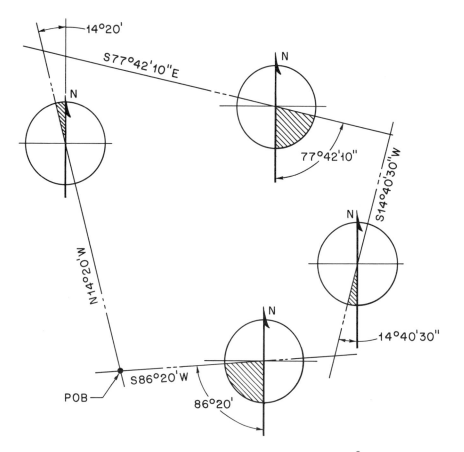

Figure 10-3 Bearing angles are always less than 90°.

✓ CHECK YOUR PROGRESS

Can you perform these tasks?

☐ Determine the length of each property boundary from the information shown on a site plan.

☐ Determine the direction of each property boundary from the information shown on a site plan.

☐ Calculate the angle formed by property boundaries.

ASSIGNMENT

Refer to the side drawings of the Lake House (in the packet) to complete this assignment.

1. What is the length of each boundary of the Lake House? Assume that the south boundary is a straight line.

2. Which of the compass points shown in Figure 10-4 corresponds with the north boundary of the Lake House? Which compass point corresponds with the east boundary?

Figure 10-4

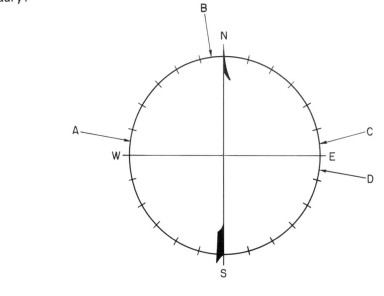

3. Write compass bearings for each part of Figure 10-5.

(A)

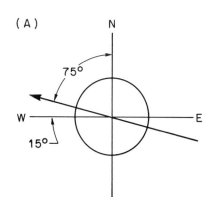

Figure 10-5

(B)

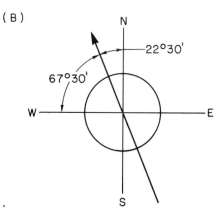

(C)

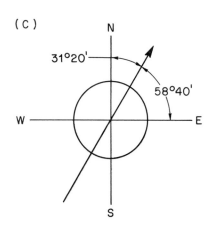

(D)

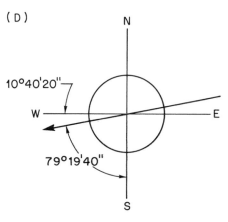

UNIT 11 Clearing and Rough Grading the Site

OBJECTIVES

After completing this unit, you will be able to perform the following tasks.

- Identify work to be included in clearing a building site according to site plans.
- Interpret grading indications on a site plan.
- Interpolate unspecified site elevations.

CLEARING THE SITE

The first step in actual construction is to prepare the site. This means clearing any brush or trees that are not to be part of the finished landscape. The architect's choice of trees to remain is based on consideration of many factors. Trees and other natural features can be an important part of architecture — not only for their natural beauty, but for energy conservation. For example, deciduous trees, which lose their leaves in the winter, can be used to effectively control the solar energy striking a house. In the winter, the sun shines through the deciduous trees on the south side of a house, Figure 11-1. In the summer, the trees shade the south side of the house, Figure 11-2. The Lake House offers a good example of the importance of the selection of trees to remain on a site. This house gets a large part of its heat from its passive-solar features. The passive-solar features are described more fully later.

Trees that are to be saved are shown on the plot plan by a symbol and a note indicating their butt diameter and species, Figure 11-3. Areas that are too densely wooded to show individual trees

Figure 11-1 The winter sun passes through deciduous trees.

Figure 11-2 The summer sun is shaded by deciduous trees.

are outlined and marked "woods," Figure 11-4. Removal of unwanted trees may require felling and stump removal, or may be accomplished with a bulldozer and dump truck. In either case, care must be exercised not to damage the trees that are to be saved.

GRADING

Grading refers to moving earth away from high areas and into low areas. Site grading is necessary to insure that water drains away from the building properly and does not puddle or run into the building. In some cases, grading may be necessary for access to the site. For example, if the site

Figure 11-3 Typical note and symbol for individual tree

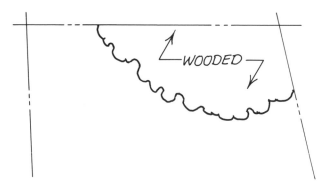

Figure 11-4 Typical note and symbol for wooded area

has a steep grade, it may be necessary to provide a more gradual slope for a driveway.

Grade is measured in vertical feet from sea level or from a fixed object such as a manhole cover. This vertical distance is called *elevation*. The term elevation to denote a vertical position should not be confused with elevation drawings that show the height of objects. The elevations of specific points are given as *spot elevations*. Spot elevations are used to establish points in a driveway, walk, or the slope of a terrace, Figure 11-5. Spot elevations are often given for trees that are to be saved.

The grade of a site is shown by *topographic contour lines*. These are lines following a particular elevation. The vertical difference between contour lines is the *vertical contour interval*. For plot plans this is usually 1 or 2 feet. When the land slopes steeply, the contour lines are closely spaced. When the slope is gradual, the contour lines are more widely spaced.

The builder must be concerned with not only the grade or contour of the existing site, but also that of the finished site. To show both contours, two sets of contour lines are included on the plot plan. Broken lines indicate natural grade (N.G.) and solid lines indicate finished grade (F.G.), Figure 11-6.

When the natural-grade elevation is higher than the finished-grade elevation, earth must be removed. This is referred to as *cut*. When the natural grade is at a lower elevation than the finished grade, *fill* is required. To determine the amount of cut or fill required at a given point, find the difference between natural grade and finished grade, Figure 11-7.

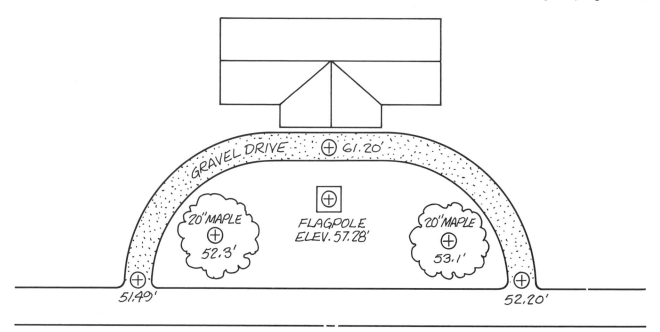

Figure 11-5 Spot elevations for specific locations

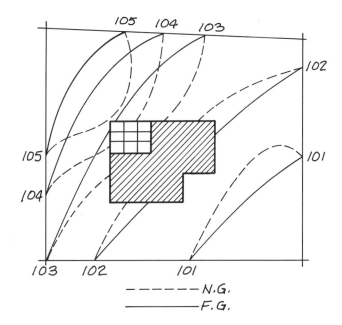

Figure 11-6 Two sets of contour lines show that this site will be graded to be more level in the area of the building.

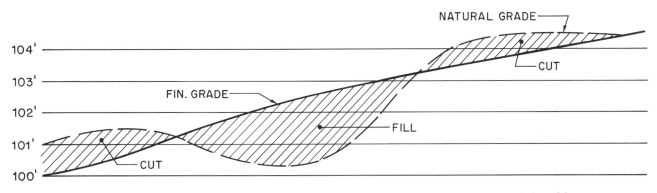

Figure 11-7 Cutting is required where NG is above FG. Fill is required where NG is below FG.

INTERPOLATING ELEVATIONS

Sometimes it is necessary to find an elevation that falls between two contour lines. This can be done by interpolation. *Interpolation* is a method of finding an unknown value by comparing it with known values.

Example: To interpolate the elevation of the tree at point A in Figure 11-8, follow the listed steps of the procedure using the information shown in the illustration and the numbers enclosed in parenthesis.

Step 1. Scale the distance between the two adjacent contour lines (12 feet).

Step 2. Scale the distance from the unknown point to the nearest contour line (4 feet).

Step 3. Multiply the contour interval by the fraction of the distance between the contour lines to the unknown point. (Contour interval = 2 feet; fraction of distance between contour lines = 4/12 = 1/3. Therefore, 2 X 1/3 = 2/3 feet.)

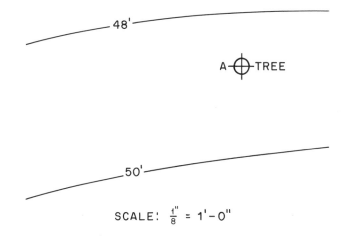

SCALE: $\frac{1}{8}"$ = 1'-0"

Figure 11-8 Interpolate the elevation of the tree.

Step 4. If the nearest contour line is below the other one, add this to it. If the nearest contour line is above the other one, subtract this amount. (Nearest contour = 48'. This is below the other contour line at 50', so 2/3' is added to 48'. 2/3' + 48' = 48.66'.)

√ CHECK YOUR PROGRESS

Can you perform these tasks?

- ☐ Identify each tree to be removed from a site.
- ☐ Give the natural grades and finished grades of all points on a site.
- ☐ Explain which direction water will naturally run from any point on a site.

ASSIGNMENT

Refer to the Lake House drawings (in the packet) to complete this assignment.

1. How many trees are indicated for removal?

2. How many trees are to remain on the site? (Do not include wooded areas.)

3. What is the finished grade elevation at the tree nearest the Lake House?

4. What was the natural-grade elevation of the most easterly tree to be saved?

5. What is the elevation of the tree to be saved nearest the lake?

6. What is the natural-grade elevation at the southwest corner of the Lake House? Do not include the deck as part of the house.

7. What is the finished-grade elevation at the southwest corner of the house?

8. How much cut or fill is required at the entrance to the garage?

9. Is cut or fill required at the southwest corner of the house? How much?

10. What is the elevation at the northeast corner of the site?

UNIT 12   *Locating the Building*

OBJECTIVES

After completing this unit you will be able to perform the following tasks:

- Lay out building lines according to a site plan.

- Use the 6-8-10 or equal-diagonals method to check the squareness of corners.

- Use a leveling instrument to measure angles and depths of excavations.

The location of a building in relation to its site is important. The architect designs the building to complement its surroundings. In fact, it may be a mistake to consider a building separately from its site. Often the site serves functional roles in the total design. Orientation to the sun, direction of prevailing winds, position of trees, and the effects of neighboring buildings can be used to achieve maximum energy efficiency and comfort.

As was pointed out in Unit 10, trees can provide valuable sun control. Deciduous trees on the south side of the house keep the heat of the sun off the house in summer. In winter when the leaves are gone, the sun shines through the trees to warm the house. However, this principle depends on the proper position of the house in relation to the trees.

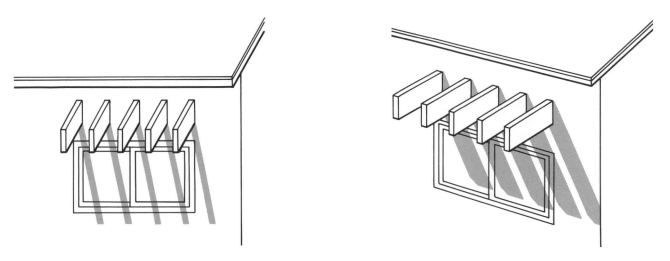

Figure 12-1 A slight change in the direction this sun screen faces makes a great difference in the sun striking the window.

It is also important that the house faces the right direction — not only for the best view, but for the best use of the heat of the sun. Changing the direction that a house faces by only a few degrees can affect the amount of solar heat received through windows and sun screens, Figure 12-1.

The location may use air currents or prevailing winds to advantage. Moving the house shown in Figure 12-2 farther up the hill could raise the outside temperature 10°F in the summer. This would increase the cost of air conditioning.

In addition to affecting energy requirements, the location of a building may be controlled by

local laws. Many communities have minimum *set-back* requirements governing how close the house can be to the street. In the locality of the Lake House, local ordinances prohibit building within 60 feet of the lake.

LAYING OUT BUILDING LINES

The position of the building is shown on the site plan. Dimensions show the distance from the street (or lake) to the building and from the side boundaries to the building. The location of one corner can easily be found by measuring with a

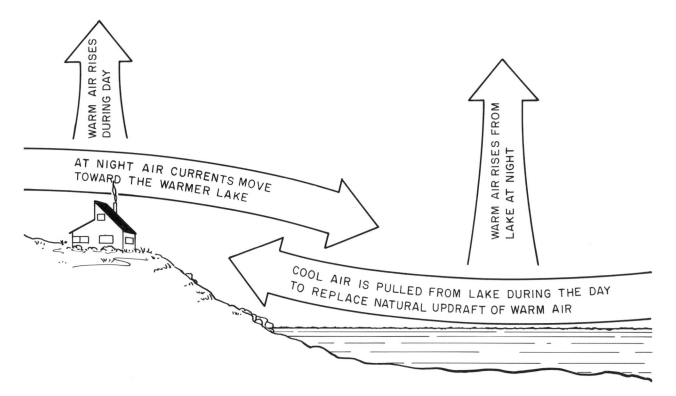

Figure 12-2 The location of a house can affect the heating and cooling load.

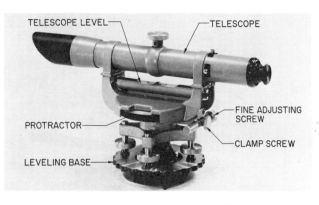

Courtesy of Keuffel & Esser Co.
Figure 12-3 Builder's level

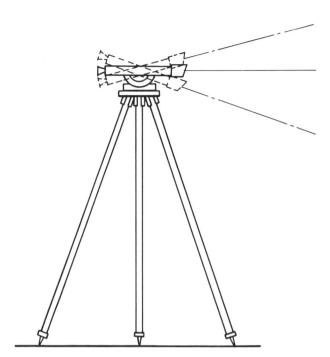

Figure 12-4 A transit is similar to a builder's level, but can be tilted to measure vertical angles.

long (100' or 200') steel tape. The most efficient way to find all remaining corners is by the use of a leveling instrument, Figure 12-3. The functions of the parts of a leveling instrument are as follows:

- *Telescope* contains the lens, focusing adjustment, and cross hairs for sighting.

- *Telescope level* is a spirit level used for leveling the instrument prior to use.

- *Clamp screw* locks the instrument in position horizontally.

- *Fine adjusting screw* makes fine adjustments in a horizontal plane.

- *Leveling base* holds the leveling screws (usually four) for leveling the instrument prior to use.

- *Protractor* is a scale graduated in degrees and minutes for measuring horizontal angles.

Two accessories are required for most operations performed with a leveling instrument. The *tripod* is a three-legged stand that provides a stable base for the instrument. A *target rod* is a separate device with a scale graduated in feet and tenths of a foot. The telescope is focused on the target rod to measure elevations. The builder's level is a device for checking the difference in elevation between two points. It can also be used for measuring angles on a horizontal plane. Another instrument, a *transit*, can be tilted to measure angles in a vertical plane, Figure 12-4. The procedure described here for laying out square corners can be used with either instrument:

Step 1. Set the tripod up over a known corner. The exact position is determined by hanging a plumb bob from the tripod.

The legs should be firmly set in the ground about three feet apart.

Step 2. Set the instrument on top of the tripod and hand tighten the clamp screw.

Step 3. Turn the leveling screws down, so they contact the tripod plate.

Step 4. Turn the telescope so that it is over one pair of leveling screws. Adjust these two screws so that the telescope is level.

Step 5. Rotate the telescope so that it is over the other pair of leveling screws. Adjust these two screws to level the telescope.

Step 6. Repeat this over each pair of leveling screws, until the telescope is level in all positions.

Step 7. Using a compass, carefully point the telescope to magnetic north. The north arrow on the drawing can point to either true north or magnetic north. The compass needle will point only to magnetic north. Because the true north pole of the earth is some distance from its magnetic north, the difference between bearings based on true north and those based on magnetic north can be several degrees. If the building is to be laid out according to true-north bearings, not magnetic compass bearings, contact the architect or surveyor to find out what correction should be made in your area.

Step 8. Set the protractor at zero degrees.

Step 9. Rotate the telescope to the bearing of one building line.

Step 10. Stretch a line (string) from the existing stake to several feet beyond the next corner.

Step 11. Have a partner hold the target rod plumb over the far end of this line. When the telescope cross hairs can be focused on the target rod, the bearing of the line is correct.

Step 12. Measure the length of this building line from the first stake and drive another stake.

Step 13. To lay out each of the remaining corners, set the tripod over the stake; line the telescope up on the existing line; and then use the protractor of the instrument to measure a 90° corner.

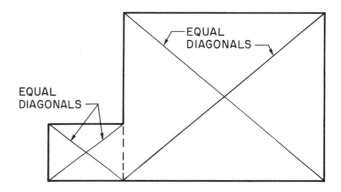

Figure 12-5 When the diagonals of a rectangle are the same length the corners are square

To check the squareness of the layout, measure the diagonals of a rectangle formed by the layout, Figure 12-5. When all four corners of the rectangle are 90°, the diagonals are equal.

Another method of checking a 90° angle is called the *6-8-10 method*, Figure 12-6. Measure 6 feet from the corner along one line. Measure 8 feet from the corner along the other line. These points should be 10 feet apart. (See Math Review 19.)

The building lines can be saved, even after the corner stakes are removed for earthwork, by erecting batter boards, Figure 12-7. *Batter boards* are sturdy horizontal boards fastened between 2x4 stakes, at least 4 feet outside the building lines. The building lines are extended and marked on the batter boards.

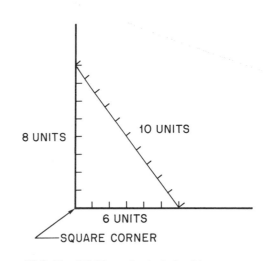

Figure 12-6 The 6-8-10 method of checking a square corner

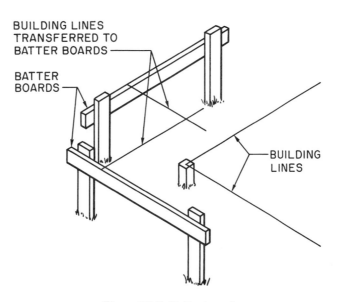

Figure 12-7 Batter boards

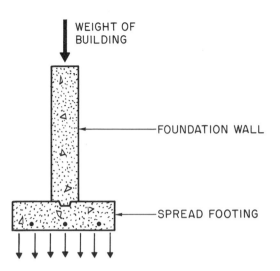

Figure 12-8 The footing spreads the weight of the building over a greater soil area.

EXCAVATING

Most buildings require some *excavation* (digging) to prepare the site for a foundation. The depth of the excavation at any point is the difference between the grade elevation and the elevation at the bottom of the excavation. If the excavation is done after the rough grading, the depth is measured from the finished grade. If the rough grading has not been done, the excavation is measured from the natural grade.

Concrete footings are placed in the bottom of the excavation to support the entire weight of the building, Figure 12-8. These footings are placed on unexcavated earth to reduce the chance of the soil compacting under them. This means that the excavation contractor must measure the depth of the excavation accurately. The footings may be *stepped*, as in Figure 12-9, to accommodate a sloping site. This requires measuring the depth at each step of the footing. Information about the footing design is found on the foundation plan and building elevations. The layout of the foundation walls and their footings is shown on the foundation plan. The foundation walls are shown by two solid lines with dimensions to indicate their sizes, Figure 12-10. A dotted line on each side of the foundation indicates the concrete footing. The size of the footing may be omitted when the plan was developed for use in several locations.

The depth of the foundation, including its footing, is shown on the elevations. To simplify calculating excavation depths, many architects

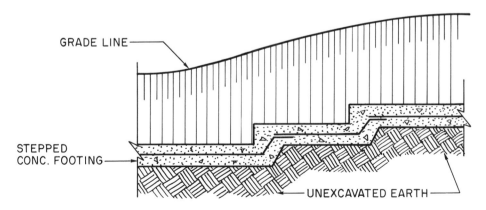

Figure 12-9 Footings can be stepped to accommodate a sloping site.

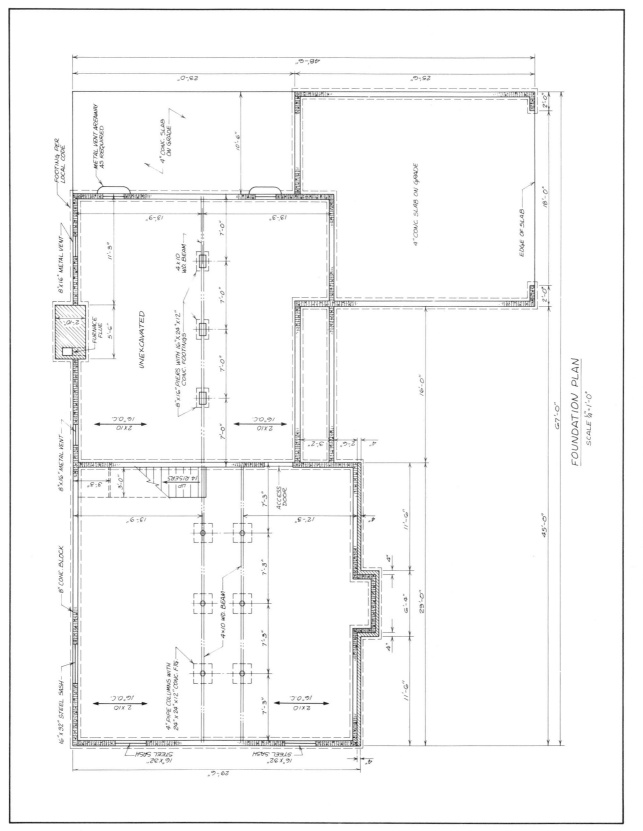

FOUNDATION PLAN
SCALE ½"=1'-0"

Figure 12-10 The foundation plan gives complete dimensions.

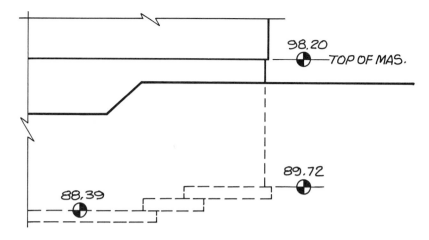

Figure 12-11 Spot elevations to show key points on footing and foundation

indicate the elevation at key points along the footings, Figure 12-11. A section view through all or part of the building may show a typical depth, but it is wise to check all of the elevations for steps in the footing. The footings may be shown on the elevation as a double or a single dotted line. Steps in footings are usually in increments of 8 inches to conform to standard concrete block sizes.

Use the following procedure to measure differences in elevation, such as the depth of an excavation:

Step 1. Set the instrument up on a tripod and level it in a convenient location.

Step 2. Have a partner hold the target rod on a known elevation, such as a bench mark or ground of known elevation, while you focus the telescope and note the reading on the target rod where the cross hairs focus.

Step 3. Take a similar reading with the target rod at the bottom of the excavation. The difference in the two readings is the depth of the excavation, Figure 12-12.

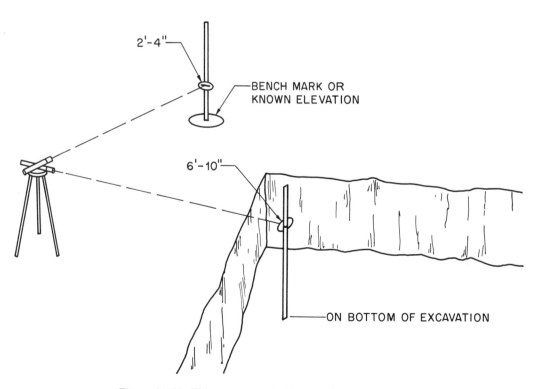

Figure 12-12 This excavation is 4'-6" deep — the difference in the two readings on the target rod.

✓ CHECK YOUR PROGRESS ────────────────────

Can you perform these tasks?

☐ Properly set up a builder's level.

☐ Use a builder's level and tape measure to lay out building lines.

☐ Check the accuracy of right angles using a tape measure only.

☐ Check the squareness of a building layout using a tape measure only.

☐ Calculate the depth excavation required for footings.

☐ Measure the depth of an excavation using a builder's level.

ASSIGNMENT ────────────────────────────

Refer to the Lake House drawings (in the packet) to complete this assignment.

1. What is the distance from the Lake House to the nearest property boundary? (Do not treat the decks as part of the house for this question.)

2. What is the distance from the Lake House to the lake?

3. What is the distance from the north property line to the garage?

4. What is the area of the basement of the Lake House, including the foundation? Ignore slight irregularities in the shape of the foundation. For ease in calculating, divide the foundation into rectangles, Figure 12-13.

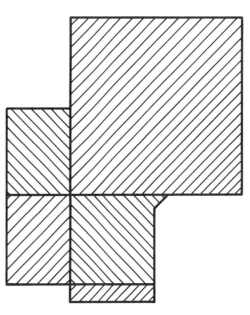

Figure 12-13 Use with assignment problem 4.

5. Find the highest and lowest natural grades meeting the house. Using the midpoint between the highest and lowest natural grades as a reference, determine the average depth of the excavation for the basement floor.

6. Approximately how many cubic yards of earth will have to be removed from the Lake House excavation? (See Math Review 28.)

7. Measuring from the natural grade, how deep is the excavation for the footing under the overhead garage door?

8. What is the elevation at the bottom of the deepest excavation for the Lake House? (Do not include the garage.)

9. Why would a row of large evergreen trees between the Lake House and the lake decrease the energy efficiency of the house?

10. What aspect of the location of a building is most often regulated by local ordinances?

UNIT 13 Site Utilities

OBJECTIVES

After completing this unit, you will be able to perform the following tasks:

- Interpret symbols and notes used to describe site utilities.
- Explain the septic system indicated on a site plan.
- Determine the pitch of drain lines.

SEWER DRAINS

The *building sewer* carries the waste to the municipal sewer or septic system, Figure 13-1. Because sewer lines usually rely on gravity flow, they are large in diameter (4 inches, minimum) and are pitched to provide flow. Because water supply lines and gas lines are pressurized, pitch is not important in their installation. Therefore, the sewer is installed first and other piping is routed around it as necessary. The size, material, and pitch of drains are usually given in a note on the site plan, Figure 13-2. The pitch of a pipe is given in fractions of an inch per foot. A pitch of 1/4 inch per foot means that for every horizontal foot, the pipe rises or falls 1/4 inch.

In some cases, sewers may have to flow uphill. This is the case with the Lake House. Uphill flow is accomplished by a *sewage pump* or *ejector*, Figure 13-3.

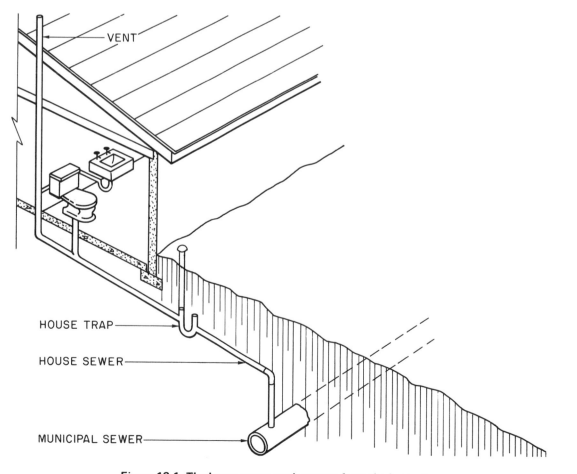

VENT

HOUSE TRAP

HOUSE SEWER

MUNICIPAL SEWER

Figure 13-1 The house sewer carries waste from the house to the municipal sewer or septic tank.

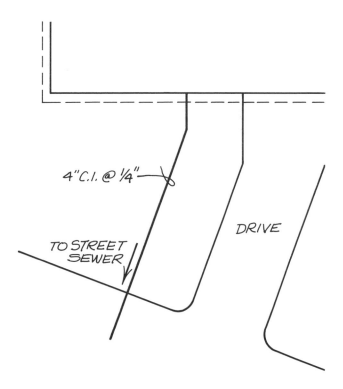

4"C.I. @ ¼"

TO STREET SEWER

DRIVE

**Figure 13-2 This note indicates cast iron pipe
pitched 1/4 inch for every foot of run.**

BUILDING SEWER

Any plumbing that is to be concealed by concrete work must be installed without fixtures or *roughed in* before the concrete is placed. Because the plumbing contractor installs all plumbing inside the building lines, this phase of construction is discussed later with mechanical systems in Section 12. However, employees of the general contractor may rough in the sewer from the building line to the street or septic system.

These workers must be able to determine the elevation at which the sewer passes through the foundation and the pitch of the line outside the building. The sewer line may be shown on plans as a solid or broken line. Although it is usually labeled, this is not always true. When the sewer is not labeled as such, it can still be recognized by its material, pitch, and ending place. In light construction, the sewer is usually the only 4-inch pipe to the building. Also, the sewer is the only line with the pitch indicated.

MUNICIPAL SEWERS AND SEPTIC SYSTEMS

In highly developed areas, the building sewer empties into a municipal sewer line in or near the street. The contractor for the new building is responsible for everything from the municipal sewer to the house.

In less developed areas, the sewer carries the sewage to a septic system. The *septic system* in-

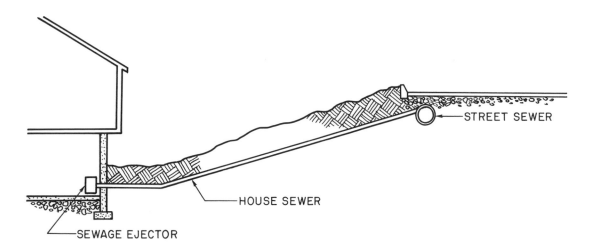

STREET SEWER

HOUSE SEWER

SEWAGE EJECTOR

Figure 13-3 A sewage ejector is a pump which moves sewage uphill.

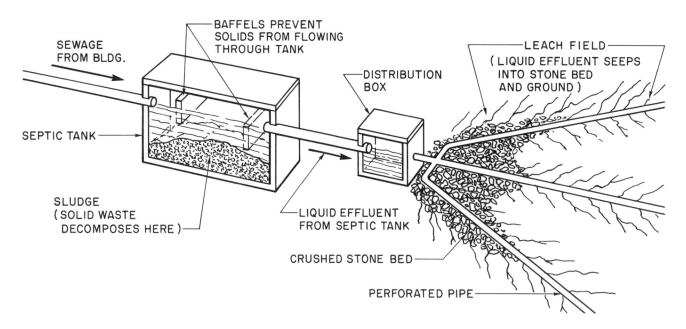

Figure 13-4 Septic system

cludes a septic tank and drain field, Figure 13-4. The septic tank holds the solid waste while it is decomposed by bacterial action. The liquids pass through the baffles and flow out of the tank to the distribution box. The distribution box (D.B.) diverts the liquid into *leach lines*. These perforated plastic or loose-fitting tile lines allow the liquid to be absorbed by the surrounding soil. The liquid gradually evaporates from or drains through the soil. The *drain field*, where the leach lines are laid, is usually made of a layer of crushed stone.

The design of septic systems is closely regulated by most health and plumbing codes. These local codes should be checked before designing or installing any septic system. The building code or health department code often requires a percolation test before the system can be approved. In a *percolation test*, holes are dug. Then a measured amount of water is poured into each hole. The amount of time required for the water to drain into the soil is an indication of how well the soil will accept water from the septic system. This ability to accept water is called *percolation*. The locations of key elements in the system are often shown on the site plan.

OTHER PIPING

Other utility piping, such as for water supply or gas, is shown on the site plan. If these lines will pass beneath the concrete footings or be concealed under a concrete slab, they must be roughed in before the concrete is placed. Water supply pipes follow the most direct route from the municipal water main or well to the main shut-off valve or pump. Gas lines run from the main to the gas meter. All supply lines on the plot plan should be labeled according to type and size of piping, Figure 13-5.

ELECTRICAL SERVICE

The electrical service is the wiring that brings electricity to the house. There are two types of electrical service: overhead and underground (or buried). Overhead service involves a cable from the utility company transformer or pole to a weatherhead on the house, Figure 13-6. The *weatherhead* is a weathertight fitting on the *mast* (usually a pipe) which serves as a conduit to the meter receptacle. In an underground service the cable is buried, Figure 13-7. Although electrical service is a site utility, it is not usually installed until the building is enclosed.

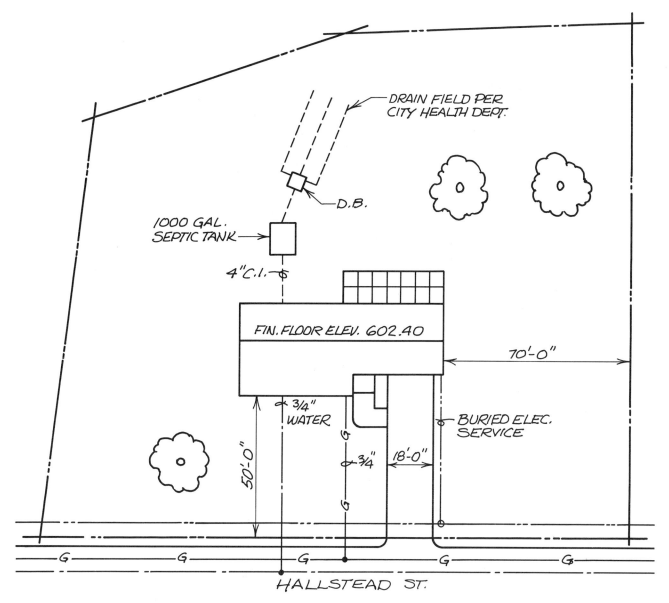

Figure 13-5 Partial site plan with utilities indicated

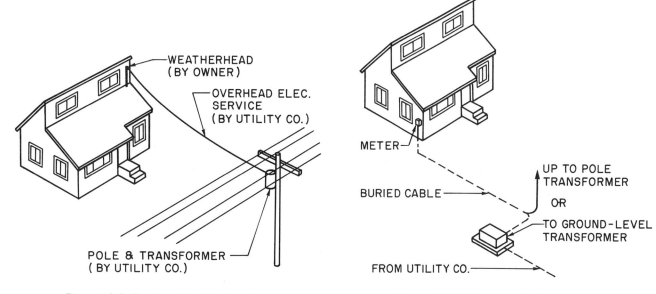

Figure 13-6 Overhead electrical service

Figure 13-7 Underground electrical service

✓ CHECK YOUR PROGRESS

Can you perform these tasks?

- ☐ Identify electrical service, water, and sewage lines shown on a site plan.
- ☐ Explain the operation of a septic system.
- ☐ Tell at what elevation sewer lines are to pass through a foundation.
- ☐ Tell what the elevation should be at each end of a pitched sewer pipe.

ASSIGNMENT

Refer to the Lake House drawings (in the packet) to complete the assignment.

1. What size is the sewer for the Lake House?

2. How many lineal feet are required from the foundation wall to the septic tank?

3. What is the rise of the sewer from the house to the septic tank?

4. Where does the sewer pass through the foundation?

5. How many lineal feet of perforated pipe are needed for the drain field?

6. How many cubic yards of crushed stone are needed for the drain field?

Foundations

Section 4

UNIT 14   Footings

OBJECTIVES

After completing this unit, you will be able to perform the following tasks:

- Find all information on a set of drawings pertaining to footing design.
- Interpret drawings for stepped footings used to accommodate changes in elevation.
- Discuss applicable building codes pertaining to building design.

All soil can change shape under force. When the tremendous weight of a building is placed on soil, the soil tends to compress under the foundation walls and allow the building to settle. To prevent settling, concrete footings are used to spread the weight of the building over more area. The footings distribute the weight of the building, so that there is less force per square foot of area.

The simplest type of footing used in residential construction is referred to as *slab-on-grade*. In this system the main floor of the building is a single concrete slab, reinforced with steel to prevent cracking. This slab supports the weight of the building, Figure 14-1. Slab-on-grade foundations are common in warm climates. This type of construction is indicated on the floor plan by a note, Figure 14-2, and on section views of the construction, Figure 14-3. If excavation is involved in the construction where a slab-on-grade is to be placed, it is very important to thoroughly compact all loose fill before placing the concrete. Tamping the fill prevents the soil from compacting under the concrete later, causing the concrete to settle or crack.

A haunch is used to further strengthen the slab where concentrated weight, such as a wall, will

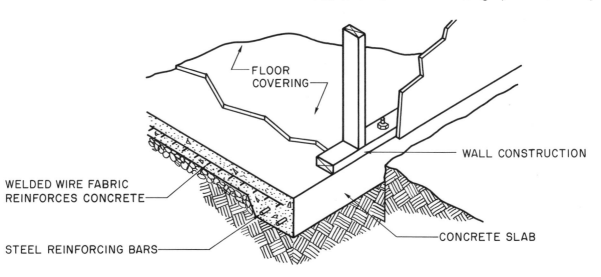

Figure 14-1 Slab-on-grade foundation

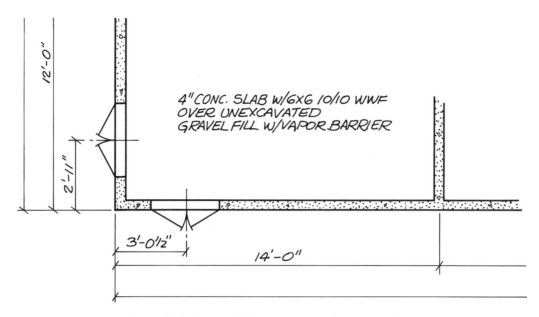

Figure 14-2 Note indicating slab-on-grade construction

be located. A *haunch* is an extra thick portion of the slab that is made by ditching the earth before the concrete is placed, Figure 14-4.

SPREAD FOOTINGS

In most sections of the country, the foundation of the house rests on a footing separate from the concrete floor. This separate concrete footing is called a *spread footing* because it spreads the force of the foundation wall over a wider area, Figure 14-5. Spread footings may be made by placing concrete inside wooden or metal forms, Figure 14-6, or placing the concrete in carefully measured ditches. In either case, the footing is shown on the foundation or basement plan by dotted lines outside the foundation wall lines, Figure 14-7.

The dimensions of the footings can be determined from the dimensions shown for the foundation. The foundation rests on the center of the footing unless otherwise specified. Therefore, if an 8-inch foundation rests on a 16-inch footing, the footing projects four inches beyond the foundation on each side. To lay out these footing lines, measure four inches from the building lines marked on the batter boards. Where footing lines cross to form a corner, suspend a plumb bob. Drive a stake under the plumb bob. Then, drive a nail in the stake to accurately mark the corner. When all corners are located in this manner, stretch a line between the nails to locate the inside of the footing forms.

A complete set of construction drawings also includes sections that show the spread footing in

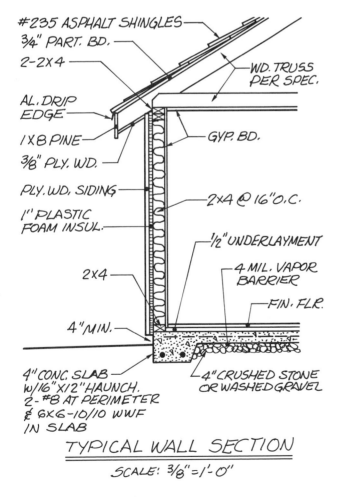

Figure 14-3 Typical wall section for slab-on-grade construction

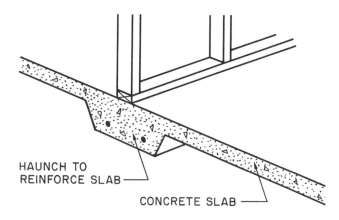

HAUNCH TO
REINFORCE SLAB

CONCRETE SLAB

Figure 14-4 A *haunch* is a thickened part of a slab to reinforce it under a load-bearing wall

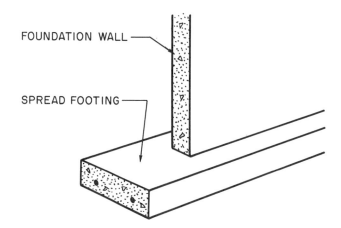

FOUNDATION WALL

SPREAD FOOTING

Figure 14-5 A spread footing is so named because it "spreads" the weight of the foundation wall over a wider soil area.

Figure 14-6 Footing forms

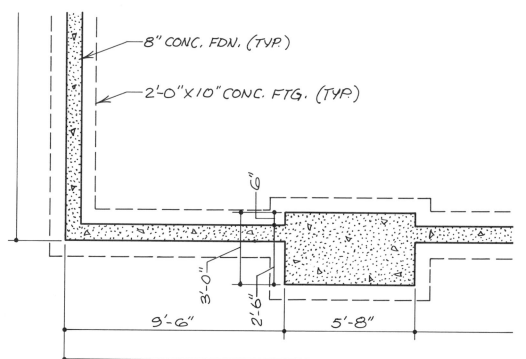

8" CONC. FDN. (TYP.)

2'-0"X10" CONC. FTG. (TYP.)

6"

3'-0"

2'-6"

9'-6"

5'-8"

Figure 14-7 The footing lines are sometimes shown on the foundation plan by broken lines around the foundation.

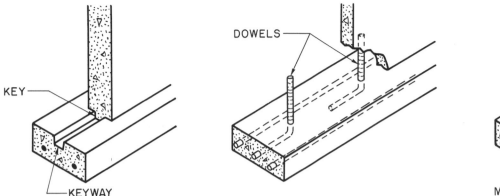
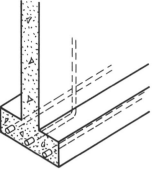

KEY

KEYWAY

DOWELS

MONOLITHIC (ONE PIECE)

Figure 14-8 Anchoring the foundation wall to the footing

SECTION 1009.0 CONCRETE FOOTINGS

1009.1 Concrete strength: Concrete in footings shall have an ultimate compressive strength of not less than 2500 psi (1.76 kg/mm^2) at 28 days.

1009.2 Design: Concrete footings shall comply with Sections 1215.0 and 1216.0 and ACI 318 listed in Appendix A.

1009.3 Thickness: The thickness of concrete footings shall comply with Sections 1009.3.1 and 1009.3.2.

1009.3.1 Plain concrete: In plain concrete footings, the edge thickness shall be not less than 8 inches (203 mm) for footings on soil: except that for buildings of Use Group R-3 and buildings less than two stories in height of Type 4 construction, the edge thickness may be reduced to 6 inches (152 mm) provided the footing does not extend beyond 4 inches (102 mm) on either side of the supported wall.

1009.3.2 Reinforced concrete: In reinforced concrete footings the thickness at the edge above the bottom reinforcement shall be not less than 6 inches (152 mm) for footings on soil, nor less than 12 inches (305 mm) for footings on piles. The clear cover on reinforcement where the concrete is cast against the earth shall not be less than 3 inches (76 mm). Where concrete is exposed to soil after it has been cast, the clear cover shall be not less than 1 1/2 inches (38 mm) for reinforcement smaller than No. 5 bars of 5/8 inch (16 mm) diameter wire, nor 2 inches (51 mm) for larger reinforcement.

1009.4 Footings on piles and pile caps: Footings on piles and pile caps shall be of reinforced concrete. The soil immediately below the pile cap shall not be considered as carrying any vertical load. The top of all piles shall be embedded not less than 3 inches (76 mm) into pile caps and the caps shall extend at least 3 inches (76 mm) beyond the edge of all piles.

1009.5 Deposition: Concrete footings shall not be poured through water unless otherwise approved by the building official. When poured under or in the presence of water, the concrete shall be deposited by approved means which insure minimum segregation of the mix and negligible turbulence of the water.

1009.6 Protection of concrete: Concrete footings shall be protected from freezing during depositing and for a period of not less than 5 days thereafter. Water shall not be allowed to flow through the deposited concrete.

Reproduced from THE BOCA BASIC BUILDING CODE with permission of Building Officials and Code Administrators International, Inc.

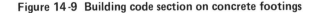

Figure 14-9 Building code section on concrete footings

greater detail. However, these drawings are often superceded by local building codes. Building codes for footings include such things as minimum permissible depth of footing, required strength of concrete for footings, the width of footings, and the use of keys or dowels, Figure 14-8. The drawings for the Lake House are drawn to satisfy the building codes that are enforced in the community where it is to be built. If the drawings were done for a plan catalog where the locality is not known in advance, they might refer the builder to local building codes and omit many dimensions. Figure 14-9 shows an example of a building code section on footings.

When reading the foundation plan for footing dimensions, pay particular attention to special features like fireplaces and pilasters. (*Pilasters* are thickened sections of the foundation wall that add strength to the wall.) The footing in these areas will probably be wider and may be deeper than under straight sections of foundation wall.

COLUMN PADS

Where steel columns, masonry piers, and wooden posts are used in the construction, a special concrete pad is indicated on the foundation plan, Figure 14-10. As with other footings, building codes for the area should be consulted for the design of these pads. In the absence of any applicable building code, the foundation plan normally includes enough information to construct the pads. These pads, as with all other footings, should rest on unexcavated or well-tamped earth.

REINFORCEMENT

Footings, column pads, and other structural concrete frequently include steel reinforcement. Footing reinforcement is normally in the form of steel reinforcement bars, commonly called *rebars*. Reinforcement bars are designated by their diameters in eighths of an inch, Figure 14-11.

DEPTH OF FOOTINGS

In many sections of North America, the moisture in the surface of the earth freezes in the winter. As this frost forms, it causes the earth to expand. The force of this expansion is so great that if the earth under the footing of a building is allowed to freeze, it either cracks the footing or moves the building. To eliminate this problem, the

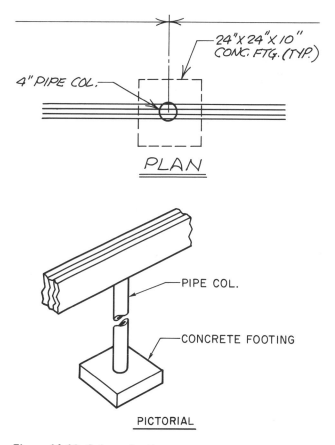

Figure 14-10 Column footings appear as a rectangle of broken lines on the plan.

REINFORCEMENT BARS	
Size Designation	Diameter in Inches
3	.375
4	.500
5	.625
6	.750
7	.875
8	1.000
9	1.128
10	1.270
11	1.410
14	1.693
18	2.257

Figure 14-11 Standard sizes of rebars

footing is always placed below the depth of any possible freezing. This depth is called the frostline, Figure 14-12.

Two methods are commonly used to indicate the elevation, or depth, of the bottom of the footings. The easiest to interpret is with the elevation shown directly referenced to the footing on the elevation drawings. Where elevations are given in

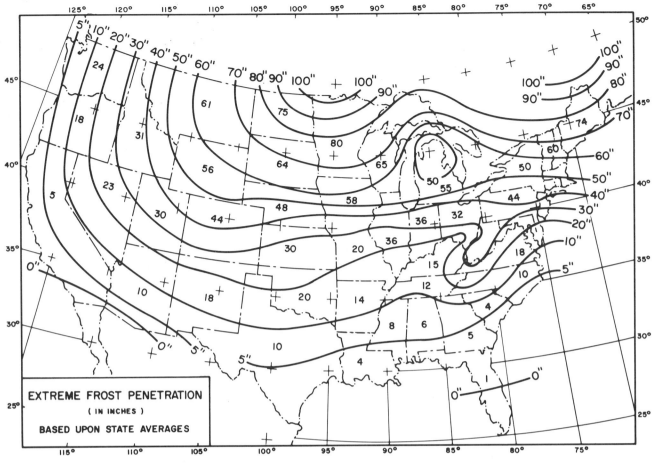

Figure 14-12 Average frost depths in the United States

this manner, the top of the footing forms are leveled with a leveling instrument, using a bench mark for reference.

The most frequently used method is to dimension the bottom of the footing from a point of known elevation. This may be the finished floor or the top of the masonry foundation, for example. These dimensions are given on the building elevations. Footings and other features marked with a reference symbol (⊕ , called a *datum* symbol) are to be used as reference points for other dimensions.

For an example, see the South Elevation 3/3 of the Lake House (included in the packet). The left end of this view shows a footing with its bottom at an elevation of 334.83 feet. This is a variation of the usual practice of showing the elevation of the top of the footing. The top of this footing is 10 inches higher or 335.66 feet. (See Math Review 15.) What room of the Lake House is this footing under?

At the right end of the South Elevation 3/3, the footing is shown to be 5' - 4" below the finished floor and masonry. This is the basement floor, which the Site Plan shows to be at 337.0 feet.

Therefore, the top of this footing is at 331' - 8" or 331.66'.

STEPPED FOOTINGS

On sloping building sites, it is necessary to change the depth of the footings to accommodate the slope. This is done by *stepping* the footings. When concrete blocks are to be used for the foundation walls, these steps are normally in increments of eight inches. This allows the concrete blocks to be laid so that the top of each footing step is even with a masonry course. Some buildings require several steps in the footing to accommodate steeply sloping sites. Steps in the footings are shown on the elevation drawings and on the foundation plan by a single line across the footing.

The Lake House has several steps in the footing. For example, see the east side of the garage. This is shown in the East Elevation 2/3. This step is also shown on the Foundation Plan. It is 8' - 0" from the north end of the garage. Notice that the two levels of the footing overlap one another. These are built in one overlapping section as shown in Figure 14-13.

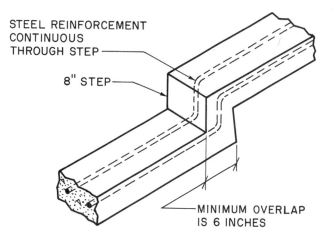

STEEL REINFORCEMENT
CONTINUOUS
THROUGH STEP

8" STEP

MINIMUM OVERLAP
IS 6 INCHES

Figure 14-13 Stepped footing

✓ CHECK YOUR PROGRESS

Can you perform these tasks?
- ☐ List the dimensions of a slab-on-grade including footings, when this type of foundation is shown on construction drawings.
- ☐ List the thickness, width, and reinforcement to be used for spread footings.
- ☐ Identify steps in footings and give the dimensions of each step.
- ☐ Give the locations and dimensions of footings for columns, posts, and other features.
- ☐ Give the dimensions and reinforcement to be used for any thickened haunches in a concrete slab-on-grade.

ASSIGNMENT

Refer to drawings of the Lake House (included in the packet) to complete this assignment.

1. What is the typical width and depth of the concrete footings for the Lake House?

2. What is the total length and width (outside dimensions) of the concrete footings for the garage of the Lake House? (Remember to allow for the footings to project beyond the foundation wall.)

3. How many concrete pads are shown for footings under columns or piers in the Lake House?

4. What are the dimensions of these pads?

5. What reinforcement is indicated for these pads?

6. What reinforcement is indicated for the spread footings under the Lake House?

7. What is indicated by the 2-inch dimension between the 18-inch round concrete footings?

8. What is the elevation of the top of the footing under the garage door?

9. What are the elevations of the tops of each section of concrete footing shown on the East Elevation 2/3?

10. How far outside the foundation walls are the typical footings?

11. Refer to the building code in your community (or the model code section shown in Figure 14-9) and list the specific differences between the Lake House footings and the minimum code requirements.

UNIT 15 Foundation Walls

OBJECTIVES

After completing this unit, you will be able to perform the following tasks:

- Determine the locations and dimensions of foundation walls indicated on a set of drawings.

- Describe special features indicated for the foundation on a set of drawings.

LAYING OUT THE FOUNDATION

When the concrete for the footings has hardened and the forms are removed, carpenters can begin erecting forms for concrete foundations or masons can begin laying blocks or bricks for masonry foundations. Although the material differs, the drawings and their interpretation for each type of foundation are similar.

In Unit 14 you referred to the dimensions on the foundation plan to lay out the footings. The same dimensions are used to lay out the foundation walls. The layout process is also similar. The outside surface of the foundation wall is laid out using previously constructed batter boards. Then the forms are erected or the masonry units are laid to these lines. The foundation plan includes overall

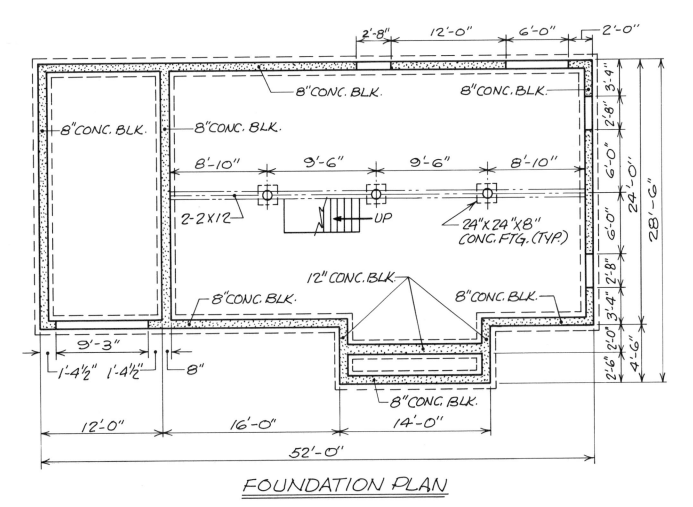

FOUNDATION PLAN

Figure 15-1 Dimensioning on a foundation plan

dimensions, dimensions to interior corners and special constructions, and dimensions of special smaller features. It is customary to place the smallest dimensions closest to the drawing. The overall dimensions are placed around the outside of the drawing, Figure 15-1.

All drawing sets include, at least, a wall section showing how the foundation is built, how it is secured to the footings, and any special construction at the top of the foundation wall, Figure 15-2. Although a typical wall section may indicate the thickness of the foundation wall, you should carefully check around the entire wall on the foundation plan to find any notes that indicate varying thicknesses of the foundation wall. For example, the wall may be 12 inches thick where it has to support brick veneer above, while it is only 8 inches thick on the back of the building where there is no brick veneer. A careful check of the foundation plan for the Lake House shows that the house foundation calls for 12-inch, 10-inch, and 6-inch concrete block and that the garage foundation calls for 8-inch block.

The details for a masonry foundation may call for reinforcement in every second or third course. This is usually prefabricated wire reinforcement to be embedded in the mortar joints. Prefabricated wire reinforcement is available in varying sizes for different sizes of concrete blocks.

The height of the foundation wall is dimensioned on the building elevations. These are the same dimensions as those used to determine the depth of the footings in the preceding unit. Just as the footing was stepped to accommodate a sloping building site, the top of the foundation wall may be stepped to accommodate varying floor levels in the *superstructure* (construction above the foundation), Figure 15-3.

The top of a masonry foundation may be built with smaller concrete blocks to form a ledge upon which later brickwork will be built. It is also common practice to use one course of 4-inch solid block as the top course of the foundation wall.

In concrete foundations, anchor bolts, are usually placed in the top of the foundation, Figure 15-4. These bolts are left protruding out of the top of the foundation so that the wood superstructure can be fastened in place later. Anchor bolts are not normally shown on the foundation plan, but a note on the wall section indicates their center-to-center or on-center spacing. On masonry walls, anchor bolts can be placed in the hollow cores of the concrete blocks. They are held in place by filling the core

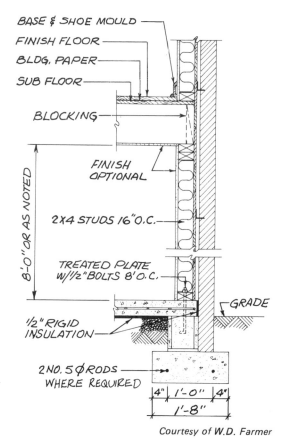

Courtesy of W.D. Farmer

Figure 15-2 Section through foundation

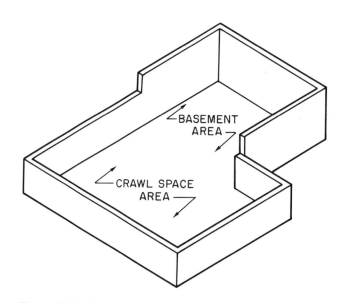

Figure 15-3 The top of the foundation may be stepped to allow for a partial basement or varying floor levels.

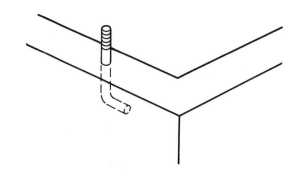

Figure 15-4 Anchor bolt

with mortar grout. *Grout* is a portland cement mixture which has high strength.

In areas where there is a threat of extremely high winds or earthquakes, additional hold-down straps may be called for, Figure 15-5. These hold-downs are normally only used with concrete foundations.

SPECIAL FEATURES

Many foundations include steel or wooden beams which act as girders to support the floor framing over long spans, Figure 15-6. When the girder is steel, it is indicated by a single line with a note specifying the size and type of structural steel. A wood girder is usually indicated by two or more lines and a note specifying the number of pieces of wood and their sizes in a built-up girder, Figure 15-7.

If the top of the girder is to be flush with the top of the foundation beam, pockets must be provided in the foundation, Figure 15-8. Beam pockets are usually shown on the details and sections of the construction drawings. The locations of these beam pockets are dimensioned on the foundation plans.

If windows are to be included in the foundation, the form carpenter or mason must provide rough openings of the proper size. The locations of windows should be dimensioned on the foundation plan. The sizes of the windows may be shown by a note or given on the window schedule. Window sizes

Figure 15-5 Anchor bolts and hold-down strap used in an earthquake zone

Courtesy of Trus Joist Corporation

Figure 15-6 The girder supports the floor framing.

COMMON SIZES OF STRUCTURAL STEEL GIRDERS

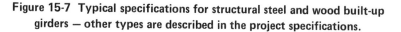

SHAPE: S = STANDARD BEAM, W = WIDE-FLANGE BEAM	NOMINAL DEPTH IN INCHES	WEIGHT PER FOOT IN POUNDS
S	10 X	35
S	8 X	23
S	8 X	18.4
W	10 X	33
W	10 X	21
W	10 X	11.5
W	8 X	31
W	8 X	28
W	8 X	17
W	8 X	13

COMMON SIZES OF WOOD BUILT-UP GIRDERS

NUMBER OF PIECES	THICKNESS EACH (NOMINAL) IN INCHES	WIDTH EACH (NOMINAL) IN INCHES
2 —	2 X	10
2 —	2 X	12
3 —	2 X	8
3 —	2 X	10
3 —	2 X	12

Figure 15-7 Typical specifications for structural steel and wood built-up girders — other types are described in the project specifications.

and window schedules are discussed later in Unit 26. It is important, however, to get the masonry opening size from the window manufacturer before forming the opening in the foundation wall. The *masonry opening* is the size of the opening required in the foundation wall to accommodate the window. This size may be different from the nominal size given in a note on the foundation plan.

The foundation may include pilasters for extra support. A *pilaster* is a thickened section of the foundation which helps it to resist the pressure exerted by the earth on the outside. The location and size of pilasters is shown on the foundation plan.

The Lake House drawing includes a special feature not commonly found on foundation plans for houses. There are four notes which read 3 1/2'' □ STD. WT. STL. COL. W/8x8x1/2 B.PL. These notes indicate a 3 1/2-inch square, standard-weight, steel column with 8 inch-by-8 inch-by-1/2 inch thick

base plates. This structural steel work is explained in more detail later, but to completely understand the foundation plan, it is necessary to know that the steel will be erected. The stress that this steel work will place on the foundation is the reason for another note on the foundation plan in the area near the 18-inch round concrete footings. That note says FILL CORE SOLID W/ GROUT 1:1:6 = PORTLAND: MAS. CEMENT:AGGREGATE. The cores in the concrete block foundation where the steelwork will be erected are filled with this grout to provide the extra strength necessary to support the steel.

ALL-WEATHER WOOD FOUNDATION

Foundations are usually constructed of concrete or concrete block. However, a type of specially treated wood foundation is sometimes used, Figure

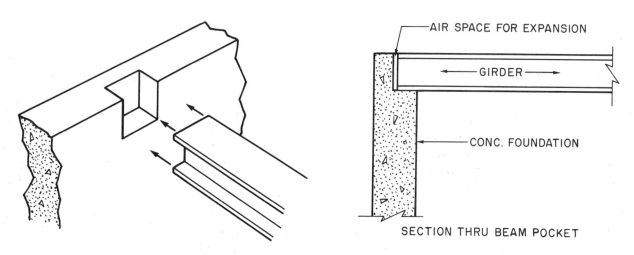

Figure 15-8 A beam pocket is a recess in the wall to hold the girder.

Courtesy of National Forest Products Association

Figure 15-9 All-weather wood foundation

15-9. These *all-weather wood foundations* do not use concrete footings.

Instead, they are built on 2x8s or 2x10s laid on gravel fill below the frostline. The foundation walls are framed with lumber that has been pressure treated to make it rot resistant and insect resistant. The framing is covered with a plywood skin and the plywood is covered with polyethylene film for complete moisture proofing, Figure 15-10.

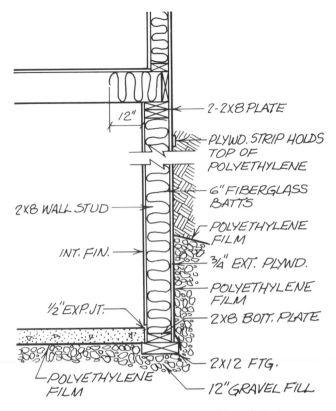

Figure 15-10 Section of wood foundation

√ CHECK YOUR PROGRESS

Can you perform these tasks?

☐ Give the length, width, and any offsets in a foundation wall.

☐ Name the material to be used for the foundation walls, including any reinforcement.

☐ Describe the locations and dimensions of all window openings, door openings, beam pockets, and other openings in the foundation walls.

☐ Tell what the elevation is at any point on the foundation wall.

☐ Describe anchor bolts and other tie-downs to be embedded in the foundation.

ASSIGNMENT

Refer to Lake House drawings (in the packet) to complete the assignment.

1. What is the typical thickness of the concrete block foundation for the Lake House?

2. How many lineal feet of block wall are included in the foundations of the Lake House and garage?

3. How thick is the south foundation of the fireplace?

4. What is the elevation of the top of the north end of the east foundation wall of the Lake House?

5. What is the highest elevation on the Lake House foundation?

6. How many courses of block are required at the highest elevation of the Lake House foundation?

7. What size anchor bolts are indicated at the top of the Lake House foundation?

8. What spacing is indicated for the anchor bolts?

9. How close (in courses) is horizontal reinforcement to be placed in the Lake House foundation?

10. What is the elevation of the top of the concrete block wall at the southwest 3 1/2"  steel column?

11. In how many places are the concrete block walls to be filled with grout?

UNIT 16   Drainage, Insulation, and Concrete Slabs

OBJECTIVES

After completing this unit, you will be able to perform the following tasks:

- Locate and explain information for control of ground water as shown on a set of drawings.

- Locate and describe subsurface insulation.

- Determine the dimensions of concrete slabs and the reinforcement to be used in concrete slabs.

DRAINAGE

After the foundation walls are erected and before the excavation outside the walls is *backfilled* (filled with earth to the finished gradeline), footing drains, if indicated, must be installed. *Footing drains* consist of either perforated plastic pipe or loose-fitting clay tile pipe placed around the footings in a bed of crushed stone, Figure 16-1. If the site has a natural slope, the footing drains can be run around

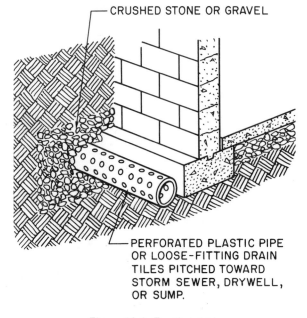

CRUSHED STONE OR GRAVEL

PERFORATED PLASTIC PIPE OR LOOSE-FITTING DRAIN TILES PITCHED TOWARD STORM SEWER, DRYWELL, OR SUMP.

Figure 16-1 Footing drain

Figure 16-2 Plastic drain pipe

the foundation wall to the lowest point, then away from the building to drain by gravity. In areas where there is no natural drainage, the drain is run to a dry well or municipal storm drain.

At one time, clay drain tile was the most common type of pipe for this purpose. However, perforated plastic pipe is used in most new construction. Plastic drain pipe is manufactured in 10-foot lengths of rigid pipe and in 250-foot rolls of flexible pipe, Figure 16-2. An assortment of plastic fittings is available for joining rigid plastic pipe. When footing drains are to be included, they are shown on a wall section or footing detail, Figure 16-3. A note on the drawing indicates the size and material of the pipe.

If floor drains are to be included in concrete slab floors, they are indicated by a symbol on the appropriate floor plan, Figure 16-4. If these floor drains run under a concrete footing, the piping had to be installed before the footing was placed (See Unit 13). However, it is better to run all piping above the footing whenever possible. The *riser* (vertical part through the floor) and drain basin are usually set at the proper elevation just prior to placing the concrete floor.

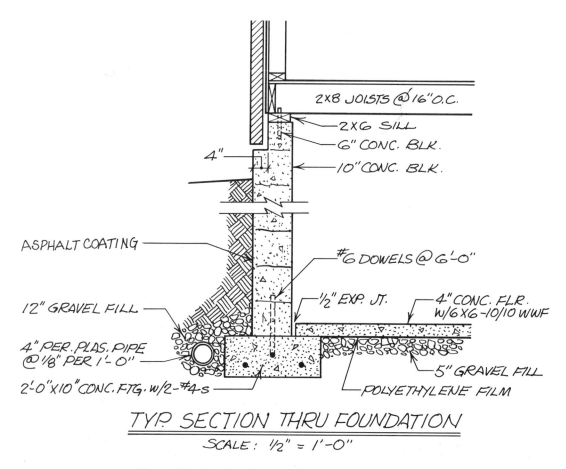

TYP. SECTION THRU FOUNDATION

SCALE: 1/2" = 1'-0"

Figure 16-3 Footing drains are shown outside the footing.

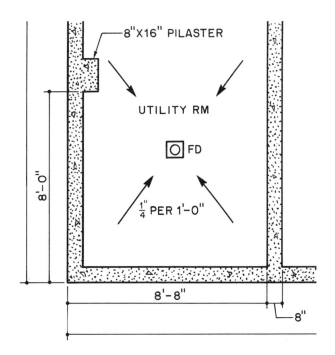

Figure 16-4 The floor drain is shown by a symbol. (Notice that the floor is pitched toward the drain.)

The floor plan may include a spot elevation for the finished drain, or it may be necessary to calculate it from information given for the pitch of the concrete slab. Pitch of concrete slabs is discussed later in this unit.

VAPOR BARRIERS

Another technique often used to prevent groundwater from seeping through the foundation is coating the foundation wall. In residential construction, the most common foundation moisture-control system is a combination of parging and coating with asphalt foundation coating. *Parging* consists of plastering the outside of the masonry wall below the grade line with a portland cement and lime plaster. The parging is then covered with asphalt-based foundation coating, Figure 16-5.

At this point, subsurface work outside the foundation wall is completed, but backfilling should not be done until the superstructure is framed. The weight and rigidity of the floor on the foundation wall helps the wall resist the pressure of the backfill. If the backfilling must be done before the framing, the foundation walls should be braced. To retard the flow of moisture from the earth through the concrete-slab floor, the drawings may call for a layer of gravel over the entire area before the concrete is placed. A polyethylene vapor barrier is laid over the gravel underfill. The thickness of polyethylene

sheeting is measured in mils. One mil equals 1/1000 of an inch. For vapor barriers, 4-mil or 6-mil polyethylene is generally used.

INSULATION

In cold climates, it is desirable to insulate the foundation and concrete slab. This insulation is

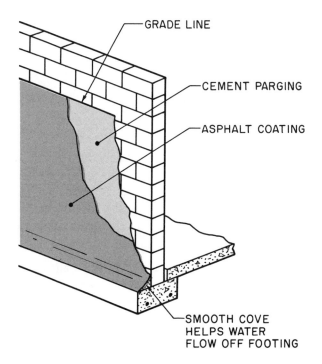

Figure 16-5 Cement parging and asphalt foundation coating are used for moisture proofing.

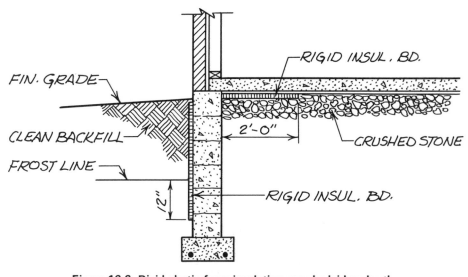

Figure 16-6 Rigid plastic-foam insulation may be laid under the perimeter of the floor or against the foundation wall.

usually rigid plastic foam boards placed against the foundation wall or laid over the gravel underfill, Figure 16-6.

Like all materials, concrete and masonry expand and contract slightly with changes in temperature. To allow for this slight expansion and contraction, the joint between the concrete slab and foundation wall should include a compressible expansion joint material. Expansion joints can be made from any compressible material such as neoprene or composition sheathing material. This expansion joint filler is as wide as the slab is thick and is simply placed against the foundation wall before the concrete is placed.

CONCRETE SLABS

When the house has a basement, the floor is a concrete slab-on-grade. The areas to be covered with concrete are indicated on the foundation plan or basement floor plan. This is usually done by an area note giving the thickness of the concrete slab and any reinforcing steel to be used. To help the concrete resist minor stresses, it is usually reinforced with welded wire fabric. The specifications for welded wire fabric are explained in Figure 16-7. Where the slab must support bearing walls or masonry partitions, it may be haunched as discussed in Unit 14.

When floor drains are included or where water must be allowed to run off, the slab is *pitched* (sloped slightly). A note on the drawings indicates the amount of pitch. One-quarter inch per foot is common. When there is any possibility of confusion about which way the slab is to be pitched, bold arrows are drawn to show the direction the water will run, Figure 16-8.

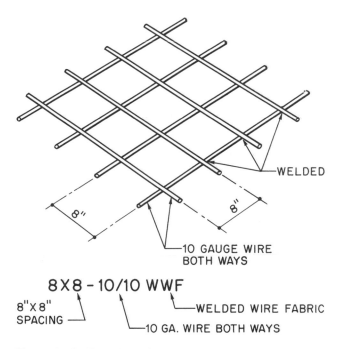

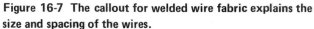

Figure 16-7 The callout for welded wire fabric explains the size and spacing of the wires.

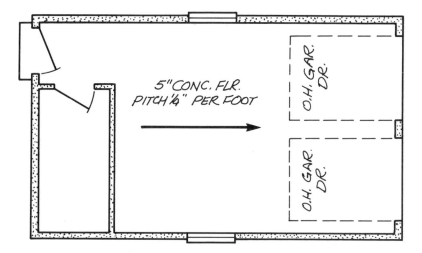

**Figure 16-8 A bold arrow indicates the way that
water will run off a pitched surface.**

When floor drains or forms are set for pitched floors, it is necessary to find the total pitch of the slab. This is done by multiplying the pitch-per-foot by the number of feet over which the slab is pitched. (See Math Review 3.) For example, if the note on a concrete apron in front of a garage door indicates a pitch of 1/2-inch per foot and the apron is 4 feet wide, the total pitch is 2 inches. The proper elevation for the form at the outer edge of the apron is 2 inches less than the finished floor elevation.

Some of the concrete work in the Lake House is of particular interest, because it is part of the passive-solar heating system the Lake House uses. Section 1/4 and the lower level floor plan 1/2 indicate that the area under the living room and dining room floors is a heat sink. A *heat sink* is a mass of dense material which absorbs the energy of the sun during the day and radiates it at night. The living room and dining room are on the south side of the Lake House. In the winter, when the leaves are off the deciduous trees, the sun shines in the large areas of glass in these rooms and warms the heat sink. At night, this heat is radiated into the house to provide additional heat when it is needed most. The floor over the heat sink is a concrete slab similar to that used in the playroom. Detail 2/5 helps explain this concrete slab.

✓ CHECK YOUR PROGRESS

Can you perform these tasks?

☐ Describe the footing drains shown on a set of construction drawings.

☐ Name the material and its thickness when foundation insulation is shown.

☐ Describe any vapor barriers or foundation coating to be applied to the foundation or under concrete slabs.

☐ List the dimensions of any concrete slab floors shown on the drawings.

☐ Describe the reinforcement to be used in concrete slabs.

☐ Explain the pitch of a concrete slab to provide for drainage.

ASSIGNMENT ⎯⎯⎯⎯⎯⎯⎯⎯⎯⎯⎯⎯⎯⎯⎯⎯⎯⎯⎯⎯⎯⎯⎯⎯⎯⎯⎯⎯⎯⎯

Refer to the Lake House drawings (in the packet) to complete this assignment.

1. What is the thickness of the concrete slab over the heat sink in the Lake House?.

2. Describe the reinforcement used in the concrete slab in the playroom of the Lake House.

3. How many square feet of two-inch rigid insulation are needed for the Lake House heat sink?

4. What prevents moisture from seeping through the concrete slab floor in the Lake House?

5. What is the finished floor elevation of the Lake House garage?

6. What is the elevation of the floor drain in the utility room of the Lake House?

7. What is the purpose of the eight-inch-thick concrete haunch in the middle of the Lake House slab?

8. How many cubic yards of concrete are required for the garage floor? The basement floor?

Framing

Section 5

| UNIT 17 | Framing Systems |

OBJECTIVES

After completing this unit, you will be able to identify each of the following types of framing on construction drawings:

- Platform
- Balloon
- Post-and-beam
- Energy-saving

PLATFORM FRAMING

Platform framing, also called *western framing,* is the type of framing used in most houses built in the last 30 years, Figure 17-1. It is called platform framing because as the rough floor is built at each level, it forms a platform on which to work while erecting the next level, Figure 17-2.

A characteristic of platform framing is that all wall *studs,* the main framing members in walls,

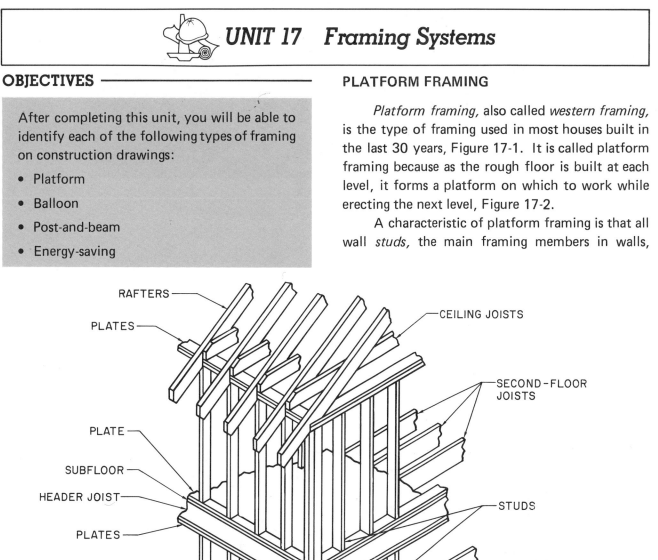
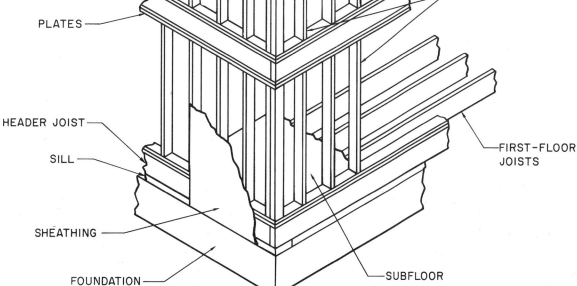

RAFTERS

PLATES

CEILING JOISTS

SECOND-FLOOR JOISTS

PLATE

SUBFLOOR

HEADER JOIST

PLATES

STUDS

HEADER JOIST

SILL

FIRST-FLOOR JOISTS

SHEATHING

FOUNDATION

SUBFLOOR

Figure 17-1 Platform or western framing

Figure 17-2 Platform or western framing provides a convenient work surface during construction.

extend only the height of one story. Interior walls, called partitions, are the same as exterior walls. The bottoms of the studs are held in position by a *bottom* (or *sole*) *plate*. The tops of the studs are held in position by a *top plate*. Usually, a *double top plate* is overlapped at the corners to tie intersecting walls and partitions together, Figure 17-3. In some construction, the second top plate is not used. Instead, metal framing clips are used to tie intersecting walls together. Upper floors rest on the top plate of the walls beneath. The framing members of the upper floors or roof are positioned over the studs of the wall that supports them.

Platform construction can be recognized on wall sections, Figure 17-4. Notice that the studs extend only from one floor to the next.

BALLOON FRAMING

Although it is not widely used now, balloon framing was once the most common framing system. In *balloon framing,* the exterior wall studs are continuous from the foundation to the top of the wall,

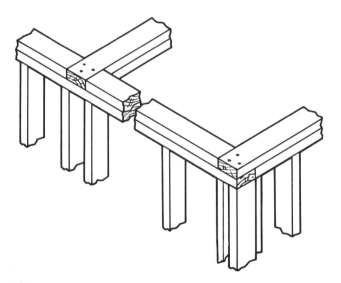

Figure 17-3 The double top plate overlaps at corners.

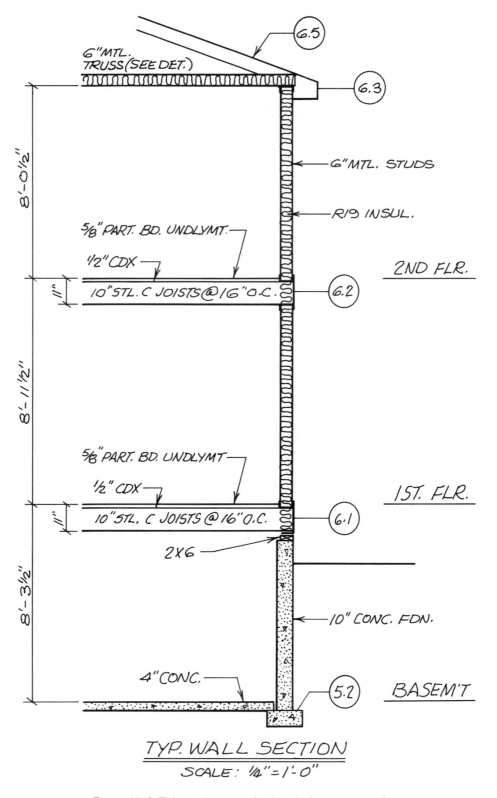

6.5

6" MTL.
TRUSS (SEE DET.)

6.3

6" MTL. STUDS

R/9 INSUL.

8'-0½"

5/8" PART. BD. UNDLYMT.

½" CDX

2ND FLR.

10" STL. C JOISTS @ 16" O.C.

6.2

11"

8'-11½"

5/8" PART. BD. UNDLYMT

½" CDX

1ST. FLR.

10" STL. C JOISTS @ 16" O.C.

6.1

11"

2X6

8'-3½"

10" CONC. FDN.

4" CONC.

5.2

BASEM'T

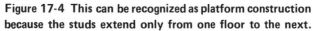

TYP. WALL SECTION

SCALE: ¼" = 1'-0"

Figure 17-4 This can be recognized as platform construction because the studs extend only from one floor to the next.

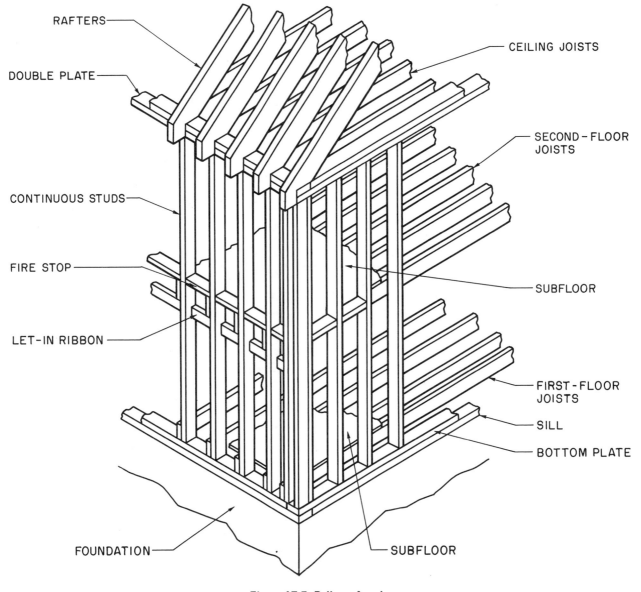

RAFTERS

DOUBLE PLATE

CONTINUOUS STUDS

FIRE STOP

LET-IN RIBBON

FOUNDATION

CEILING JOISTS

SECOND-FLOOR JOISTS

SUBFLOOR

FIRST-FLOOR JOISTS

SILL

BOTTOM PLATE

SUBFLOOR

Figure 17-5 Balloon framing

Figure 17-5. Floor framing at intermediate levels is supported by *let-in ribbon boards.* This is a board that fits into a notch in each joist and forms a support for the joists.

In both platform-frame and balloon-frame construction, the structural frame of the walls is covered with sheathing. Sheathing encloses the structure and, if a structural grade is used, prevents wracking of the wall. *Wracking* is the tendency of all of the studs to move, as in a parallelogram, allowing the wall to collapse to the side, Figure 17-6. There are two ways to prevent wracking. Plywood or other structural sheathing at the corners of the building

prevents this movement. Also, diagonal braces can be attached to the wall framing at the corners to prevent wracking, Figure 17-7.

Seismic codes (building codes for earthquake protection) often call for interior shear walls. These are framed walls covered with structural sheathing, Figure 17-8. Interior shear walls make the building stronger.

POST-AND-BEAM FRAMING

Platform framing and balloon framing are characterized by closely spaced, lightweight framing

Figure 17-6 Wracking

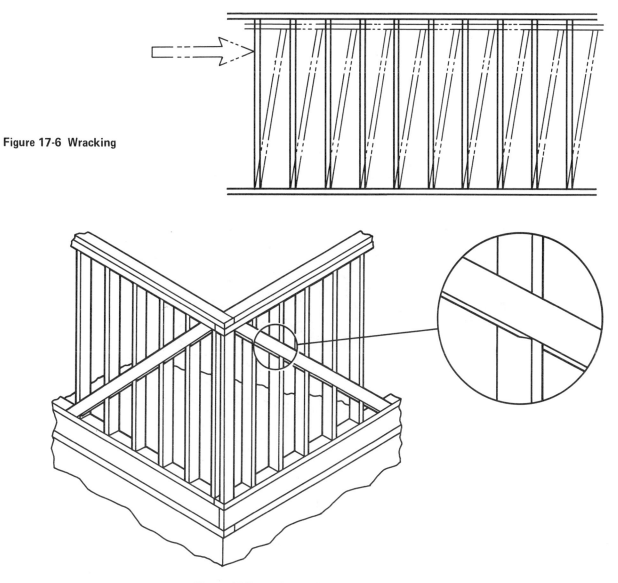

Figure 17-7 Let-in wind bracing prevents wracking.

Figure 17-8 Interior shear walls may be used in earthquake zones.

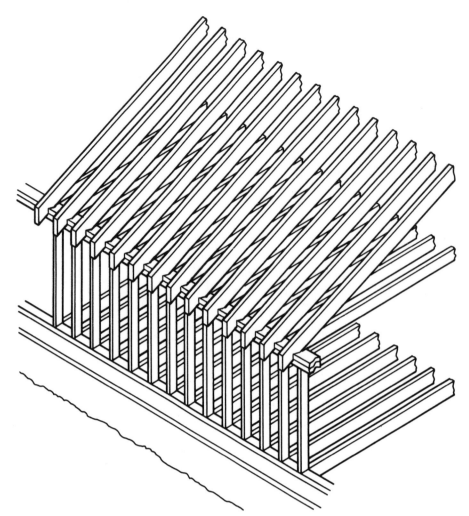

Figure 17-9 Conventional framing, 16" O.C.

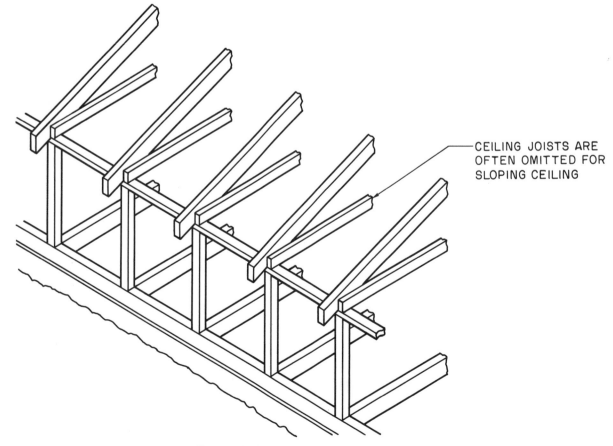

CEILING JOISTS ARE
OFTEN OMITTED FOR
SLOPING CEILING

Figure 17-10 Post and beam framing, 4' O.C.

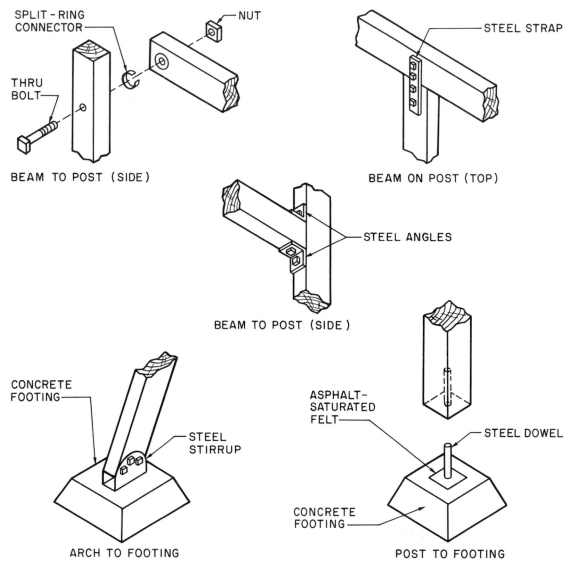

Figure 17-11 Common timber fastenings

members, Figure 17-9. *Post-and-beam* framing uses heavier framing members spaced further apart, Figure 17-10. These heavy timbers are joined or fastened with special hardware, Figure 17-11. Because post-and-beam framing uses fewer pieces of material, it can be erected more quickly. Also, although the framing members are large (ranging from 3 inches by 6 inches to 5 inches by 8 inches), their wider spacing results in a savings of material. However, to span this wider spacing, floor and roof decking must be heavier. Post-and-beam framing is sometimes left exposed to create special architectural effects, Figure 17-12.

The structural core of the Lake House uses posts and beams, detail 6/6. However, this is not purely post-and-beam construction. The posts are 3 1/2-inch square steel tubing and the beams are plywood box beams. These are properly called *beams* because they carry a load without continuous support from below. These beams are supported only in the beam pockets on the steel posts,

Courtesy of Weyerhaeuser

Figure 17-12 The framing in this cottage is exposed to create a special architectural effect.

Figure 17-13. In the Lake House the post-and-beam construction does not include exterior walls. In pure post-and-beam construction, the exterior walls have widely spaced posts and the space between is filled in with nonload-bearing curtain walls. These curtain walls are merely panels that fill in the space between the structural elements — the posts and beams.

In areas where hurricanes and high tides are a threat, some houses are built as pole structures. Pole buildings are a variation of post-and-beam construction. Poles, which are treated to be insect and rot resistant, are set several feet in the ground and 8 to 12 feet apart. A *band joist,* or *header,* is then bolted to these poles, Figure 17-14. The floors, walls, and roof are framed within the pole structure, Figure 17-15. Pole buildings are strong; they resist severe winds. Pole construction allows buildings to be kept above damaging flood waters.

ENERGY-SAVING TECHNIQUES

Approximately 11 percent of all energy consumed in the United States is used for heating homes. As energy resources dwindle, new techniques for saving energy are developed. One of these techniques is the *Arkansas Energy Saving System* of framing. This system reduces the amount of heat that is allowed to escape through the walls of a house.

When wall framing is done with 2x4s spaced 16 inches on centers, up to 25 percent of the wall is solid wood. Wood conducts heat out of the building. Only the space between the solid wood framing can be filled with insulation. By using 2x6 studs

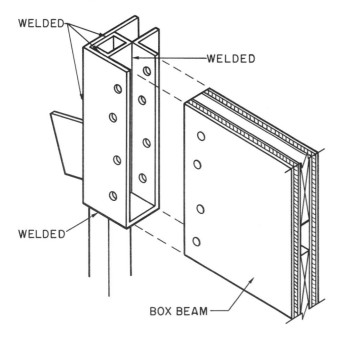

Figure 17-13 Beam pocket for Lake House

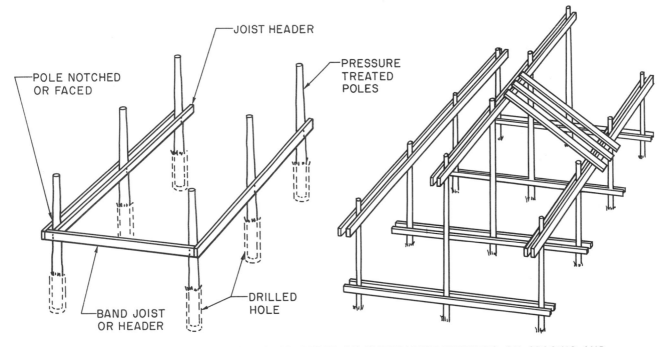

EMBEDMENT AND ALIGNMENT OF POLES. DEPTH OF EMBEDMENT DEPENDS ON SPACING AND SIZE OF POLES, WIND LOADS, AND SO FORTH, AND MAY VARY FROM 5 TO 8 FEET.

Courtesy of American Wood Preservers Institute

Figure 17-14 Basic elements of a pole building

spaced 24 inches on centers, the area of solid wood is reduced to less than 20 percent of the wall. The amount of framing material is the same. Not only does this reduce the amount of wood exposed to the surface of the wall, but it also allows for 2 inches more insulation.

The area of exposed wood is further reduced by special corner construction. In conventional framing, three pieces are used to frame the corner of a wall, Figure 17-16. In the Arkansas Energy Saving System only two pieces are used, Figure 17-17. The third piece, which normally provides a nailing surface for the interior drywall, is replaced by metal clips.

Examination of the first floor plan and the detail drawings of the Lake House shows that the walls are framed for maximum efficiency. Several details indicate that the studs are 2x6s @ 24 O.C. This allows room for more insulation in the wall. Also notice that the house is sheathed with 3/4-inch insulating sheathing. Interior partitions do not need insulation, so they are framed with 2x4s.

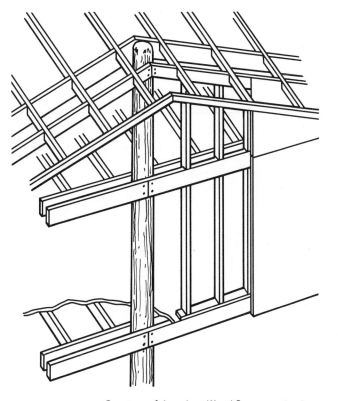

Courtesy of American Wood Preservers Institute

Figure 17-15 Framing in a pole building

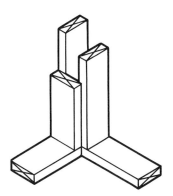
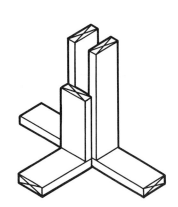
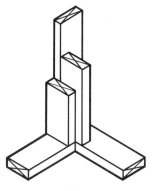

Figure 17-16 Conventional corner posts for 2x4 framing

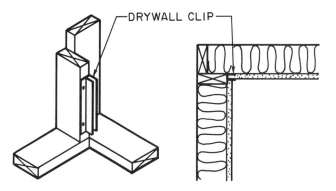

DRYWALL CLIP

Figure 17-17 Corner construction for Arkansas Energy Saving System

✓ CHECK YOUR PROGRESS

Can you perform these tasks?

☐ Identify platform framing shown on drawings and describe how it differs from other types of framing.

☐ Identify balloon framing shown on drawings and describe how it differs from other types of framing.

☐ Identify post-and-beam framing shown on drawings and describe how it differs from other types.

☐ Identify the following framing members on a wall or building section: joists, bottom or sole plate, top plate, stud, ribbon (to support joists), posts, beams, and sheathing.

☐ Explain what prevents a building frame from wracking.

☐ Describe at least three techniques for reducing the amount of heat lost through the solid wood in a building frame.

ASSIGNMENT

1. Identify *a* through *h* in Figure 17-18.
2. What kind of framing is shown in Figure 17-18?

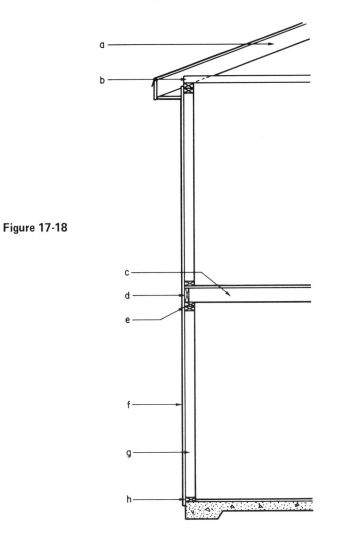

Figure 17-18

3. Identify *a* through *c* in Figure 17-19.

4. What kind of framing is shown in Figure 17-19?

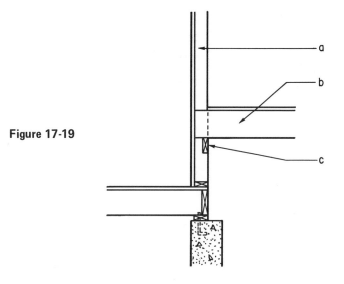

Figure 17-19

5. Identify *a* through *c* in Figure 17-20.

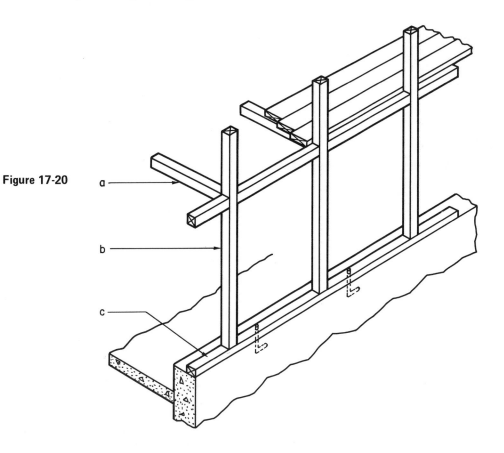

Figure 17-20

6. Sketch a plan view of a conventional corner detail. Include drywall and sheathing.

7. Sketch a plan view of an energy-efficient corner detail. Include drywall and sheathing.

    Refer to Lake House drawings (in the packet) to complete the rest of the assignment.

8. Which of the types of framing discussed in this topic is used for the Lake House?

9. What supports the west ends of the floor joists in bedroom #1?
10. What supports the east ends of the kitchen rafters?
11. How are the box beams fastened to the 3 1/2-inch square posts?
12. How are the steel posts anchored?
13. What does the north-west square steel post rest on?
14. What supports the north edge of the living room floor?
15. List the dimensions (thickness x depth x length) of all box beams. (Note: The length can be found by subtracting the outside dimension of the posts from the centerline spacing shown on the plan views.)
16. List the length of each piece of 3 1/2-inch steel.

UNIT 18   Columns, Piers, and Girders

OBJECTIVES

After completing this unit, you will be able to perform the following tasks:

• Locate columns and piers and describe each from drawings.
• Locate and describe the girders which support floor framing.
• Determine the lengths of columns and the heights of piers.

The most common system of floor framing in residential construction involves the use of joists and girders. *Joists* are parallel beams used in the framing of most floors, Figure 18-1. Usually buildings are too wide for continuous joists to span the full width. In this case, the joists are supported by one or more *girders* (beams) running the length of the building. The girder is supported at regular intervals by wood or metal posts or by masonry or concrete piers.

COLUMNS AND PIERS

Metal posts called *pipe columns* are the most common supports for girders. However, masonry or concrete piers may be specified. The locations of

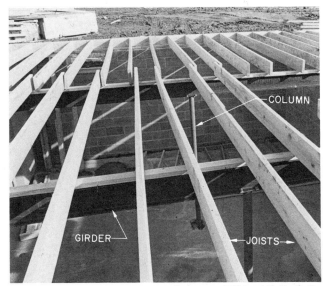

Courtesy of Richard T. Kreh, Sr.

Figure 18-1   Joist-and-girder floor framing

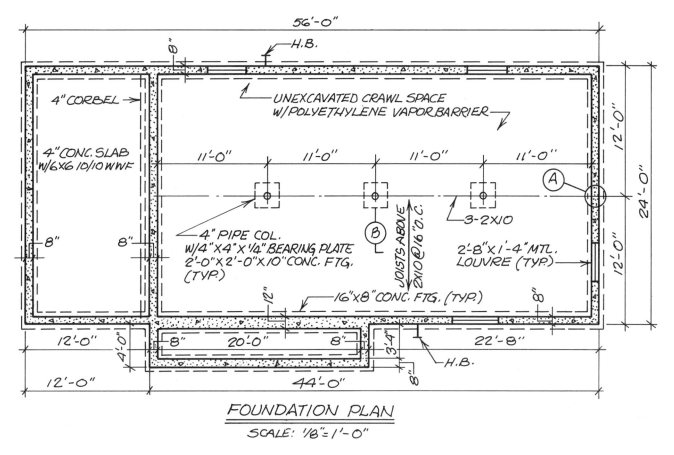

Figure 18-2 Foundation plan with note for column and footing

columns, posts, or piers are given by dimensions to their centerlines. When metal or wooden posts are indicated, the only description may be a note on the foundation plan, Figure 18-2. This note may give the size and material of the posts only, or it may also specify the kind of bearing plates to be used at the top and bottom of the post, Figure 18-3. A *bearing plate* is a steel plate that provides a flat surface at the top or bottom of the column or post. If the girder is supported by masonry or concrete piers, a special detail may be included to give dimensions and reinforcement details.

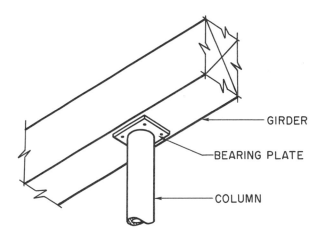

Figure 18-3 Bearing plate

GIRDERS

Before the total length of the columns or height of the piers can be calculated, it is necessary to determine the size of the girder and its relationship to the floor joist. There are three types of girders commonly used in residential construction. Steel beams are often used where strength is a critical factor. Built-up wood girders are constructed on the site. These consist of two, three, or more pieces of 2-inch lumber nailed together with staggered joints to form larger beams. The sizes and specifications for built-up wood girders and structural steel girders are discussed in Unit 15. Plywood box beams, Figure 18-4,

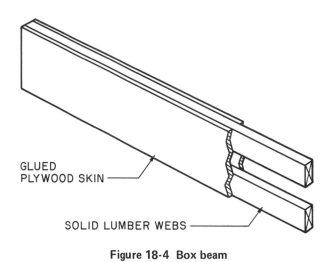

Figure 18-4 Box beam

are made up of a solid wood frame covered with plywood. The plywood skin adds rigidity to the box beam. Box beams use a relatively small amount of material and are light in weight. The notes commonly found on a foundation plan to indicate the type of girder to be used are shown in Figure 18-2.

DETERMINING THE HEIGHTS OF COLUMNS AND PIERS

The length of the columns or height of the piers depends on how the joists will be attached to the girder. The floor joists may rest directly on top of the girder or may be butted against the girder so that the top surface of the floor joist is flush with the top surface of the girder, Figure 18-5.

To find the height of the columns or piers, first determine the dimension from the basement floor to the finished first floor. Then subtract from this dimension the depth of the first floor including all of the framing and the girder. Then add the distance from the top of the basement floor to the bottom of the column. The result equals the height of the column or pier. (See Math Reviews 5 and 6.) For example, the following shows the calculation of the height of the steel column in Figure 18-6:

* Dimension from finished basement floor to finished first floor = 8'-10 1/2"

    Allowance for finished floor = 1"
    Nominal 2x8 joists = 7 1/4"
    2x4 bearing surface on girder = 1 1/2"
    W8 x 31 = 8"
    Total floor framing = 17 3/4" or 1'-5 3/4"

* 8'-10 1/2" minus 1'-5 3/4" = 7'-4 3/4"
    Add thickness of concrete slab

* 7'-4 3/4" plus 4" = 7'-8 3/4"

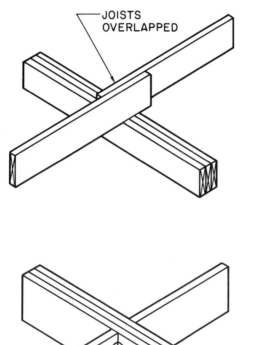

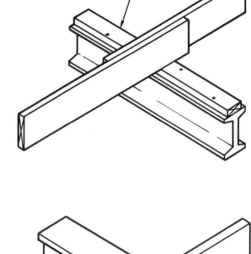

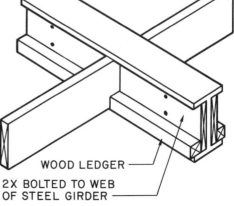

Figure 18-5 Several methods of attaching joists to girders

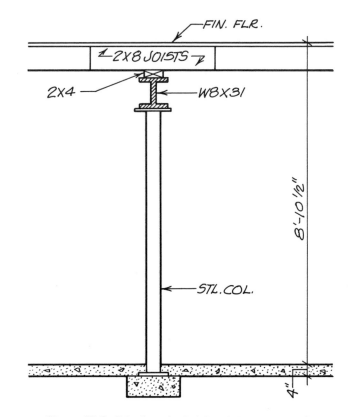

Figure 18-6 Calculate the height of the steel column.

✓ **CHECK YOUR PROGRESS**

Can you perform these tasks?

☐ Give the dimensions to locate columns or piers to support floor framing.

☐ Describe columns, including their material, diameter, and bearing plates.

☐ Calculate the length or height of columns and piers.

☐ Describe the girders to be used in a building, including their material, depth, weight or thickness, and length.

☐ Explain how the floor joists bear on the girders.

ASSIGNMENT

Questions 1 through 5 refer to Figure 18-7.

1. What is the length of the girder?
2. Describe the material used to build the girder including the size of material.
3. What supports the girder? (Include material and cross sectional size.)
4. How many posts, columns, or piers support the girder?
5. What is the height of the column or pier supporting the girder including bearing plates?

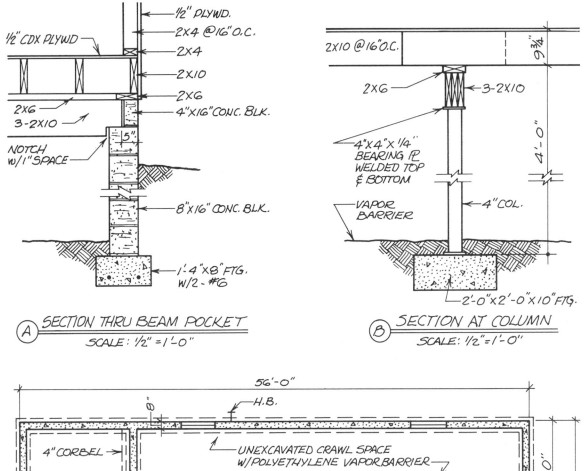

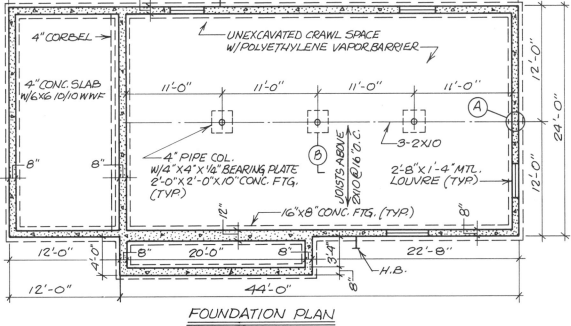

Figure 18-7 Use assignment questions 1-5.

Questions 6 through 10 refer to Figure 18-8.

6. What is the length of the girder?

7. Describe the material used to build the girder including the size of material.

8. What supports the girder? (Include material and cross sectional size.)

9. How many posts, columns, or piers support the girder?

10. What is the height of the column or pier supporting the girder, including bearing plates?

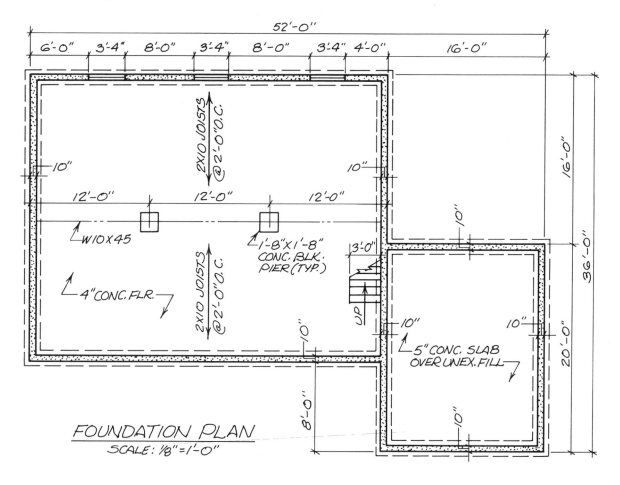

FOUNDATION PLAN
SCALE: 1/8"=1'-0"

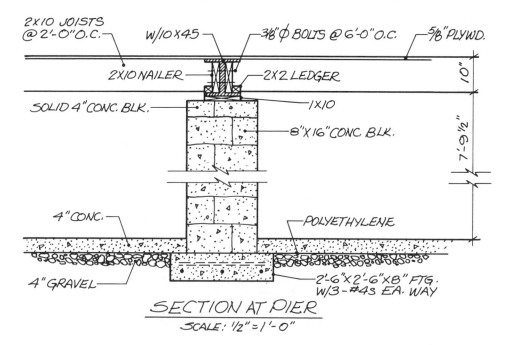

SECTION AT PIER
SCALE: 1/2"=1'-0"

Figure 18-8 Use with assignment questions 6-10.

UNIT 19 Floor Framing

OBJECTIVES

After completing this unit, you will be able to perform the following tasks:

- Describe the sill construction shown on a set of drawings.

- Identify the size, direction, and spacing of floor joists according to a set of drawings.

- Describe the floor framing around openings in a floor.

SILL CONSTRUCTION

Where the framing rests on concrete or masonry foundation walls, the piece in contact with the foundation is called the *sill plate*, Figure 19-1. The sill plate is the piece through which the anchor bolts pass to secure the floor in place. To prevent the sill plate from coming in direct contact with the foundation, and to seal any small gaps, a *sill sealer* is often included. This is a compressible, fiberous material that acts like a gasket in the sill construction.

The entire construction of the floor frame at the top of the foundation is called *sill construction* or the *box sill*. The box sill is made up of the sill sealer, sill plate, joist, and joist header, Figure 19-2. The sizes of materials are given on a wall section or sill detail. For areas where termites are present, a termite shield is included in the sill construction, Figure 19-3. A *termite shield* is a continuous metal shield that prevents termites from getting to the wood superstructure.

FLOOR JOISTS

Floor joists are the parallel framing members that make up most of the floor framing. Until recently, joists in residential construction were 2-inch framing lumber. However, recent advances in the use of materials have produced several types of engineered joists, Figure 19-4. Although the materials in each type are different, their use is essentially the same.

Figure 19-1 Sill plate

114

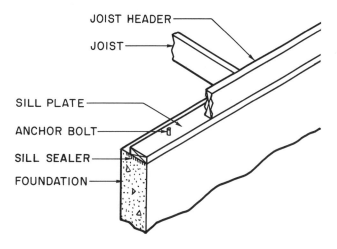

JOIST HEADER

JOIST

SILL PLATE

ANCHOR BOLT

SILL SEALER

FOUNDATION

Figure 19-2 Box sill

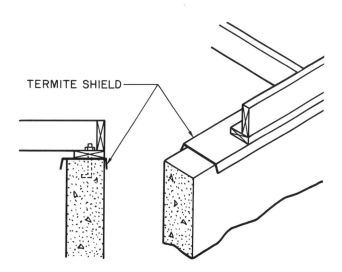

TERMITE SHIELD

Figure 19-3 Termite shield

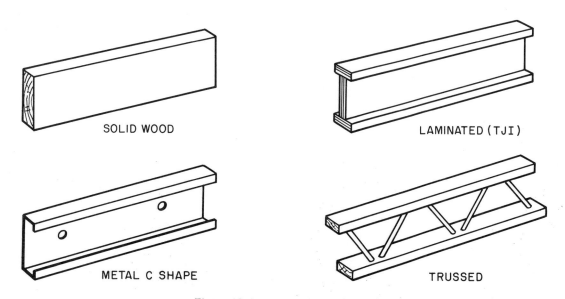

SOLID WOOD

LAMINATED (TJI)

METAL C SHAPE

TRUSSED

Figure 19-4 Several styles of joists

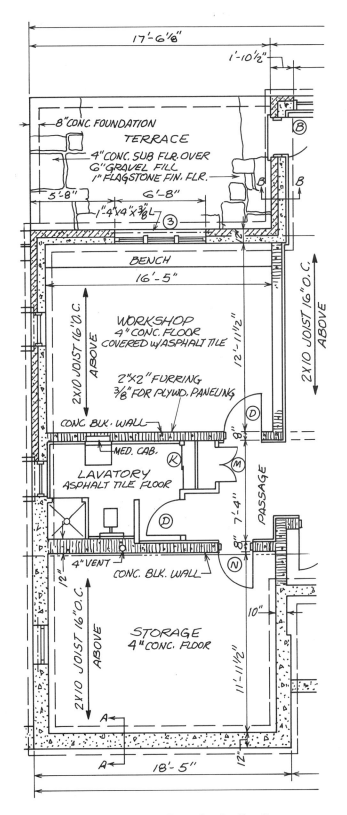

Figure 19-5 Joist callouts for the first floor
are shown on the foundation plan.

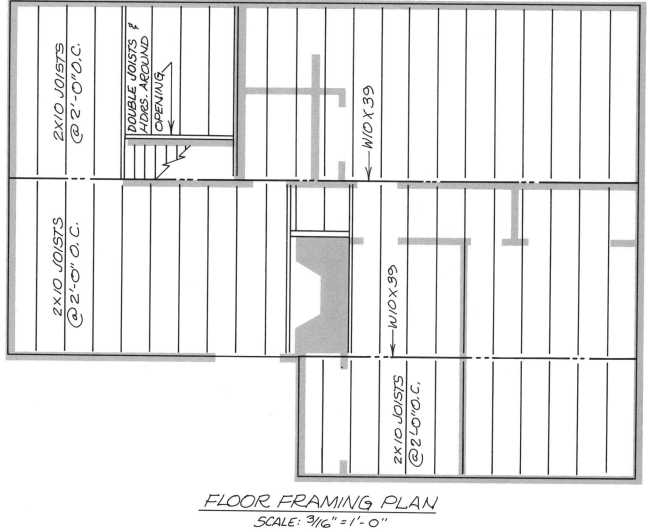

FLOOR FRAMING PLAN
SCALE: 3/16" = 1'-0"

Figure 19-6 A simple floor framing plan

Notes on the floor plans indicate the size, direction, and spacing of the joists in the floor above, Figure 19-5. For example, notes on the foundation plan give the information for the first floor framing. When the arrangement of framing members is complicated, a framing plan may be included, Figure 19-6. On most framing plans, each member is represented by a single line.

In the simplest building, all joists run in the same direction and are supported between the foundation walls by a single girder. However, irregularities in building shapes require that joists run in different directions, Figure 19-7. As the building de-

sign becomes more complex, more variations in floor framing are necessary. In a building such as the Lake House, which has floors at varying levels, the joists are supported by a combination of girders, beams, and load-bearing walls.

Lake House Floor Framing

The floor framing for the Lake House is shown on Framing Plan 1/6 (included in the packet). Notice that the framing plan is made up of a simplified floor plan and single lines to represent floor joists, beams, and joist headers. A more elaborate type of

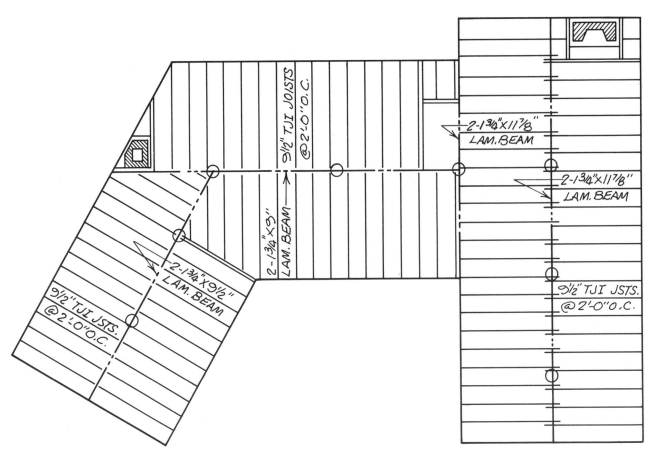

Figure 19-7 Floor framing plan for irregular-shaped house

framing plan uses double lines to represent the thickness of each member, Figure 19-8. The framing plan shows the location and direction of framing members, but for more detail it will be necessary to refer to Floor Plan 2/2 also.

The floor framing in the Lake House can be studied most easily if it is viewed as having four parts: the kitchen, bedroom #1, bedroom #2 and bathrooms, and the loft. There is no floor framing for the living room and dining room. These two areas form the heat sink for the passive-solar features discussed in earlier units. The floors in these rooms are concrete to absorb and radiate solar heat. As each area is framed, the carpenter must identify the following:

- Joist headers (locate the outer ends of joists)
- Bearing for inner ends of joists (beams, walls, etc.)

- Size and type of framing materials
- Length of joists
- Spacing
- Framing at openings

The joist headers are easily identified on the framing plan, but their exact position should be checked on the detail drawings. They may be set back to create a brick ledge, to accommodate wall finish, or to be flush with the foundation, Figure 19-9.

The *bearing* (support) for the inner ends of some of the joists in the Lake House is the structural-steel-and-box-beam core, Detail 6/6. This core consists of four 3 1/2" square steel posts with wood box beams and steel-channel beams. The positions of the posts are shown on all plan views of the house. The C8 x 11.5 steel acts as a beam to support the floor in bedroom #2. Two 2x10s support the loft. The

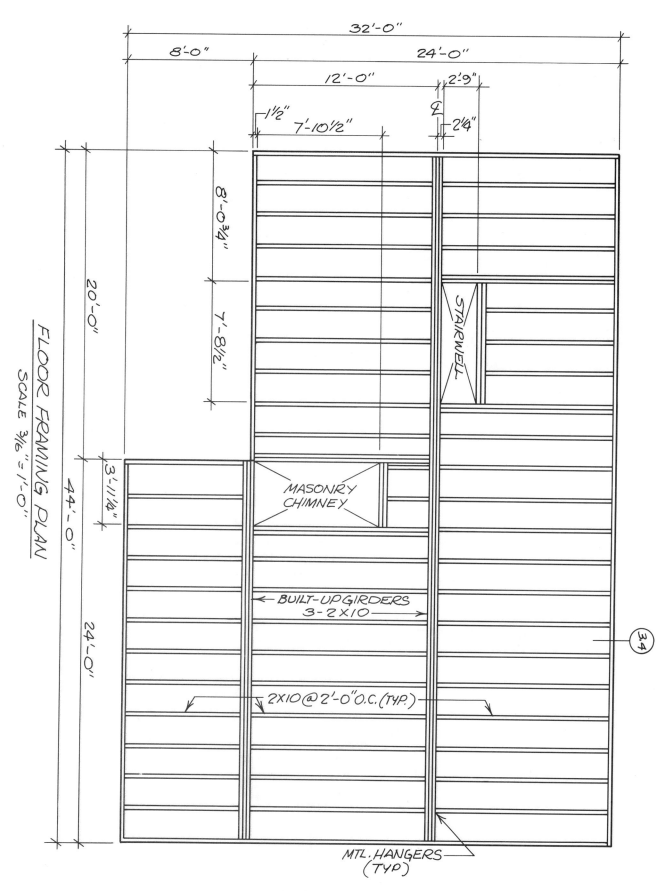

Figure 19-8 Double-line framing plan

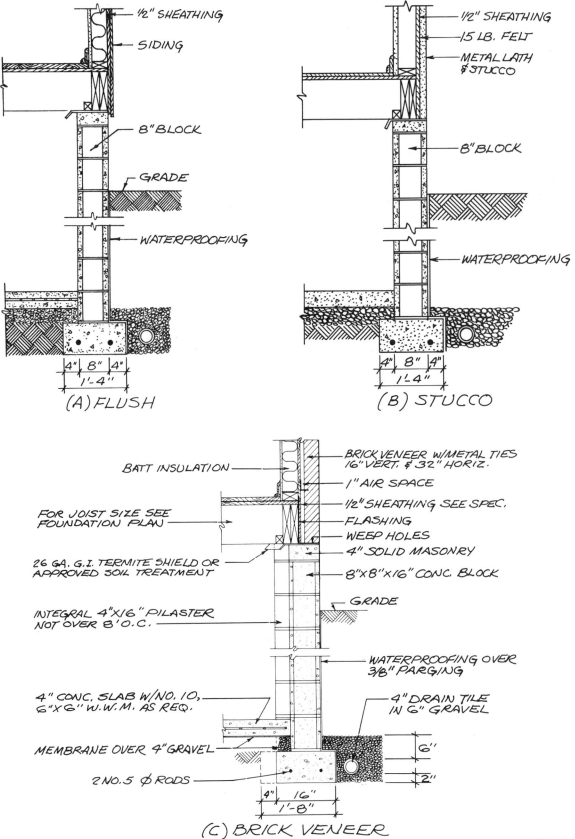

½" SHEATHING
SIDING
8" BLOCK
GRADE
WATERPROOFING

4" | 8" | 4"
1'-4"
(A) FLUSH

½" SHEATHING
15 LB. FELT
METAL LATH & STUCCO
8" BLOCK
WATERPROOFING

4" | 8" | 4"
1'-4"
(B) STUCCO

BATT INSULATION
FOR JOIST SIZE SEE FOUNDATION PLAN
26 GA. G.I. TERMITE SHIELD OR APPROVED SOIL TREATMENT
INTEGRAL 4"X16" PILASTER NOT OVER 8' O.C.
4" CONC. SLAB W/NO. 10, 6"X6" W.W.M. AS REQ.
MEMBRANE OVER 4" GRAVEL
2 NO. 5 ⌀ RODS

BRICK VENEER W/METAL TIES 16" VERT. & 32" HORIZ.
1" AIR SPACE
½" SHEATHING SEE SPEC.
FLASHING
WEEP HOLES
4" SOLID MASONRY
8"X8"X16" CONC. BLOCK
GRADE
WATERPROOFING OVER 3/8" PARGING
4" DRAIN TILE IN 6" GRAVEL
6"
2"

4" | 16"
1'-8"
(C) BRICK VENEER

Courtesy of W.D. Farmer

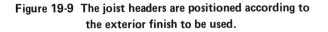

Figure 19-9 The joist headers are positioned according to the exterior finish to be used.

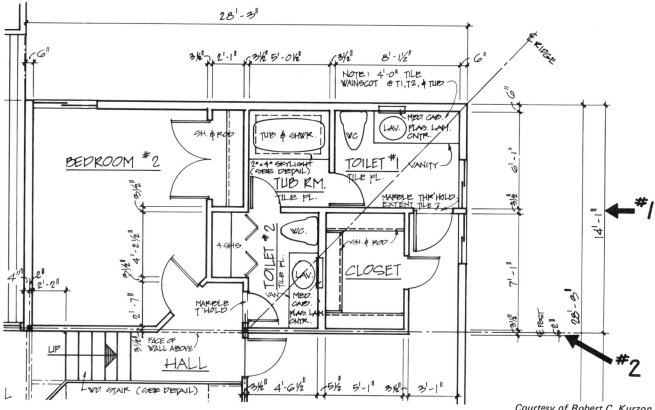

Figure 19-10 Part of Lake House Floor plan

Courtesy of Robert C. Kurzon

plywood box beams form part of the roof framing and are discussed in Unit 23. The double 2x10 beam rests in beam pockets that are welded to the posts. The structural steel channel (C8 x 11.5) is bolted directly to the posts.

The inner ends of the remaining floor joists in the Lake House are supported by bearing walls in the lower level. For example, some of the kitchen floor joists are supported by the west wall of the play-room, near the fireplace; and some, by the 2x10 header that spans the distance from the playroom wall to the foundation in the crawl space.

The size and spacing of material to be used are given on the framing plan by a note. They can also

be found on the wall section. Lengths of framing members are usually not included on framing plans. However, these lengths can be found easily by referring to the floor plan that shows the location of the walls or beams on which the joists rest. For example, refer to Figures 19-10 and 19-11 and find the length of the floor joists in bedroom #2 as follows:

1. The dimension from the outside of the north wall to the centerline of the 3 1/2" ⊡ post is 14'-1".
2. The dimension from the centerline of the 3 1/2" ⊡ post is 2", so the overall dimension of the bedroom floor is 14'-3".

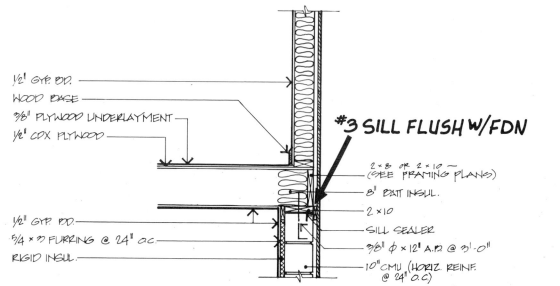

Courtesy of Robert C. Kurzon

Figure 19-11 Part of Lake House wall section

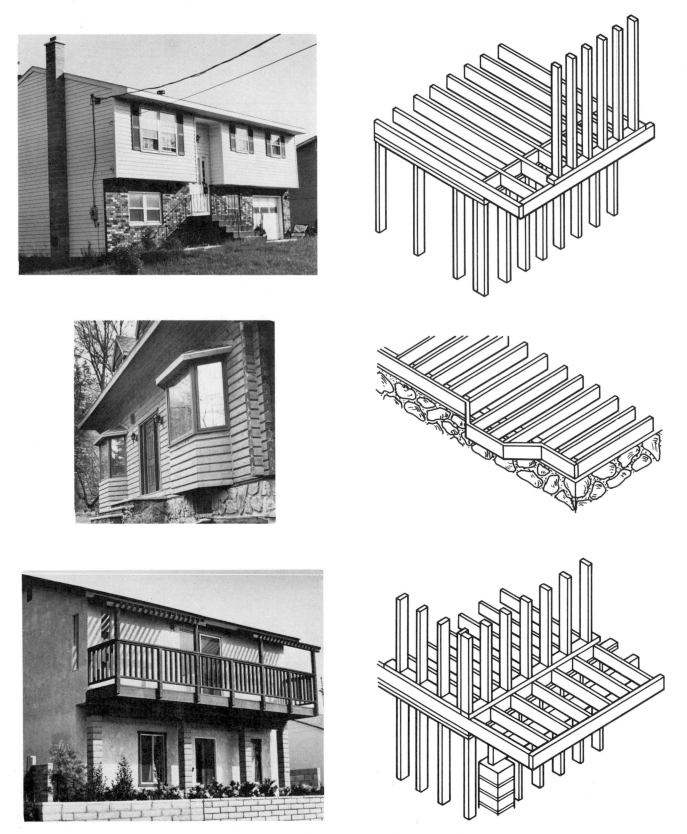

Figure 19-12 Typical uses of cantilevered framing

3. According to Wall Section 3/4, the joist header is flush with the north foundation wall, so subtract 1 1/2" (the thickness of the joist header) from each end.
4. 14'-3" minus 3" (1 1/2" at each end) equals 14'-0" (the length of the joists).

Some floor framing is cantilevered to create a seemingly unsupported deck. *Cantilevered* framing consists of joists that project beyond the bearing surface to create a wide overhang. This technique is used extensively for balconies, under bay windows, and for garrison-style houses, Figure 19-12.

FRAMING AT OPENINGS

Where stairs and chimneys pass through the floor frame, some of the joists must be cut out to form an opening. The ends of these joists are supported by double headers. The full joists at the sides of the opening must carry the extra load of the shortened joists, so they are also doubled, Figure 19-13.

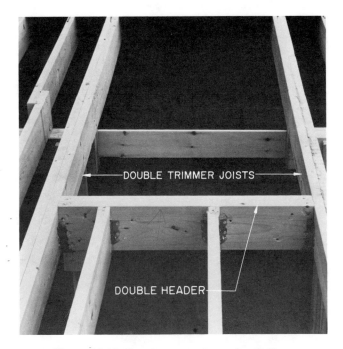

Figure 19-13 Framing around opening in floor

✓ CHECK YOUR PROGRESS

Can you perform these tasks?
- ☐ Describe the box sill shown on a set of drawings, including size of materials, anchoring, and setback from the face of the foundation.
- ☐ List the sizes of all the floor joists in a building.
- ☐ Describe the direction and spacing of all of the floor joists in a building.
- ☐ Describe the framing around openings in the floor.

 ## ASSIGNMENT

Refer to the Lake House drawings to complete the assignment.
1. What size lumber is used for the floor framing in bedroom #2?
2. What size lumber is used for the floor joists in the loft?
3. What size lumber is used for the joist headers in the loft?
4. How long are the joists in the loft?
5. How long are the floor joists in bedroom #1?
6. What supports the west end of the floor joists in bedroom #1?
7. What supports the south end of the floor joists in bedroom #2?
8. How long are the floor joists in bedroom #2?
9. How many floor joists are needed for bedroom #2 and the adjacent closets?
10. What size material is used for the sill?
11. What does the double 2x10 beam in the structural core of the Lake House support?
12. How long are the headers that support the loft floor joists?
13. Is the box sill of the Lake House flush with the foundation wall or set back?
14. When an opening in a floor is framed, why are the joists at the sides of the opening doubled?

UNIT 20 Laying Out Walls and Partitions

OBJECTIVES

After completing this unit, you will be able to perform the following tasks:

- Describe the layout of a house from its floor plans.
- Find specific dimensions given on floor plans.

When the deck (framing and subfloor or concrete slab) is completed, the framing carpenter lays out the location of walls and partitions. The size and location of each wall is indicated on the floor plans. Drawings 1/2 and 2/2 are the floor plans for the Lake House.

Of all the sheets in a set of construction drawings, the floor plans often contain the most information. Before looking for specific details on floor plans, it may help to mentally walk through the house. Start at the main entrance to the lowest floor and visualize each room as if you were walking through the house.

VISUALIZING THE LAYOUT OF WALLS AND PARTITIONS

The lowest floor with frame walls in the Lake House is the basement. Start at the $6^0 \times 6^8$ SGD (sliding glass door) on the east wall of the playroom. This large L-shaped room covers most of the basement floor. Some plans list overall dimensions for each room. The plans for the Lake House give this information by conventional dimension lines only.

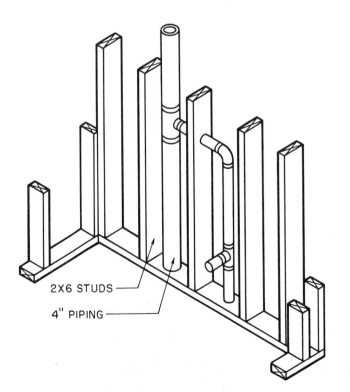

Figure 20-1 Six-inch plumbing wall

2X6 STUDS

4" PIPING

The north-south dimension of the playroom is 28'-3". The east-west dimension of the north part of the playroom is 14'-1". The section of the play-room with the fireplace is 13'-8" by 13'-10". It may be helpful to notice the overall dimensions of each room as you visualize its shape and relationship to other rooms. It will be necessary to refer to these dimensions many more times. At the north end of the playroom is a small closet. South of the closet is a hall leading to the bathroom and utility room.

The bathroom has a shower, water closet (toi-let), and lavatory. Notice that although most interi-or partitions are 3 1/2 inches thick, the wall behind the water closet and lavatory is 5 1/2 inches thick. This thicker wall is called a *plumbing wall.* Its extra thickness allows room for plumbing, Figure 20-1.

The purpose of the utility room beyond the bathroom is to house mechanical equipment such as water heater, furnace or boiler, and water pump. In the southwest corner of the utility room is a small opening into the crawl space on the west side of the house. Notice that the concrete ends here.

Walking back through the playroom to the stairs, you will see some interesting features, Figure 20-2. A broken line in this area indicates the edge of a floor above. These floors can be seen more easily on the first floor plan. Also notice the location of the fireplace which will extend up through the upper floor. The L-shaped stairs lead up to the first floor. It is obvious that this is not a full story higher be-cause only four steps are shown. This can be seen more clearly in the Sections on sheet 4. The L-shaped stairs lead up to the living room, Figure 20-3. Against the west and south walls is a plywood plat-form or built-in bench.

Another set of stairs in the northwest corner of the living room leads to the kitchen and dining room. These rooms are separated only by a peninsula of kitchen cabinets. The kitchen is separated from a hall by a free-standing closet and enclosure for the refrigerator and oven, Figure 20-4. This closet and enclosure cannot be recognized as free-standing (meaning it does not reach the ceiling) on the floor plan. However, this can be seen on Section 2/4. On the north side of the kitchen is a door to the deck outside. The east side of the hall has a railing that continues up another set of stairs to the upper hall. Beyond this railing the floor is open to the playroom below.

Above the open area of the playroom is a loft that provides storage or extra sleeping space, Figure 20-5. Access to the loft is by a ladder in the upper hall. The loft is shown on a separate plan on Sheet

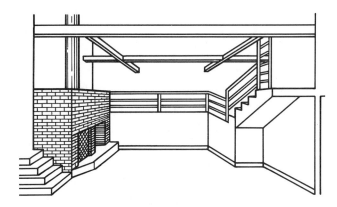

Figure 20-2 Lake House playroom

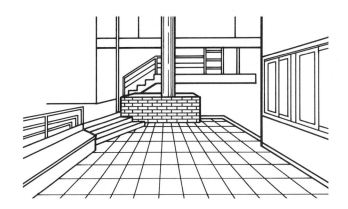

Figure 20-3 Lake House living room

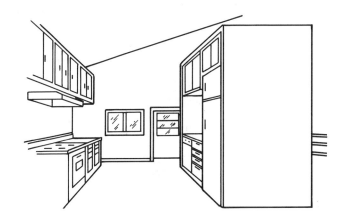

Figure 20-4 Lake House kitchen

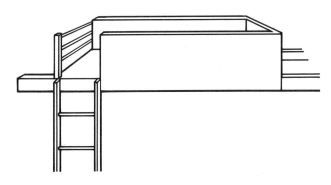

Figure 20-5 Lake House loft

4. The loft is suspended on two beams built-up of three 2x10s each. The north and south sides of the loft are enclosed by a wall 2'-6" high. The east and west ends have a railing.

One of the doors in the upper hall opens into bedroom #1. On the east side of this bedroom is a hall that leads past a large (7'-1" x 5'-1") closet to toilet #1 in the northeast corner of the house. This toilet room has only a lavatory and water closet. A tub in the next room also serves toilet #2 to the east. Toilet #2 has a closet with shelves for linen storage. This toilet room can also be entered from the upper hall. The remaining room is bedroom #2.

The garage is attached to the house only by the wood deck and the roof. The garage is a rectangular building with 4-inch walls, an overhead door, and a walk-through door.

FINDING DIMENSIONS

When you understand the relationships of the rooms to one another, you are ready to look for more detailed information. Frame walls are dimensioned in one of three ways, Figure 20-6. Exterior walls are usually dimensioned to the face of the studs or the face of the sheathing. Interior walls may be dimensioned either to the face of the studs or to their centerlines.

When walls are dimensioned to their centerlines, one half of the wall thickness must be subtracted to find the face of the studs. For example, in Figure 20-7, the end walls are 12'-4" on centers. However, the plates for the side walls are 12'-0 1/2" long (12'-4" minus 3 1/2", the width of the studs, equals 12'-0 1/2").

Dimensions are usually given in a continuous string where practical, Figure 20-8. Overall dimensions and major wall locations are given outside the view. Minor partitions and more detailed features are dimensioned either on or off the view, whichever is most practical.

The Lake House includes an angled wall. The length of such walls can be accurately found by trigonometry. However, the accuracy obtained by measuring with an architects' scale should be adequate for normal estimating.

As each wall is laid out, the plates are cut to length and the postions of all openings and intersecting walls are marked. The wall frame is usually assembled flat on the deck. Its position is marked on the deck with a chalkline, then the assembly is tipped up and slid into place.

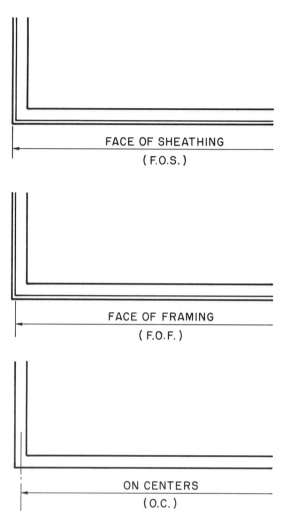

FACE OF SHEATHING
(F.O.S.)

FACE OF FRAMING
(F.O.F.)

ON CENTERS
(O.C.)

Figure 20-6 Three types of dimensioning

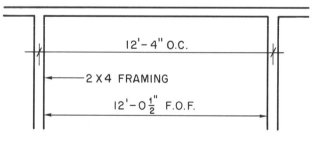

12'-4" O.C.

2 X 4 FRAMING

12'-0 $\frac{1}{2}$" F.O.F.

Figure 20-7 It is important to consider the size of the framing material when working with O.C. dimensions.

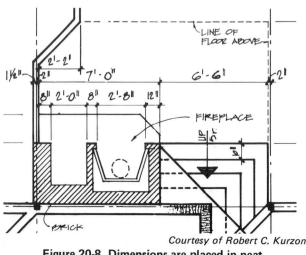

Courtesy of Robert C. Kurzon

Figure 20-8 Dimensions are placed in neat strings as much as possible.

√ CHECK YOUR PROGRESS

Can you perform these tasks?

☐ List the dimensions for locating all of the walls and partitions in a house.

☐ Explain whether wall dimensions are to face of sheathing, face of framing, or on centers.

☐ Identify plumbing walls shown on floor plans.

ASSIGNMENT

Refer to the Lake House drawings (in the packet) to complete the following assignment.

1. What are the inside dimensions of each room? Disregard slight irregularities in room shape, but remember to allow for wall thicknesses.

Room	N-S dimension	E-W dimension
Utility	X	
Basement Bath	X	
Living room	X	
Bedroom #1	X	
Bedroom #2	X	
Toilet #1	X	
Tub	X	
Toilet #2	X	
Loft	X	

2. How many lineal feet of 2x4 frame wall are there in the basement?

3. How many lineal feet of 2x6 frame wall are there in the basement?

4. What is between the dining room and the living room?

5. What is directly below the loft?

6. What is directly below the tub room?

7. How many lineal feet of 2x6 frame wall are there on the Upper Level Floor Plan?

UNIT 21 *Framing Openings in Walls*

OBJECTIVES

After completing this unit, you will be able to perform the following tasks:

- Describe typical rough openings.
- Locate and interpret specific information for framing openings.

Two types of dimensions must be known before window and door openings can be framed: location and size. Opening locations are given by dimensioning to their centerlines in a string of dimensions outside the floor plan, Figure 21-1. Such dimensions are usually given from the face of the studs. One-half the rough opening is then allowed on each side of the centerline.

DIMENSIONS OF ROUGH OPENINGS

The size of the *rough opening* (R.O.), or opening in the framing, is often listed on the door and

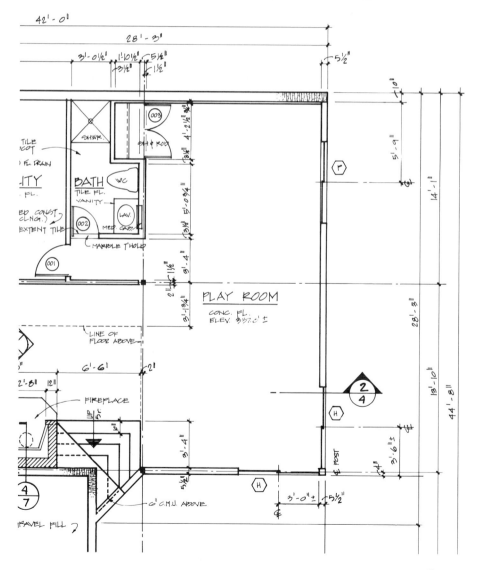

Courtesy of Robert C. Kurzon

Figure 21-1 The locations of openings in framed walls are usually given to their centerlines. The ± on the sliding doors allows the builder to place the doors next to the corner post.

window schedule. This is a list of all doors and windows in the house, Figure 21-2. Doors and windows are identified on the floor plans by a *mark* — a letter or number. All doors or windows of a certain size and type have the same mark. Each mark is listed on the schedule with the information for the doors or windows of that type and size.

Occasionally the rough opening dimensions are not given on the drawings. In this case they should be obtained from the manufacturer. Window manu-

SCHEDULE OF WINDOWS

◯	SIZE	TYPE	GLASS	COMMENTS
A	3'-0" × 3'-0"	THERMAL BREAK ALUMINUM	5/8" INSULATED	
B	3'-0" × 5'-0"	D.O.	D.O.	
C	4'-0" × 4'-0"	D.O.	D.O.	
D	4'-0" × 5'-0"	D.O.	D.O.	
E	5'-0" × 5'-0"	D.O.	D.O.	
F	6'-0" × 4'-0"	D.O.	1" INSULATED	
G	6'-0" × 5'-0"	D.O.	1" INSULATED	
H	6'-0" × 6'-8"	D.O.	1" TEMP. INSUL.	SLIDING GLASS DOORS

SCHEDULE OF DOORS

◯	SIZE		THK	DOOR MAT'L	DOOR TYPE	FRAME MAT'L	FRAME TYPE	GLASS	LOCKSET	COMMENTS
001	2'-8" × 6'-8"		1 3/8	WD.	C	WD.	A		CLASSROOM	
002	2'-0" × 6'-8"		1 3/8	WD.	C	WD.	A		PRIVACY	
003	PR 2'-0" × 6'-8"		1 3/8	WD.	D	WD.	A		PASSAGE R.H. DUMMY TRIM L.H.	
101	3'-0" × 6'-8"		1 3/4	H.M.	A	WD.	A	1" INSUL.	ENTRY	1'-8" SIDELIGHT
102	3'-0" × 6'-8"		1 3/4	H.M.	B	WD.	A		ENTRY	
103	PR 2'-4" × 6'-8"		1 3/8	WD.	D	WD.	B		NONE	BI-PASSING HDWARE
104	2'-6" × 6'-8"		1 3/8	WD.	C	WD.	B		PRIVACY	
105	PR 2'-6" × 6'-8"		1 3/8	WD.	D	WD.	B		PASSAGE R.H. DUMMY TRIM L.H.	
106	2'-0" × 6'-8"		1 3/8	WD.	C	WD.	B		PRIVACY	
107	(4) 1'-0" × 6'-8"		1 3/8	WD.	E	WD.	B		NONE	BI-FOLDING HDWARE
108	2'-0" × 6'-8"		1 3/8	WD.	C	WD.	B		PRIVACY	
109	2'-0" × 6'-8"		1 3/8	WD.	C	WD.	B		PRIVACY	
110	2'-6" × 6'-8"		1 3/8	WD.	C	WD.	B		PRIVACY	
111	2'-0" × 6'-8"		1 3/8	WD.	C	WD.	B		PASSAGE	
112	2'-6" × 6'-8"		1 3/8	WD.	C	WD.	B		PRIVACY	
113	2'-6" × 6'-8"		1 3/8	WD.	C	WD.	B		PASSAGE	

Courtesy of Robert C. Kurzon

Figure 21-2 Window and door schedules list all of the windows and doors with their sizes.

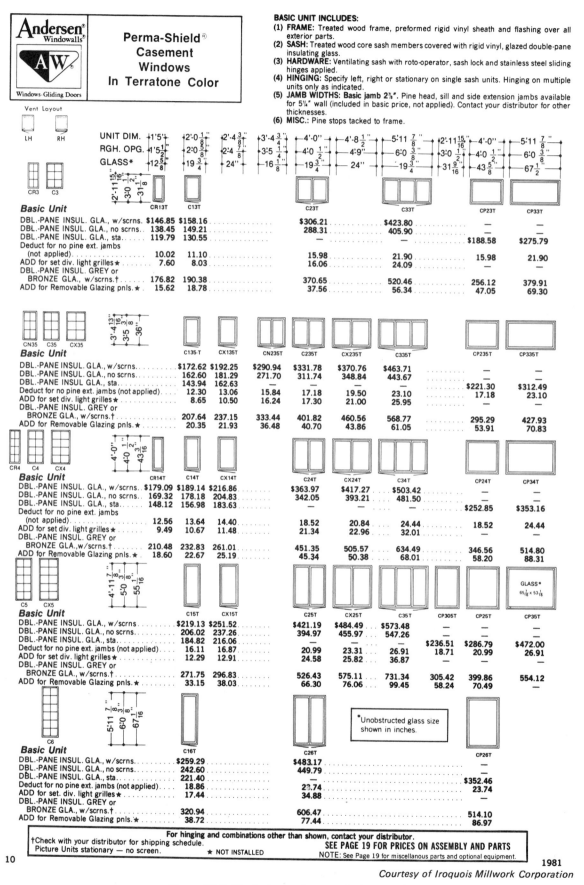

BASIC UNIT INCLUDES:
(1) **FRAME:** Treated wood frame, preformed rigid vinyl sheath and flashing over all exterior parts.
(2) **SASH:** Treated wood core sash members covered with rigid vinyl, glazed double-pane insulating glass.
(3) **HARDWARE:** Ventilating sash with roto-operator, sash lock and stainless steel sliding hinges applied.
(4) **HINGING:** Specify left, right or stationary on single sash units. Hinging on multiple units only as indicated.
(5) **JAMB WIDTHS:** Basic jamb 2⅝". Pine head, sill and side extension jambs available for 5¼" wall (included in basic price, not applied). Contact your distributor for other thicknesses.
(6) **MISC.:** Pine stops tacked to frame.

Andersen® Windowalls®
AW
Windows·Gliding Doors

Perma-Shield®
Casement
Windows
In Terratone Color

Basic Unit (CR13T / C13T ... C23T / C33T ... CP23T / CP33T)

	CR13T	C13T		C23T	C33T	CP23T	CP33T
DBL.-PANE INSUL. GLA., w/scrns.	$146.85	$158.16		$306.21	$423.80	—	—
DBL.-PANE INSUL. GLA., no scrns.	138.45	149.21		288.31	405.90	—	—
DBL.-PANE INSUL. GLA., sta.	119.79	130.55		—	—	$188.58	$275.79
Deduct for no pine ext. jambs (not applied)	10.02	11.10		15.98	21.90	15.98	21.90
ADD for set div. light grilles ★	7.60	8.03		16.06	24.09	—	—
DBL.-PANE INSUL. GREY or BRONZE GLA., w/scrns.†	176.82	190.38		370.65	520.46	256.12	379.91
ADD for Removable Glazing pnls. ★	15.62	18.78		37.56	56.34	47.05	69.30

Basic Unit (CN35 / C35 / CX35 ...)

	C135-T	CX135T	CN235T	C235T	CX235T	C335T	CP235T	CP335T
DBL.-PANE INSUL. GLA., w/scrns.	$172.62	$192.25	$290.94	$331.78	$370.76	$463.71	—	—
DBL.-PANE INSUL. GLA., no scrns.	162.60	181.29	271.70	311.74	348.84	443.67	—	—
DBL.-PANE INSUL GLA., sta.	143.94	162.63	—	—	—	—	$221.30	$312.49
Deduct for no pine ext. jambs (not applied)	12.30	13.06	15.84	17.18	19.50	23.10	17.18	23.10
ADD for set div. light grilles ★	8.65	10.50	16.24	17.30	21.00	25.95	—	—
DBL.-PANE INSUL. GREY or BRONZE GLA., w/scrns.†	207.64	237.15	333.44	401.82	460.56	568.77	295.29	427.93
ADD for Removable Glazing pnls. ★	20.35	21.93	36.48	40.70	43.86	61.05	53.91	70.83

Basic Unit (CR4 / C4 / CX4 ...)

	CR14T	C14T	CX14T	C24T	CX24T	C34T	CP24T	CP34T
DBL.-PANE INSUL. GLA., w/scrns.	$179.09	$189.14	$216.86	$363.97	$417.27	$503.42	—	—
DBL.-PANE INSUL. GLA., no scrns.	169.32	178.18	204.83	342.05	393.21	481.50	—	—
DBL.-PANE INSUL. GLA., sta.	148.12	156.98	183.63	—	—	—	$252.85	$353.16
Deduct for no pine ext. jambs (not applied)	12.56	13.64	14.40	18.52	20.84	24.44	18.52	24.44
ADD for set div. light grilles ★	9.49	10.67	11.48	21.34	22.96	32.01	—	—
DBL.-PANE INSUL. GREY or BRONZE GLA., w/scrns.†	210.48	232.83	261.01	451.35	505.57	634.49	346.56	514.80
ADD for Removable Glazing pnls. ★	18.60	22.67	25.19	45.34	50.38	68.01	58.20	88.31

Basic Unit (C5 / CX5 ...)

GLASS* 65⅝ × 53 1/16

	C15T	CX15T	C25T	CX25T	C35T	CP305T	CP25T	CP35T
DBL.-PANE INSUL. GLA., w/scrns.	$219.13	$251.52	$421.19	$484.49	$573.48	—	—	—
DBL.-PANE INSUL. GLA., no scrns.	206.02	237.26	394.97	455.97	547.26	—	—	—
DBL.-PANE INSUL. GLA., sta.	184.82	216.06	—	—	—	$236.51	$286.79	$472.00
Deduct for no pine ext. jambs (not applied)	16.11	16.87	20.99	23.31	26.91	18.71	20.99	26.91
ADD for set div. light grilles ★	12.29	12.91	24.58	25.82	36.87	—	—	—
DBL.-PANE INSUL. GREY or BRONZE GLA., w/scrns.†	271.75	296.83	526.43	575.11	731.34	305.42	399.86	554.12
ADD for Removable Glazing pnls. ★	33.15	38.03	66.30	76.06	99.45	58.24	70.49	—

*Unobstructed glass size shown in inches.

Basic Unit (C6 ...)

	C16T	C26T	CP26T
DBL.-PANE INSUL. GLA., w/scrns.	$259.29	$483.17	—
DBL.-PANE INSUL. GLA., no scrns.	242.60	449.79	—
DBL.-PANE INSUL. GLA., sta.	221.40	—	$352.46
Deduct for no pine ext. jambs (not applied)	18.86	23.74	23.74
ADD for set div. light grilles ★	17.44	34.88	—
DBL.-PANE INSUL. GREY or BRONZE GLA., w/scrns.†	320.94	606.47	514.10
ADD for Removable Glazing pnls. ★	38.72	77.44	86.97

For hinging and combinations other than shown, contact your distributor.
†Check with your distributor for shipping schedule.
Picture Units stationary — no screen.
★ NOT INSTALLED

SEE PAGE 19 FOR PRICES ON ASSEMBLY AND PARTS
NOTE: See Page 19 for miscellaneous parts and optional equipment.

10

1981

Courtesy of Iroquois Millwork Corporation

Figure 21-3 Typical page from a window catalog

facturers list glass size, sash opening, rough opening, and unit dimensions in their catalogs, Figure 21-3. If the finished doorway, the *jamb,* is to be built on the site by a carpenter, the rough opening size will not be available. In this case the rough opening for swing doors can be built 2 inches wider and 1 1/2 inches higher than the door. If the door is another type, you must first determine what the finished opening is to be. Sliding door panels should overlap 1 inch so the finished opening is 2 inches narrower than the width of two panels. Sliding doors, bifold doors, and accordian doors all need an extra inch at the top for hardware; build these rough openings 2 1/2 inches higher than the door itself.

Sizes of doors and windows are given with the width first and height second. To further simplify dimensioning, they are often listed as feet/inches. For example, a 2^6x6^8 door is 2 feet 6 inches wide by 6 feet 8 inches high.

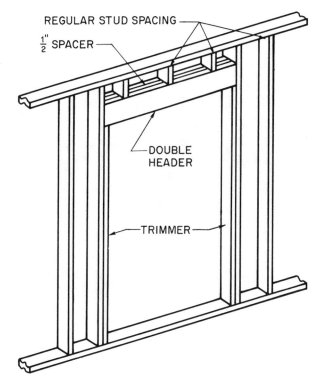

Figure 21-4 The trimmer studs support the header.

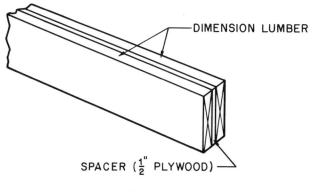

Figure 21-5 Solid wood header

The information in this unit is intended only to help you determine rough opening sizes. More information about windows and doors can be found in Unit 26.

FRAMING OPENINGS

In stud-wall construction, it is usually necessary to cut off or completely eliminate one or more studs where windows and doors are installed. The load normally carried by these studs must be transferred to the sides of the opening. The construction over an opening which transfers this load to the sides of the opening is called the *header.* Additional studs are installed at the sides to carry this load, Figure 21-4.

A set of construction drawings includes details showing the type of rough opening construction intended. The simplest type of wood construction uses two 2x10s or 2x12s with plywood spacers between to form the header, Figure 21-5. A flat 2x4 may be nailed to the bottom of the header to reduce the height of the top of the opening.

To conserve lumber in nonbearing walls, the header may be two 2x4s. The area over the header is framed with cripple studs. The area below a window, also, is framed with *cripple studs,* Figure 21-6. These cripple studs are installed on the normal spacing for wall studs. This may require a cripple within

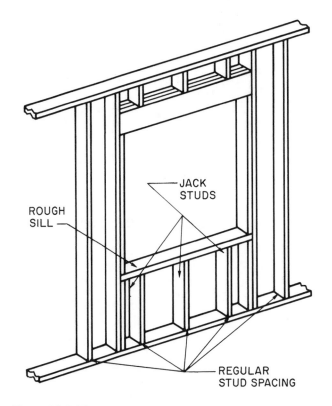

Figure 21-6 The spaces above and below the window opening are framed with cripple studs.

a few inches of a side trimmer, but is necessary to provide a nailing surface for the sheathing. To conserve material and reduce the area of solid wood exposed to the exterior, trussed headers may be used, Figure 21-7. The spaces in the trussed header also allow for more insulation.

Several systems of metal framing are available for light construction. The basic elements of light-gauge metal framing are the same as wood framing, Figure 21-8. These systems use top and bottom wall plates, studs, and floor joists. To make light-gauge framing compatible with wood, the metal members are made in common sizes for wood framing. The greatest difference is that metal framing is joined with screws instead of nails.

Headers over openings in light-gauge metal framing are usually very similar to those in wood framing. However, the system designed by the metal framing manufacturer should always be followed.

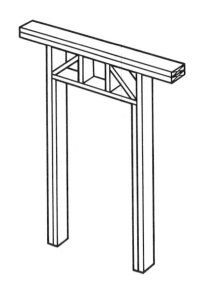

Figure 21-7 Trussed header

Courtesy of Zinc Institute, Inc.

Figure 21-8 The parts of this steel frame are the same as a wood frame.
Notice the window header and cripple studs.

✓ CHECK YOUR PROGRESS

Can you perform these tasks?
- ☐ Identify each door or window listed on a schedule of doors and windows according to its location on the floor plan.
- ☐ Determine the rough opening for a window from the information in a manufacturer's catalog.
- ☐ Describe the framing for a window opening.
- ☐ Describe the framing for a door opening.
- ☐ Determine the rough opening for a door.
- ☐ Explain the difference between a solid wood header and a trussed header.

ASSIGNMENT

Refer to the Lake House drawings (in the packet) to complete the assignment.

1. What type of header should be used over the door from the hall to bedroom #1?

2. What type of header should be used over the door from the deck to the kitchen?

3. Why should these two headers be made differently?

4. What is the length of the header over the garage overhead door? Allow for one trimmer at each side and 1 1/4" for jambs at each side.

5. What are the R.O. dimensions for the door from the deck into the garage?

6. Name the location and give the R.O. dimensions for each interior door on the Upper Level Floor Plan.

7. According to Figure 21-3, what are the R.O. dimensions for the windows in bedroom #2?

8. How long is the header over the window in bedroom #2?

9. How many cripple studs are needed beneath the windows in the south wall of the living room?

10. What is the length of the cripple studs beneath the window in bedroom #2? (Assume the bottom of the header is 6'-8 1/2" from the top of the subfloor.)

Roof Construction

Section 6

UNIT 22   Roof Construction Terms

OBJECTIVES

After completing this unit, you will be able to perform the following tasks:

- Identify common roof types.

- Define the terms used in laying out and constructing a roof.

- State roof pitches as a ratio of rise to run or as a fraction.

TYPES OF ROOFS

Several types of roofs are commonly used in residential construction, Figure 22-1. Variations of these roof types may be used to create certain architectural styles.

The *gable roof* is one of the most common types used on houses. The gable roof consists of two sloping sides which meet at the ridge. The triangle formed at the ends of the house between the top plates of the wall and roof is called the *gable*.

The *gambrel roof* is similar to the gable roof. On this roof, the sides slope very steeply from the

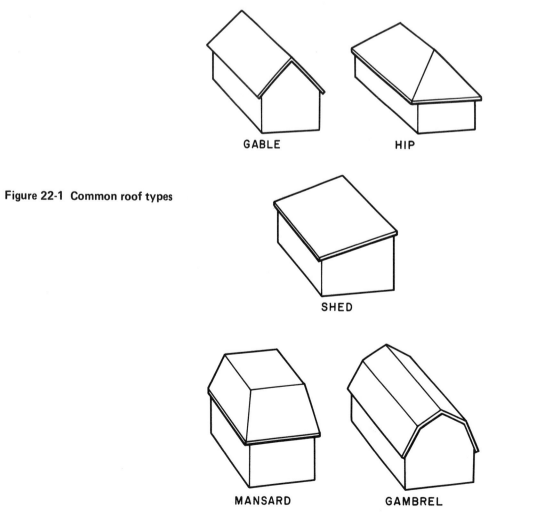

GABLE   HIP

SHED

MANSARD   GAMBREL

Figure 22-1 Common roof types

walls to a point about halfway up the roof. Above this point, they have a more gradual slope.

The *hip roof* slopes on all four sides. The hip roof has no exposed wall above the top plates. This means that all four sides of the house are equally protected from the weather.

The *mansard roof* is similar to the hip roof, except the lower half of the roof has a very steep slope and the top half is more gradual. This roof style is used extensively in commercial construction — on stores, for example.

The *shed roof* is a simple sloped roof with no ridge. A shed roof is much like one side of a gable roof. This type of roof is used in modern architecture and for additions to existing buildings.

ROOF CONSTRUCTION TERMS

The roof construction terms defined in the following list are illustrated in Figure 22-2.

- *Span* is the distance between the outsides of the walls covered by a roof.

- *Run* is the horizontal distance covered by one rafter. Run does not include any part of the rafter that extends beyond the wall. On a common two-sided roof, the run is one-half the span.

- *Rise* is the vertical distance from the top of the wall to the measuring line of the ridge board.

- The *measuring line* is an imaginary line along which all roof dimensions are taken. The measuring line of a rafter is a line parallel to

its edges and passing through the deepest part of the bird's mouth.

- A *bird's mouth* is a notch cut in the lower edge of the rafter to fit around the wall plate.

- The *ridge board* is the horizontal framing member to which the upper ends of the rafters are connected.

- The *overhang* is the horizontal distance covered by the roof outside the walls.

- The *tail* is the portion of the rafter that is outside the walls. The rafter tail is measured along the measuring line and forms the overhang.

- The *pitch* of a roof is a way of indicating how steep the roof is. Pitch can be given as a fraction or as the amount of rise per foot of run.

In the fractional method of indicating pitch the rise is the numerator (top of the fraction) and the span the denominator (bottom of the fraction), Figure 22-3. This fraction is always reduced to the lowest possible terms. (See Math Review 1.)

In the rise-per-foot-of-run method, the pitch is given as the number of units (feet or inches) of rise for every twelve units of run, Figure 22-4. The rise per foot of run is given on drawings with the symbol ▱. The horizontal leg of this triangle represents the run. The vertical leg represents the rise, Figure 22-5. Notice that Figure 22-5 shows two roof pitches. These pitches are written as 10 in 12 and 5 in 12.

When the dimensions are given for a run of other than 12 feet, the rise per foot of run can be

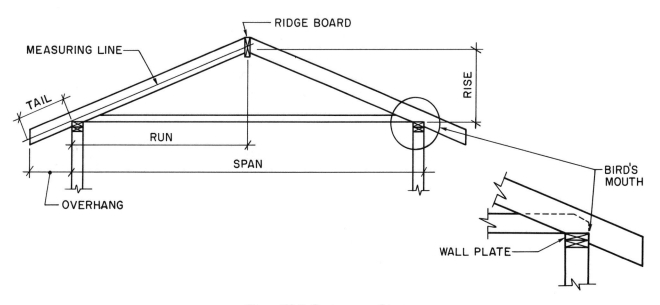

Figure 22-2 Common roof terms

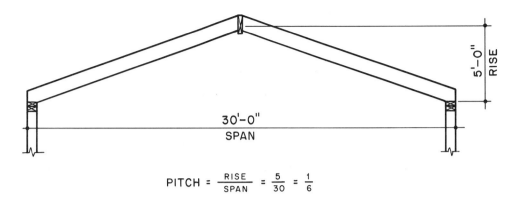

$$\text{PITCH} = \frac{\text{RISE}}{\text{SPAN}} = \frac{5}{30} = \frac{1}{6}$$

Figure 22-3 Fractional pitch

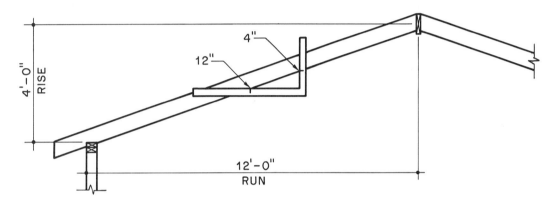

Figure 22-4 Rise per foot of run: This roof has 4 in 12 pitch.

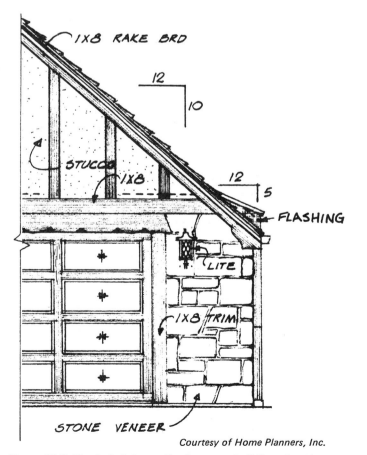

Courtesy of Home Planners, Inc.

Figure 22-5 Roof pitch is usually shown on building elevations.

calculated: Divide 12 by the actual run to find the proper ratio of rise. Multiply this result by the actual rise to find the rise per foot of run. For example, the run is 14'-6" and the rise is 5'-0"; what is the rise per foot of run? 12 ÷ 14'-6" (14.5') = 0.83. 0.83 × 5 = 4.15. The result (4.15) is close enough to 4 that rafter calculations can be based on 4 in 12.

✓ CHECK YOUR PROGRESS

Can you perform these tasks?

☐ Identify the following roof types from information given on construction drawings: gable, gambrel, hip, mansard, and shed.

☐ List each of the following from information given on construction drawings: span, run, rise, overhang, rafter tails, pitch.

ASSIGNMENT

1. Give the following information for the roof shown in Figure 22-6:
   a. Span
   b. Run
   c. Rise
   d. Overhang
   e. Length of the rafter tails
   f. Fractional pitch
   g. Rise per foot of run

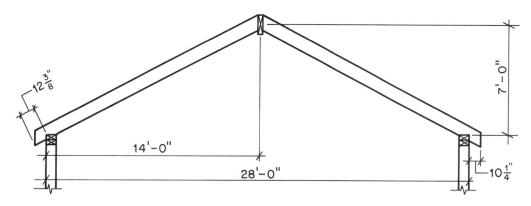

Figure 22-6

Refer to Lake House drawings (in the packet) to complete rest of assignment.

2. What style roof is used for most of the Lake House?
3. What is the span of the rafters over the Lake House garage?
4. What is the rise per foot of run of the rafters over the Lake House garage?
5. What is the run of the rafters over the Lake House bedroom #1?
6. What is the rise per foot of run of the rafters over the Lake House bedroom #1.
7. What is the overhang of the rafters over the Lake House bedroom #1?

UNIT 23 Common Roof Framing

OBJECTIVES

After completing this unit, you will be able to perform the following tasks:

- Find information about roof construction on drawings.
- Calculate the length of common rafters.
- Calculate the length of rafter tails when overhang is given.
- Describe gable framing.

ROOF CONSTRUCTION

The roof is designed to support weight and give protection from the weather. In a common frame roof, the rafters and ridge board are the structural members. They are sized and spaced to support the weight of the roof itself plus any snow that can be expected. The protection from weather is provided by sheathing (or roof decking) and roofing material, Figure 23-1.

The size and spacing of the rafters varies depending on their length, pitch, and *load* (weight they must support). If the rafters span a great distance or are spaced far apart, they must be deep to support their load. (The vertical dimension of rafters and joists is called *depth*.) The size and spacing of the rafters is shown on a section view, Figure 23-2. The ridge board is also usually shown on a section view. The ridge board should be made of stock two inches deeper than the rafters. The greater depth is needed because an angled cut across the rafters is longer than a square cut across the rafters, Figure 23-3.

When the roof framing is complicated, a separate roof framing plan may be included, Figure

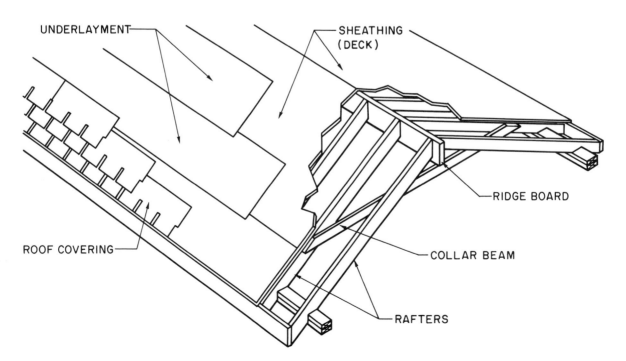

Figure 23-1 Elements of a roof

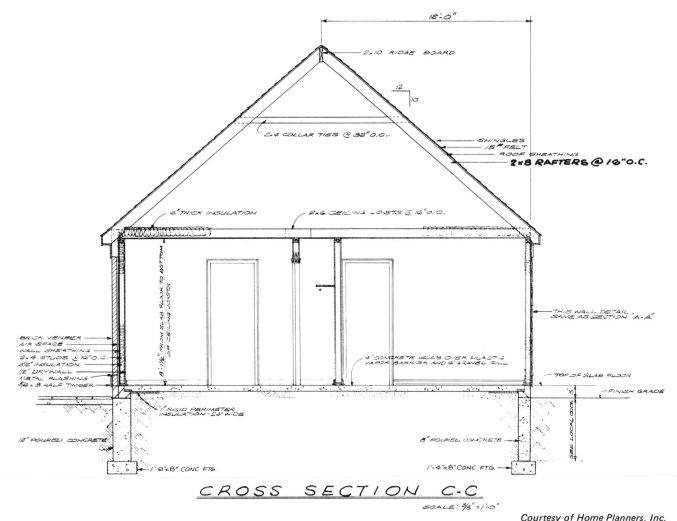

CROSS SECTION C-C

SCALE: 3/8" = 1'-0"

Courtesy of Home Planners, Inc.

Figure 23-2 A section view of the roof shows the size and spacing of rafters.

23-4. Roof framing plans are used only to show the general arrangement of the framing. Therefore, unless specific dimensions are included, framing plans should not be relied upon for the sizes of the framing members.

The roof frame may also include *collar beams*. Collar beams are usually made of one-inch (nominal thickness) lumber. They are normally included on every second or third pair of rafters. This information should be shown on the roof section if collar beams are planned.

The pitch of the roof may be shown on any section view that shows the rafter size. Pitch is usually also shown on the building elevations.

GABLE AND RAKE FRAMING

The triangle formed by the top plate of the end walls and the rafters in a gable roof is the *gable*. The gable is framed like the lower parts of the house walls. Unless a detail drawing shows differently, the gable is framed with the same size and spacing of studs as the rest of the walls, Figure 23-5.

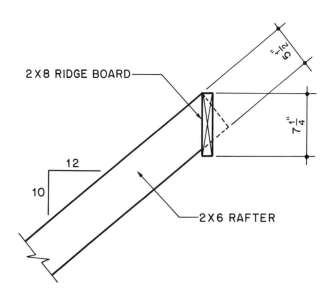

Figure 23-3 The ridge board must be made of wider lumber than the rafters to allow for the angled plumb cut.

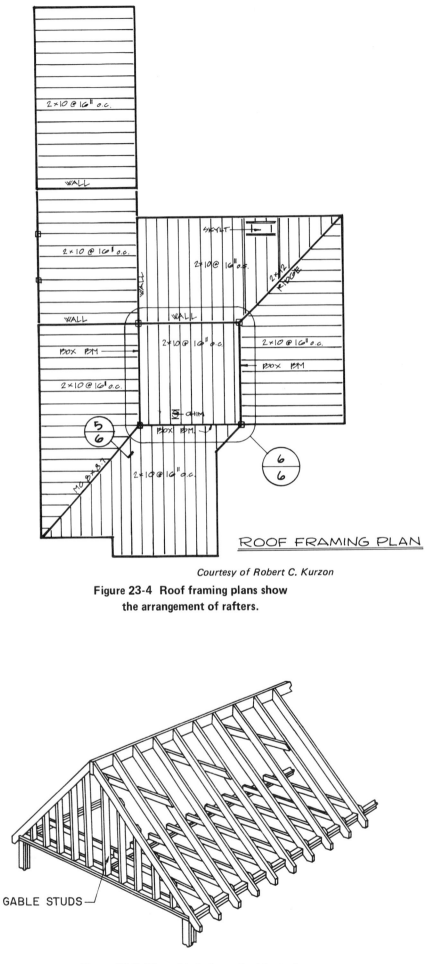

2×10 @ 16" o.c.

WALL

2×10 @ 16' o.c.

WALL

WALL

BOX BM

2×10 @ 16' o.c.

5
6

2×10 @ 16" o.c.

SKYLT

2×10 @ 16" o.c.

2×12
RIDGE

2×10 @ 16" o.c.

2×10 @ 16" o.c.

BOX BM

CHIM

BOX BM

6
6

ROOF FRAMING PLAN

Courtesy of Robert C. Kurzon

**Figure 23-4 Roof framing plans show
the arrangement of rafters.**

GABLE STUDS

Figure 23-5 The gable is framed with studs.

The end of a gable roof is called the *rake*. If the rake is *tight* (does not overhang), the gable studs are notched to fit the rake rafters, Figure 23-6. If there is a rake overhang, the gable framing includes a plate. The rake rafters are then supported by *lookouts* nailed to the last rafter and the gable plate, Figure 23-7. The overhang of the lookouts should not be more than their length inside the gable frame.

ROOF COVERING

The most common roof decking material for houses is plywood. However, in post-and-beam construction where the extra strength is needed to span the distance between the rafters, dimensional lumber is used for the roof deck.

The material used to provide weather protection is often chosen for its architectural style, Figure 23-8. Some of the most common materials are asphalt shingles, wood shingles, and terra-cotta tiles. The material to be used is shown on the section view of the roof and usually on the building elevations. The roofing should be applied over a layer of asphalt-saturated building paper — sometimes called *slater's felt*.

Asphalt roofing materials, including felt and shingles, are sold by their weight per hundred square feet of coverage. One hundred square feet of roof is called a *square*. If enough shingles to cover one square weigh 235 pounds, they are called 235-lb shingles. The felt used under roofing is typically 15-lb weight.

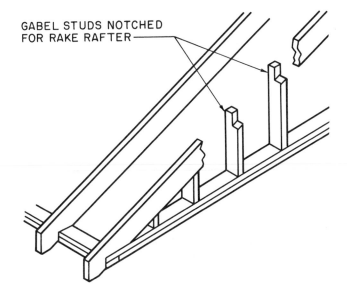

GABEL STUDS NOTCHED
FOR RAKE RAFTER

Figure 23-6 Gable framing for tight rake

Figure 23-8 Tile roofs are popular in
some parts of the country.

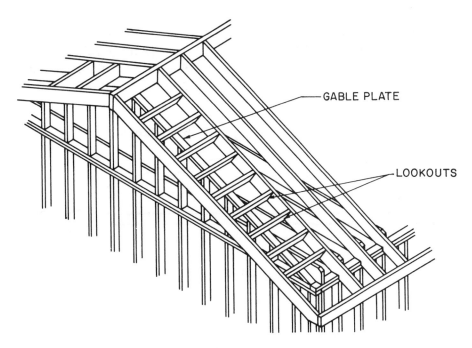

GABLE PLATE

LOOKOUTS

Figure 23-7 Framing for an overhang at the gable end

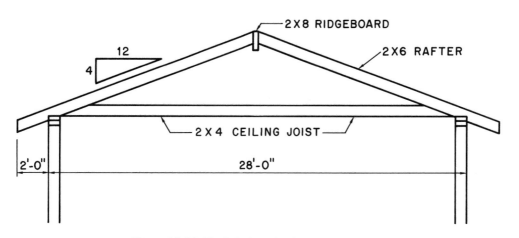

(Top rafter table — readings under each inch graduation)

	23	22	21	20	19	18	17	16	15	14	13	12
LENGTH COMMON RAFTERS PER FOOT RUN					21 63	20 81	20	19 21	18 44	17 69	16 97
" HIP OR VALLEY " " "					24 74	24 02	23 32	22 65	22	21 38	20 78
DIFF IN LENGTH OF JACKS 16 INCHES CENTERS					28 7/8	27 3/4	26 11/16	25 5/8	24 9/16	23 9/16	22 5/8
" " " 2 FEET JACKS					43 1/4	41 5/8	40	38 7/16	36 7/8	35 3/8	33 15/16
SIDE CUT OF JACKS USE					6 11/18	6 15/16	7 3/16	7 1/2	7 13/16	8 1/8	8 1/2
" " HIP OR VALLEY					8 1/4	8 1/2	8 3/4	9 1/16	9 3/8	9 5/8	9 7/8

(Lower rafter table — readings under each inch graduation)

12	11	10	9	8	7	6	5	4	3	2	1
16 97	16 28	15 62	15	14 42	13 89	13 42	13	12 65	12 37	12 16	
20 78	20 22	19 70	19 21	18 76	18 36	18	17 69	17 44	17 23	17 09	
22 5/8	21 11/16	20 13/16	20	19 1/4	18 1/2	17 7/8	17 5/16	16 7/8	16 1/2	16 1/4	
33 15/16	32 9/16	31 1/4	30	28 7/8	27 3/4	26 13/16	26	25 5/16	24 3/4	24 5/16	
8 1/2	8 7/8	9 1/4	9 5/8	10	10 3/8	10 3/4	11 1/16	11 3/8	11 5/8	11 13/16	
9 7/8	0 1/8	10 3/8	10 5/8	10 7/8	11 1/16	11 5/16	11 1/2	11 11/16	11 13/16	11 15/16	

Figure 23-9 Rafter table on the face of a square: The top line is the length of common rafters per foot of run.

FINDING THE LENGTH OF COMMON RAFTERS

Carpenters use a rafter table to find the length of rafters. These tables are available in handbooks and are printed on the face of a framing square, Figure 23-9. To find the length of a common rafter, you must know the run of the rafter and its rise per foot of run. The run can be found on the floor plan of the building. The rise per foot of run is shown on the building elevations. The length of the common rafter is then found by following these steps:

Step 1. Find the number of inches of rise per foot of run at the top of the table. These numbers are the regular graduations on the square.

Step 2. Under this number, find the length of the rafter per foot of run. A space between the numbers indicates a decimal point.

Step 3. Multiply the length of the common rafter per foot of run (the number found in Step 2) by the number of feet of run.

Step 4. Add the length of the tail and subtract one-half the thickness of the ridge board. The result is the length of the common rafter as measured along the measuring line.

Note: If the overhang is given on the working drawings, it can be added to the run of the rafter instead of adding the length of the tail.

Example: Find the length of a common rafter for the roof in Figure 23-10.

1. Rise per foot of run = 4"
2. Length of common rafter per foot of run = 12.65"
3. Run of one rafter including overhang = 16'-0"
4. 16 × 12.65" = 202.40" (round off to 202 1/2")
5. Subtract 1/2 the thickness of the ridge board: 202 1/2" – 3/4" = 201 3/4"

2 X 8 RIDGEBOARD

2 X 6 RAFTER

12

4

2 X 4 CEILING JOIST

2'-0"   28'-0"

Figure 23-10 Find the length of a common rafter.

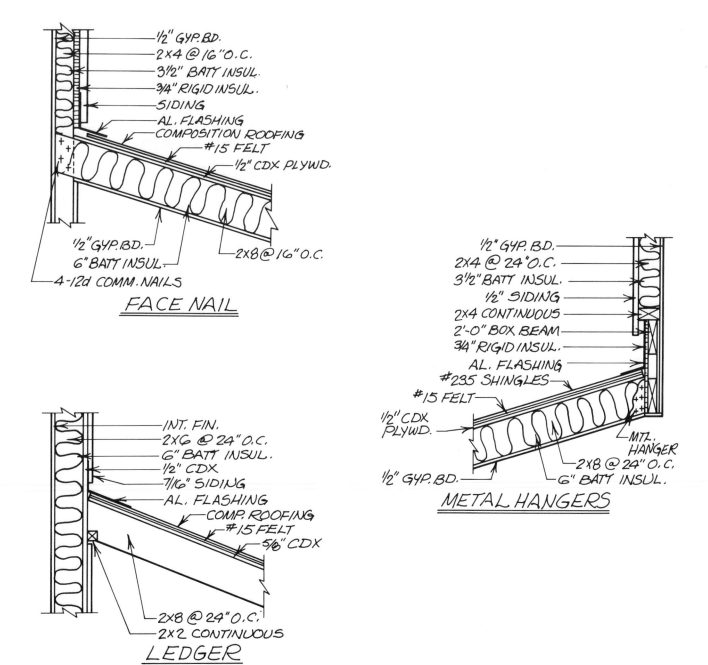

Figure 23-11 Details of shed roof to wall

SHED ROOF FRAMING

Most of the Lake House has shed roof construction. That is, the common rafters cover the full span of the area they cover in a single slope. To find the length of shed rafters, treat the entire span of the rafters as *run*. For example, the total width of the garage is 13'-9". There is no overhang on either side of the garage, so the span of the garage rafters is 13'-9". Use 13'-9", or 13.75 feet, as the run in calculating the length of common rafters. (See Math Review 20.)

Where shed rafters butt against a wall or other vertical surface, the drawings should include a detail to show how they are fastened. Figure 23-11 shows three methods of fastening rafters to a vertical surface.

ROOF OPENINGS

It is often necessary to frame openings in the roof. Where chimneys, skylights, or other features require openings through the rafters; headers and double framing members are used. This method of framing openings is similar to that used in floor framing.

✓ CHECK YOUR PROGRESS ────────────────────────────

Can you perform these tasks?

☐ List the size and spacing of rafters, ridge board, and collar beams.

☐ Describe the framing of a gable and rake overhang.

☐ List the materials to be used for roof covering.

☐ Use a rafter table to find the length of a common rafter including the rafter tail.

ASSIGNMENT ──────────────────────────────────

Give the following information for the rafters of the Lake House:

Rafter Location	Thickness X Depth	Run	Rise per Foot Run	Length	O.C.Spacing
1. Kitchen					
2. Bedroom #2 & Loft					
3. Living room					
4. Bedroom #1					
5. Garage					

UNIT 24   Hip and Valley Framing

OBJECTIVES

After completing this unit, you will be able to perform the following tasks:

- Calculate the length of hip rafters.
- Calculate the length of valley rafters.
- Calculate the length of hip and valley jack rafters.

HIP RAFTERS

Hip rafters run from the corner of the building to the ridge at a 45° angle, Figure 24-1. The length of hip rafters can be found by using a table found on most framing squares, Figure 24-2. The second line of this table is used to calculate the length of hip rafters. This table is based on the unit-run-and-rise method for finding the length of common rafters, explained earlier in Unit 23.

To calculate the length of a hip rafter, you must know the run of the common rafters in the

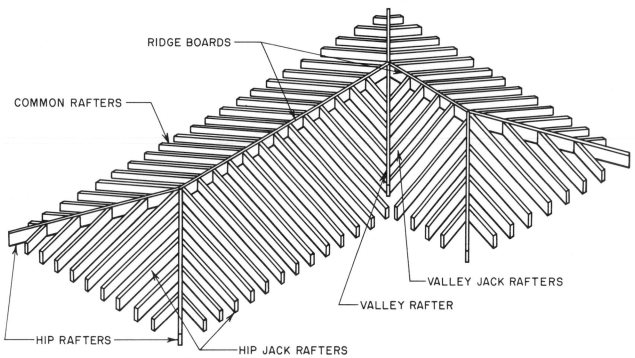

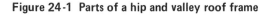

Figure 24-1 Parts of a hip and valley roof frame

					23	22	21	20	19	18	17	16	15	14	13	12
LENGTH	COMMON	RAFTERS	PER FOOT	RUN						21 63	20 81	20	19 21	18 44	17 69	16 97
"	HIP OR	VALLEY	"	"						24 74	24 02	23 32	22 65	22	21 38	20 78
DIFF	IN LENGTH	OF JACKS	16 INCHES	CENTERS						28 7/8	27 3/4	26 11/16	25 5/8	24 9/16	23 9/16	22 5/8
"	"	"	2 FEET	"						43 1/4	41 5/8	40	38 7/16	36 7/8	35 3/8	33 15/16
SIDE	CUT	OF	JACKS	USE						6 11/18	6 15/16	7 3/16	7 1/2	7 13/16	8 1/8	8 1/2
"	"	HIP OR	VALLEY	"						8 1/4	8 1/2	8 3/4	9 1/16	9 3/8	9 5/8	9 7/8

12	11	10	9	8	7	6	5	4	3	2	1
16 97	16 28	15 62	15	14 42	13 89	13 42	13	12 65	12 37	12 16	
20 78	20 22	19 70	19 21	18 76	18 36	18	17 69	17 44	17 23	17 09	
22 5/8	21 11/16	20 13/16	20	19 1/4	18 1/2	17 7/8	17 5/16	16 7/8	16 1/2	16 1/4	
33 15/16	32 9/16	31 1/4	30	28 7/8	27 3/4	28 13/16	26	25 5/16	24 3/4	24 5/16	
8 1/2	8 7/8	9 1/4	9 5/8	10	10 3/8	10 3/4	11 1/16	11 3/8	11 5/8	11 13/16	
9 7/8	0 1/8	10 3/8	10 5/8	10 7/8	11 1/16	11 5/16	11 1/2	11 11/16	11 13/16	11 15/16	

Figure 24-2 Rafter table on the face of a square

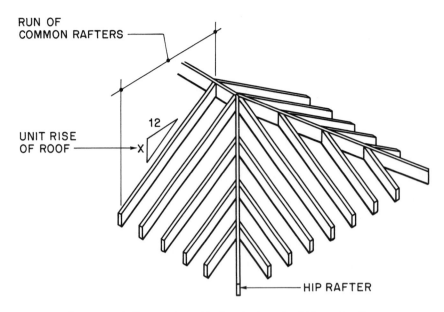

Figure 24-3 To use the table for the length of hip and valley rafters use the run of the common rafters in the roof.

roof and the unit rise of the roof, Figure 24-3. The length of the hip rafters is then found by using the table for length of hip and valley rafters in the same way the table for the length of common rafters was used in Unit 23.

Step 1. Find the unit rise (number of inches of rise per foot of run) at the top of the table. These numbers are the regular graduations on the square.

Step 2. Under this number, find the length of the hip rafter per foot of run of the common rafters.

Step 3. Multiply the length of the hip rafter per foot of common-rafter run (the number found in Step 2) by the number of feet of run of the common rafters (1/2 the width of the building).

Step 4. Subtract the ridge allowance. Because the hip rafter meets the ridge board at a 45° angle, the ridge allowance is one half the 45° thickness of the ridge board, Figure 24-4. The *45° thickness* is the length of a 45° line across the thickness of the ridge board. The 45° thickness of a 1 1/2-inch (2-inch nominal) ridge board is 2 1/8 inches. Therefore, the ridge allowance

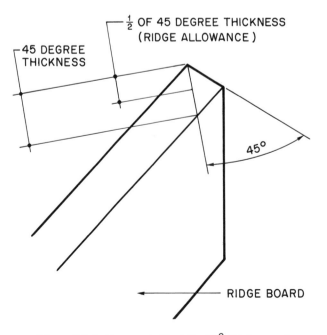

Figure 24-4 Use one-half of the 45° thickness of the ridge as a ridge allowance for hip rafters.

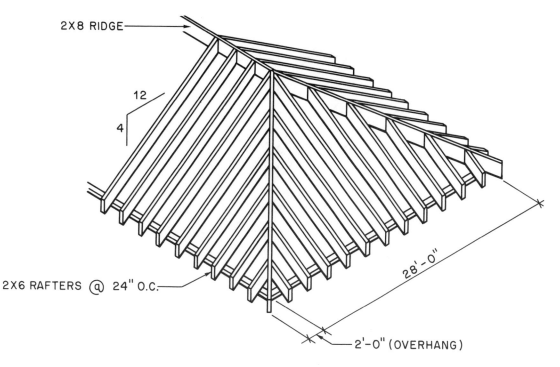

Figure 24-5 Find the length of a hip rafter.

for a hip rafter on a 1 1/2-inch ridge board is 1 1/16 inches.

Note: If the hip rafter includes an overhang, add the overhang of the common rafters to the run of the common rafters.

Example: Find the length of a hip rafter for the roof shown in Figure 24-5.

1. Rise per foot of run = 4"
2. Length of hip rafter per foot of common-rafter run = 17.44"
3. Common-rafter run including overhang = 16'-0"

4. 16 × 17.44" = 279.04" (round off to 279 1/16")
5. Subtract 1 1/16" ridge allowance: 279 1/16"- 1 1/16" = 278"

The rafters that butt against the hip rafter (called *hip jack rafters*) are cut at an angle, Figure 24-6. This angled cut produces a surface which is longer than the width of the lumber from which the rafter is cut. Therefore, the hip rafters are made of wider lumber than the common rafters and jack rafters.

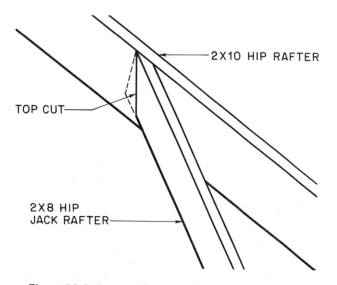

Figure 24-6 Because the top of the hip jack is cut on an angle, the hip rafter must be wider.

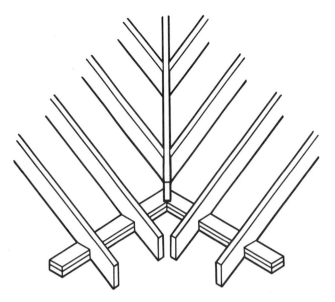

Figure 24-7 On some roofs the common rafter tails are close enough so no tail is needed on the valley rafters.

VALLEY RAFTERS

The line where two pitched roofs meet is called a *valley*. The rafter that follows the valley is a *valley rafter*, as shown in Figure 24-1. It is most common for both roofs to have the same pitch. This results in the valley rafter being at a 45° angle with both ridges — the same angle as a hip rafter. Because the angles of the hip and valley rafters are the same and the pitch is the same, the same table can be used to compute their lengths.

All steps of the procedure given earlier for hip rafters can be followed to find the length of valley rafters. However, the valley rafters often have no tail even though the roof has an overhang, Figure 24-7. In this case, the total length of the valley rafters is computed on the basis of the run of the common rafters excluding the overhang.

When both roofs have the same span and rise, the valley extends from the eave to the ridge, Figure 24-8. When one roof has a greater span than the other, the valley does not reach the ridge, Figure 24-9. In this case, the valley is framed in one of two ways. One valley rafter can extend to the ridge, and the other can butt against the first, Figure 24-10. The other method is to install a common rafter on the wider roof in line with the ridge of the narrower roof. Both valley rafters can then butt against this common rafter, Figure 24-11.

The length of the valley rafters is based on the run of the common rafters that have their upper

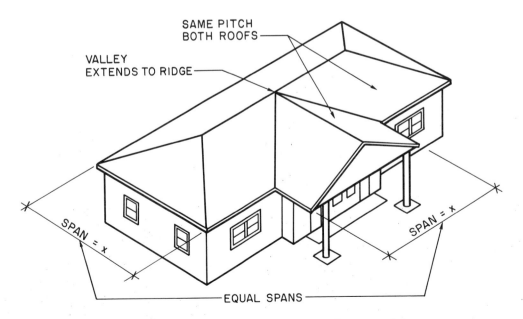

Figure 24-8 When both spans are the same distance and the same pitch, the valley goes all the way to the ridge.

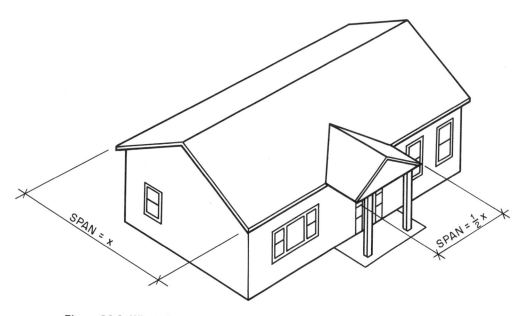

Figure 24-9 When the spans are not equal, the valley does not reach the ridge.

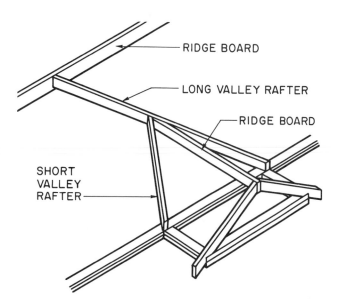

Figure 24-10 Framing valleys with long and short valley rafters

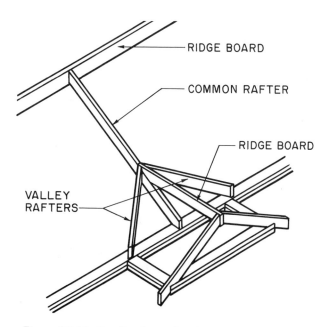

Figure 24-11 A valley framed against a common rafter.

(ridge) ends at the same level as the valley rafter. In other words, when the long and short valley rafters are used, the length of the long valley rafter is based on the run of the common rafters in the wider roof. The length of the short valley rafter is based on the run of the common rafters in the narrower roof. When both valley rafters butt against a common rafter of a higher roof, their lengths are based on the run of the common rafters in the narrower roof.

JACK RAFTERS

Rafters that extend from the wall plate to a hip rafter are called *hip jack rafters*, as was shown in Figure 24-1. Those that extend from a valley rafter to the ridge board are called *valley jack rafters*, also shown in Figure 24-1.

The third and fourth lines of the rafter table on most framing squares, shown in Figure 24-2, are used to calculate the length of jack rafters. The length of each jack rafter in a roof varies from the length of the one next to it by the same amount, Figure 24-12. The amount of this variance depends on the spacing of the rafters and the pitch of the roof. The third line of the rafter table is used when the rafters are spaced 16 inches O.C.; the fourth line is used when they are spaced 24 inches O.C. As with the other lines of the rafter table, the inch numerals at the top of the square are used to indicate the unit rise of the roof. For example, if a roof has a 6 in 12 slope and the roof framing is 16 inches O.C., the difference in the length of jack rafters is 17 7/8 inches.

The first jack rafter should be a full space (16 inches or 24 inches) from the bottom of the hip or the top of the valley. Therefore, the first jack rafter should be the length shown on the rafter table.

The length of the tail must be added to the theoretical length from the table. The length of the tail on hip jacks is the same as the length of the tail on common rafters. Therefore, the length of the tail can be found by using the table for the length of common rafters. Simply treat the tail as a very short common rafter. If the overhang is not in even feet, divide the length of common rafter per foot of run on the table by twelve. This gives you the length of the common rafter (or jack rafter tail) per inch of run.

Example: Find the length of the tails of the jack rafters in Figure 24-13.

1. Unit rise (rise per foot of run) = 4''
2. Length of common rafter per foot of run = 12.65''

3. Length of common rafter per inch of run =
$$\frac{12.65''}{12} = 1.05''$$

4. Run of rafter tail (overhang) = 8''
5. Length of rafter tail = 8 × 1.05'' = 8.4'', approximately 8 7/16''.

Hip jack rafters are also shortened at the top to allow for the thickness of the hip rafter they butt against. This allowance is one-half the 45° thickness of the hip rafter. Valley jack rafters are shortened at the bottom to allow for the thickness of the valley rafter. This allowance is one-half the 45° thickness of the valley rafter. The valley jack rafters are also shortened at the top to allow for the thickness of the ridge board. The ridge board allowance for valley jacks is the same as the ridge board allowance for common rafters — one-half the actual thickness of the ridge board.

Example: Find the length of the hip jack rafters (A, B, and C) and the valley jack rafters (D, E, and F) in Figure 24-13.

Hip Jack Rafters

1. Rise per foot of run = 4''
2. Spacing of rafters = 24'' O.C.
3. Difference in the length of jacks = 25 5/16''
4. Theoretical length of hip jack rafter A = 0 + 25 5/16'' = 25 5/16''
5. Add tail as found in earlier example: 25 5/16'' + 8 7/16'' = 33 3/4''
6. Subtract one half the 45° thickness of the hip rafter 33 3/4''- 1 1/16'' = 32 11/16'' (actual length of A)
7. Actual length of hip jack B - 32 11/16'' (length of A) + 25 5/16'' (from rafter table) = 58''
8. Actual length of hip jack C = 58'' (length of B) + 25 5/16'' = 83 5/16''

Valley Jack Rafters

9. Theoretical length of valley jack rafter D = 0 + 25 5/16'' = 25 5/16''
10. Subtract one-half the 45° thickness of the valley rafter: 25 5/16''- 1 1/16'' = 24 1/4''
11. Subtract one-half the actual thickness of the ridge board: 24 1/4''- 3/4'' = 23 1/2'' (actual length of D)
12. Actual length of valley jack E = 23 1/2'' (length of D) + 25 5/16'' (from rafter table) = 48 13/16''
13. Actual length of valley jack F = 48 13/16'' (length of E) + 25 5/16'' = 74 1/8''

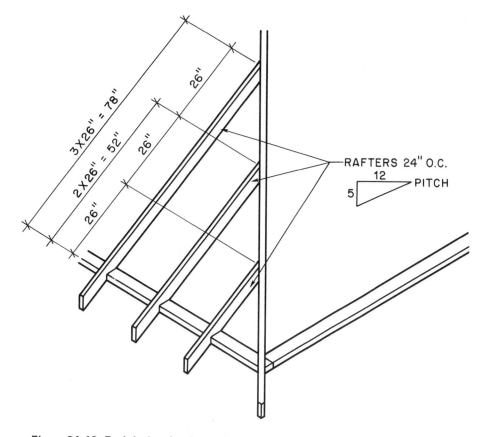

Figure 24-12 Each jack rafter in a string varies from the next by the same amount.

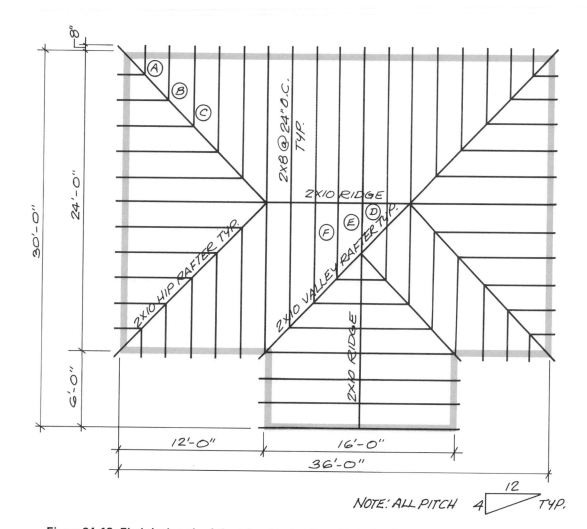

Figure 24-13 Find the length of the tails of jack rafters, hip jack rafters, and valley jack rafters.

✓ CHECK YOUR PROGRESS

Can you perform these tasks?
- ☐ Use a rafter table to find the length of hip rafters.
- ☐ Use a rafter table to find the length of valley rafters.
- ☐ Use a rafter table to find the length of hip jack rafters and valley jack rafters.

ASSIGNMENT

A.   Refer to Figure 24-14 to complete questions 1-11.

1.   What is the run of the common rafters at A?
2.   How much overhang does the roof have?
3.   What is the actual length of the common rafters at A?
4.   What is the actual length of the hip rafter at B?
5.   What is the run of the common rafters at C?
6.   What is the actual length of the common rafters at C?
7.   What is the length of the short valley rafter?
8.   What is the actual length of the shortest hip jack rafter?
9.   What is the actual length of the second shortest hip jack rafter?
10.   What is the actual length of the shortest valley jack rafter?
11.   What is the actual length of the second shortest valley jack rafter?

Figure 24-14

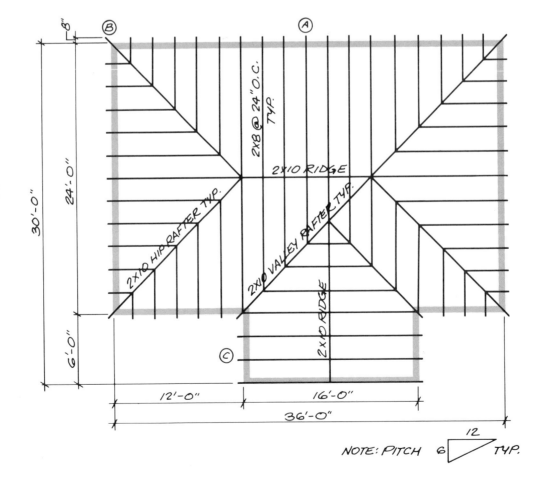

B.   Refer to Lake House drawings (in the packet) to complete this part of the assignment.

What is the length of the structural steel hip rafter over the dining room?

Notes: • This hip rafter is a steel channel shown on details 5/6 and 6/6, and marked as MC8X8.7

• Remember to allow for the distance from the column centerline and the end of the rafter as dimensioned on the detail drawing.

• This roof has an unusual pitch of 2.96 in 12. This is close enough to use 3 in 12 for calculating rafter lengths.

OUACHITA TECHNICAL COLLEGE

Exterior Trim

Section 7

UNIT 25 Cornices

OBJECTIVES

After completing this unit, you will be able to perform the following tasks:

- Describe the cornice construction shown on a set of drawings.
- List the sizes of the individual parts of the cornice shown on a set of drawings.
- Describe the provisions for attic or roof ventilation as shown on a set of drawings.

TYPES OF CORNICES

The *cornice* is the construction at the place where the edge of the roof joins the sidewall of the building. On hip roofs, the cornice is similar on all four sides of the building. On gable and shed roofs, the cornice follows the pitch of the end (*rake*) rafters. The cornice on the ends of a gable or shed roof is sometimes called simply the rake, Figure 25-1. The three main types of cornice are the *box cornice*, the *open cornice*, and the *close cornice*.

Box Cornice

The box cornice boxes the rafter tails. This type of cornice includes a fascia and soffit, Figure 25-2. The *fascia* covers the ends of the rafter tails. The *soffit* covers the underside of the rafter tails. There are three types of box cornices. These types vary in the way the soffit is applied.

Sloping Box Cornice. In the sloping box cornice, the soffit is nailed directly to the bottom edge of the rafter tails. This causes the soffit to have the same slope or pitch as the rafter, Figure 25-3.

Narrow Box Cornice. In the narrow box cornice, the rafter tails are cut level. The soffit is nailed to this level-cut surface, Figure 25-4.

Figure 25-1 The cornice is the construction at the place where the roof and the sidewall meet.

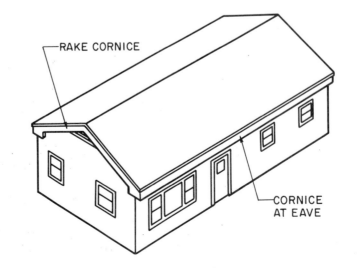

RAKE CORNICE

CORNICE AT EAVE

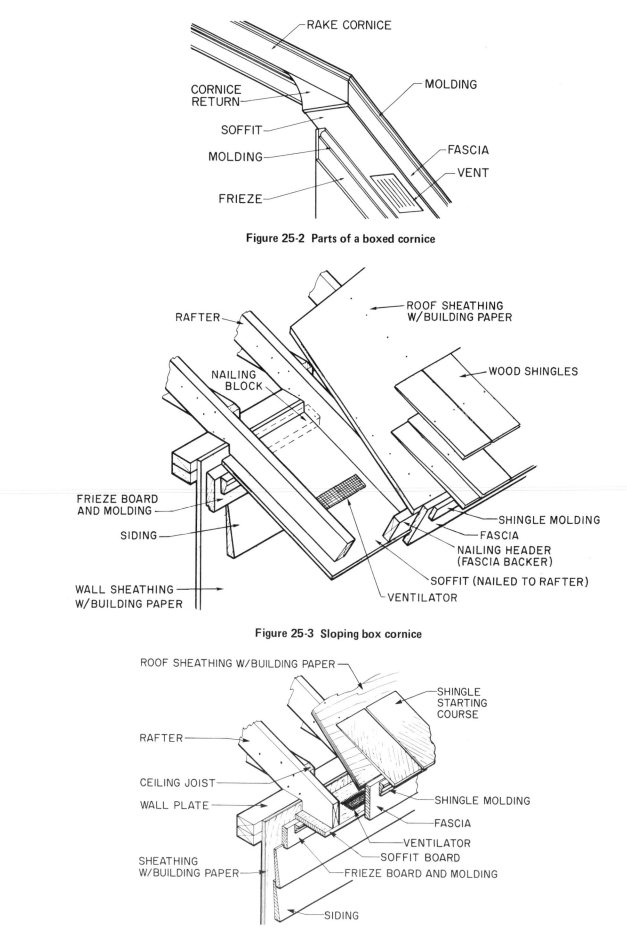

Figure 25-2 Parts of a boxed cornice

Figure 25-3 Sloping box cornice

Figure 25-4 Narrow box cornice

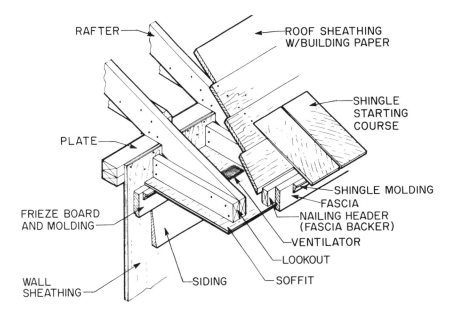

RAFTER

ROOF SHEATHING
W/BUILDING PAPER

SHINGLE
STARTING
COURSE

PLATE

SHINGLE MOLDING
FASCIA
NAILING HEADER
(FASCIA BACKER)
VENTILATOR
LOOKOUT
SOFFIT

FRIEZE BOARD
AND MOLDING

WALL
SHEATHING

SIDING

Figure 25-5 Wide box cornice (with horizontal lookouts)

Wide Box Cornice. In a wide box cornice, the overhang is too wide for a level cut on the rafter tails to hold the full width of the soffit. In conventional wood framing, *lookouts* are installed between the rafter ends and the sidewall. The lookouts provide a nailing surface for the soffit, Figure 25-5. For a metal soffit, special metal channels fastened to the sidewall and back of the fascia hold the soffit, Figure 25-6.

Open Cornice

In an open cornice, the underside of the rafters is left exposed, Figure 25-7. Blocking is

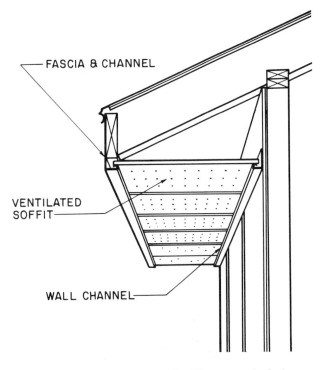

FASCIA & CHANNEL

VENTILATED
SOFFIT

WALL CHANNEL

**Figure 25-6 Metal and vinyl soffit systems include
ventilated soffits and channels.**

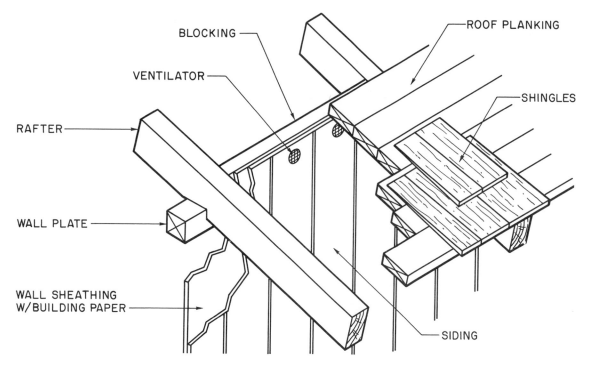

BLOCKING

VENTILATOR

RAFTER

WALL PLATE

WALL SHEATHING
W/BUILDING PAPER

ROOF PLANKING

SHINGLES

SIDING

Figure 25-7 Open Cornice

installed between the rafters and above the wall plate to seal the cornice from the weather. An open cornice may or may not include a fascia.

Close Cornice

In a close cornice, the rafters do not overhang beyond the sidewall, Figure 25-8. The interior

may be sealed by the sheathing and siding or by a fascia. In either case, there must be some provision for ventilation.

CORNICE RETURNS

Any of the types of construction described for cornices can be used for the rake. When a

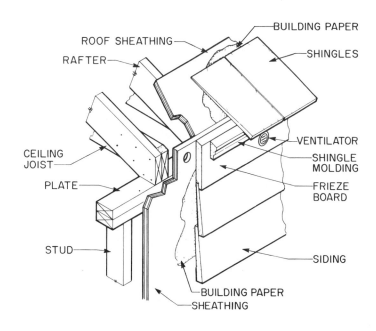

ROOF SHEATHING

RAFTER

CEILING
JOIST

PLATE

STUD

BUILDING PAPER

SHINGLES

VENTILATOR

SHINGLE
MOLDING

FRIEZE
BOARD

SIDING

BUILDING PAPER

SHEATHING

Figure 25-8 Close cornice

Courtesy of Richard T. Kreh, Sr.

Figure 25-9 Cornice return

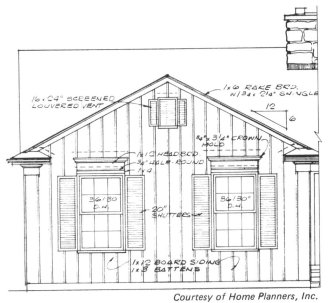

Courtesy of Home Planners, Inc.

Figure 25-10 The cornice returns can be seen on the building elevations.

sloping cornice is used, the fascia and soffit follow the line of the roof up the rake. When a level box cornice is used, a *cornice return* is necessary. This is the construction that joins the level soffit and fascia of the eave with the sloping rake, Figure 25-9.

The style of cornice return is shown on the building elevations, Figure 25-10. Although good architectural drafting practice requires details of all special construction, many architects do not include details of cornice returns. The carpenter is expected to know how to achieve the desired results. Figures 25-11 and 25-12 show the construction of two popular types of cornice returns.

VENTILATION

The cornice usually allows for ventilation of the attic or roof. Attic or roof ventilation is necessary in both hot and cold weather. Without ventilation, the air in the attic becomes stagnant because it is trapped and unable to circulate.

In hot weather, this stagnant air builds up heat and makes the house warmer. Hot, stagnant air can hold a large amount of moisture. When this moisture-laden air comes in contact with the cooler roof, the moisture condenses. The condensation can reduce the effectiveness of the insulation. Condensation can also cause the wood in the attic to rot.

Attic or roof ventilation also helps prevent ice buildup in cold climates. Without ventilation, the heat from the building melts the snow that falls on the roof. As the melted snow reaches the overhang of the roof, it refreezes. Eventually an ice dam may build up. The ice dam can back

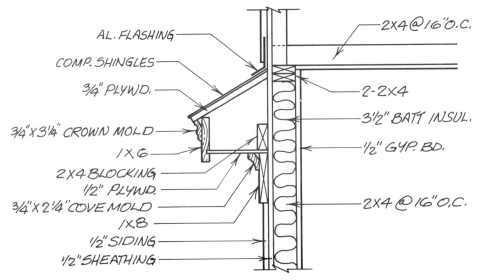

Figure 25-11 Section through cornice return shown in Figure 25-10

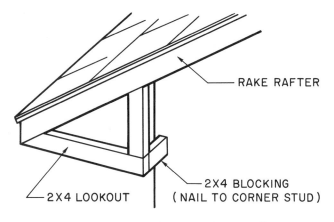

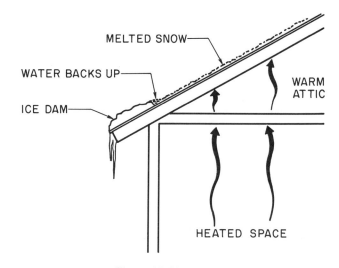

Figure 25-12 Framing for the return shown in Figure 25-9

Figure 25-13 Ice dam

up newly melted snow, causing it to seep under the shingles, Figure 25-13.

Ventilation is created by allowing cool air to enter through the cornice and exit at the ridge or through special ventilators, Figure 25-14. A section view of the sidewalls or a special roof and cornice detail shows the construction of the cornice including any ventilators. Notice that each of the soffits in Figures 25-3, 25-4, and 25-5 includes a ventilator.

The heated air can be allowed to exit in one of three ways. Some buildings have a ventilated, metal ridge cap. This may be shown on the building elevations or on a separate detail. When a ventilated ridge is used, an opening is left in the roof decking at the ridge. Metal roof ventilators can be installed in the roof. Ventilators can be installed in the gables. Metal roof ventilators and gable ventilators are usually shown on the building elevations only.

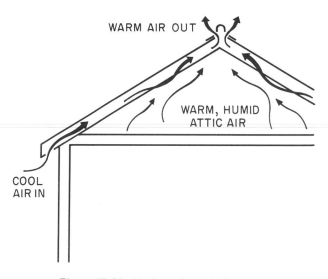

Figure 25-14 Air flow through the attic

✓ CHECK YOUR PROGRESS

Can you perform these tasks?

☐ Describe the construction of a typical sloping box cornice.

☐ Describe the construction of a typical narrow box cornice.

☐ Describe the construction of a typical wide box cornice.

☐ Describe the construction of a typical open cornice.

☐ Describe the construction of a typical close cornice.

☐ Describe the construction of a return for each of the above cornice types.

☐ Trace the flow of ventilating air from the point it enters a cornice to the point it exits the attic.

ASSIGNMENT

Refer to the Duplex and the Lake House drawings (in the packet) to complete the assignment.

1. Which type of cornice does the Duplex have?
2. What material is used for the Duplex cornice?
3. How wide is the Duplex soffit?
4. The Duplex fascia is made of two parts. What are they?
5. What provision does the Duplex cornice have for ventilation?
6. How does attic air exit from the Duplex?
7. Sketch the Lake House cornice and show where air enters for ventilation.
8. There are two ways that air can escape from the Lake House roof. Describe one.

UNIT 26   Windows and Doors

OBJECTIVES

After completing this unit, you will be able to perform the following tasks:

- Interpret information shown on window and door details.
- Find information in window and door manufacturers' catalogs.

WINDOW CONSTRUCTION

Most windows are supplied by manufacturers as completely assembled units. However, the carpenters who install windows often have to refer to window details for information. Some special installations require knowledge of the construction of the window unit. Also, when a special window is required, the carpenter may build parts of it on the construction site.

Wood Windows

The major types of windows were briefly discussed in Unit 21. All these windows include a frame and sash. The *sash* is the glass and the wood (or metal) that holds the glass. The sash is made of *rails* (horizontal parts) and *stiles* (vertical parts), Figure 26-1. The sash may also include muntins. *Muntins* are small strips that divide the glass into smaller panes. The glass is sometimes called the *lite*.

The window frame is made of the *side jambs*, the *head jamb*, and the *sill*. Stop molding is applied to the inside of the jambs to hold the sash in place. Factory-built windows also come with the exterior casing installed. The *casing* is the molding that goes against the wall around the frame. The interior casing and the apron, if one is included, are applied after the window is installed.

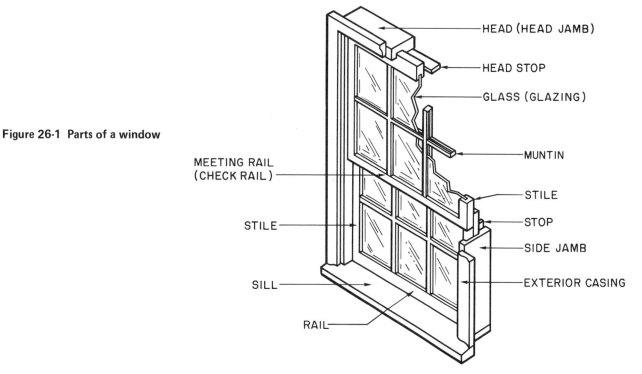

Figure 26-1 Parts of a window

Metal Windows

In recent years many buildings have been designed with metal windows. Improvements in the design of metal windows have made them competitive with wood windows in both cost and energy efficiency. The most important of these design improvements has been the development of thermal-break windows. *Thermal-break* windows use a combination of air spaces and materials that do not conduct heat easily to separate the exterior from the interior, Figure 26-2.

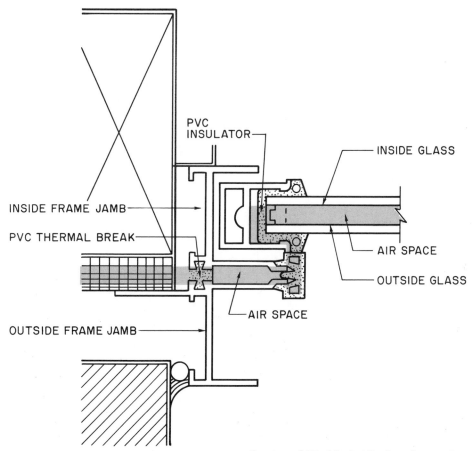

Courtesy of Ethyl Capitol Products Corporation

Figure 26-2 Thermal-break windows use insulating materials and air spaces to separate the interior from the exterior

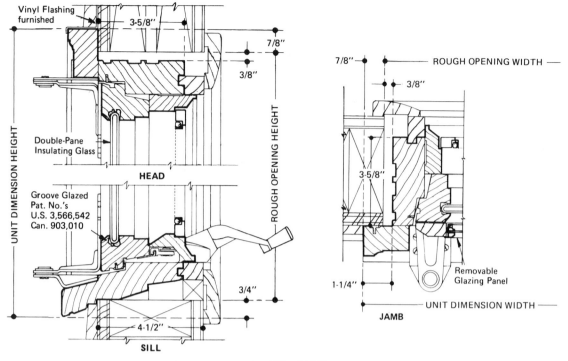

scale: 3" = 1'-0"

Courtesy of Andersen Corporation, Bayport, Minnesota 55003

Figure 26-3 Typical window detail drawings

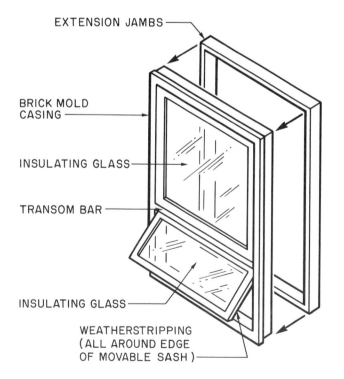

Figure 26-4

The basic parts of a metal window are similar to those of a wood window. The sash consists of stiles, rails, and glazing. The frame is made up of side jambs, head jamb, and sill. However, the trim (casing) is not included as part of the window. Often the window frame itself is the only trim used on the exterior. The frame includes a nailing fin for attaching the window to the building framing.

WINDOW DETAILS

All windows include the parts discussed so far. However, to show the smaller parts which vary from one window style to another, architects and manufacturers use detail drawings. The most common type of window detail is a section, Figure 26-3. All of the parts can be shown in section views of the head, sill, and one side jamb. These sections also usually show the wall framing around the window.

Some of the parts that can be found on window sections are defined here. Find each of the parts on the sections and illustrations in Figures 26-4, 26-5, and 26-6.

- *Weather stripping* is used on windows that open and close. It forms a weather-tight seal around the sash.

- The *transom bar* is the horizontal part of a window frame that separates the upper and lower sash.

- *Meeting rails* or check rails are the rails that meet in the middle of a double-hung window.

- *Insulating glass* is a double layer of glass, creating a dead-air space. The dead air acts as an insulator.

- *Extension jambs* are fastened to standard jambs when the window is installed in a thicker than normal wall.

- A *mullion* is a vertical section of the frame that separates side-by-side sash. If the mul-

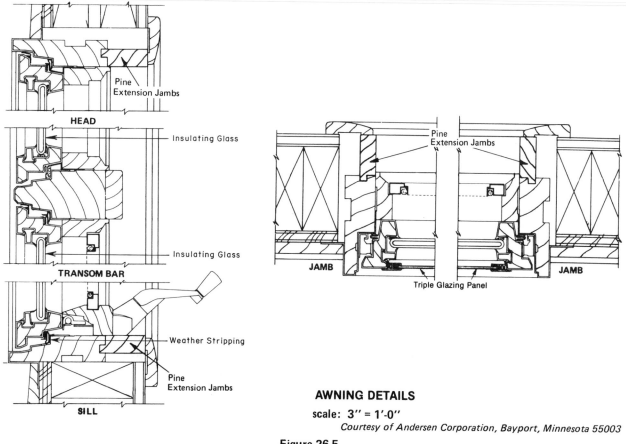

AWNING DETAILS
scale: 3" = 1'-0"
Courtesy of Andersen Corporation, Bayport, Minnesota 55003

Figure 26-5

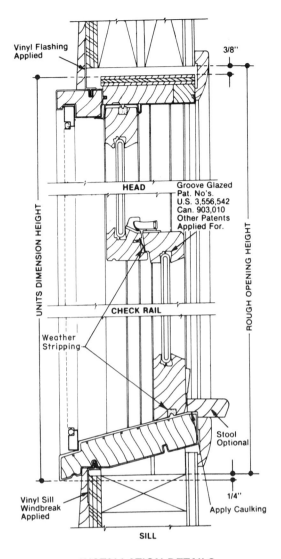

INSTALLATION DETAILS
scale: 3'' = 1'-0''

Courtesy of Anderson Corporation, Bayport, Minnesota 55003

Figure 26-6

lion is formed by butting two windows together, it is called a narrow mullion. If the mullion is built around a stud or other structural support, it is called a support mullion.

• *Triple glazing* is a third layer of glass installed with insulating glass. Triple glazing improves the insulating quality of the window. Triple glazing is often called storm sash.

Details for metal windows are often drawn as simplified sections of the frame only. Figure 26-7 shows typical details for the thermal-break window in Figure 26-2.

DOOR CONSTRUCTION

Doors include many of the same basic parts as windows, Figure 26-8. A door frame consists of side jambs and a head jamb with stop and casing. Exterior door frames also include a sill. Many doors are made of a framework with panels, Figure 26-9. The parts of panel doors are named similarly to the parts of a window. The vertical parts are *stiles*, and the horizontal parts are *rails*. Doors with glass or louvers are variations of panel doors. The framework is made of rails and stiles, and the glass or louvers replace the panels. Several manufacturers make molded doors. The most common type of

Figure 26-7 Typical metal window details

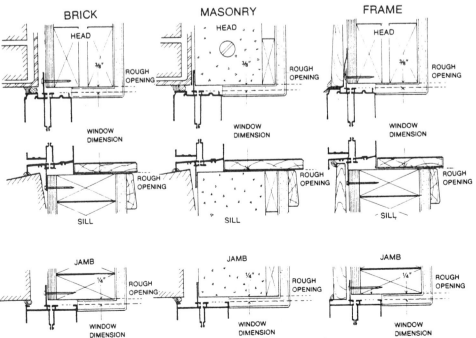

Courtesy of Ethyl Capitol Products Corporation

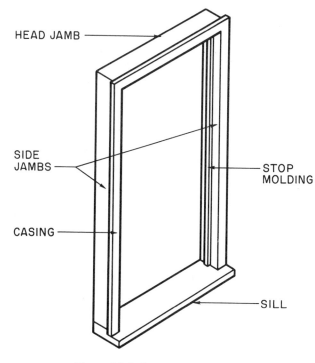

Figure 26-8 Parts of a door frame

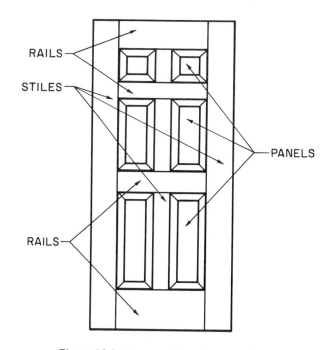

Figure 26-9 Construction of a panel door

molded door is made of hardboard which is man- ufactured in molds that contour the surface to look like panel doors, Figure 26-10. *Flush doors* consist of an internal frame with flat "skin" ap- plied to each side, Figure 26-11. Some exterior doors have insulation between the two outer skins.

These insulated doors result in considerable heat- ing and cooling savings.

DOOR DETAILS

Door details are usually less complex than window details. Carpenters rarely make doors, so

Figure 26-10 Molded hardboard door

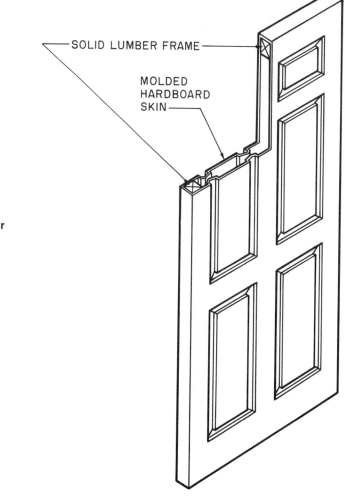

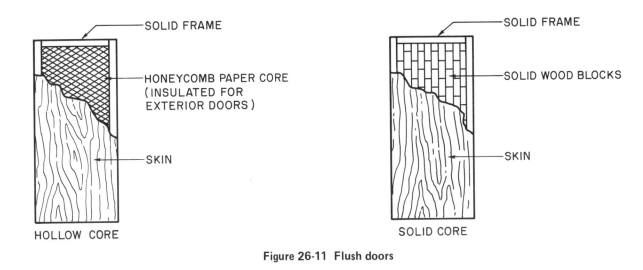

SOLID FRAME

HONEYCOMB PAPER CORE
(INSULATED FOR
EXTERIOR DOORS)

SKIN

HOLLOW CORE

SOLID FRAME

SOLID WOOD BLOCKS

SKIN

SOLID CORE

Figure 26-11 Flush doors

all that is needed are simple details of the door frame and its trim, Figure 26-12.

Many doors are sold as prehung units. In these units the frame is assembled, including the trim, and the door is hung in the frame. A section view of the jambs shows how the door is installed. For example, the door detailed in Figure 26-12 is made with two-piece jambs. These split jambs are pulled apart; each side is then slid into the opening for installation. The stop can be either applied or integral. *Applied stop* is molding which is applied to the jambs with finish nails. *Integral stop* is milled as a part of the jamb when the jamb is manufactured.

READING CATALOGS

It is often necessary to find specific information about windows or doors in the manufacturer's catalog. Usually the catalog has a table of contents listing the types of windows and doors shown. Figure 26-13 shows typical pages reprinted from a

manufacturer's catalog. A careful reading of these sample pages will help you to find the information you need in manufacturers' catalogs. For each type of window or door, you will find some or all of the listed information:

- A brief description of the window type and some of the features the manufacturer wants to highlight — a little advertising
- Installation detail drawings
- sizes available — This usually consists of drawings of the various sizes and arrangements, with dimensions for glass size, stud or rough opening, and unit dimensions. (Notice that the nominal size in the catalog sample is written as a four-digit number. The first two digits are the approximate width in feet and inches. The last two digits are the approximate height in feet and inches. For example, a 6030 window is roughly 6 feet wide by 3 feet high.)
- Additional information, such as optional equipment available

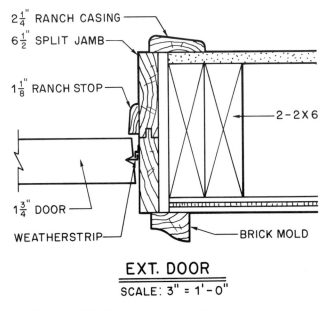

$2\frac{1}{4}"$ RANCH CASING

$6\frac{1}{2}"$ SPLIT JAMB

$1\frac{1}{8}"$ RANCH STOP

2-2X6

$1\frac{3}{4}"$ DOOR

WEATHERSTRIP

BRICK MOLD

EXT. DOOR

SCALE: 3" = 1'-0"

Figure 26-12 Typical jamb detail for an exterior door

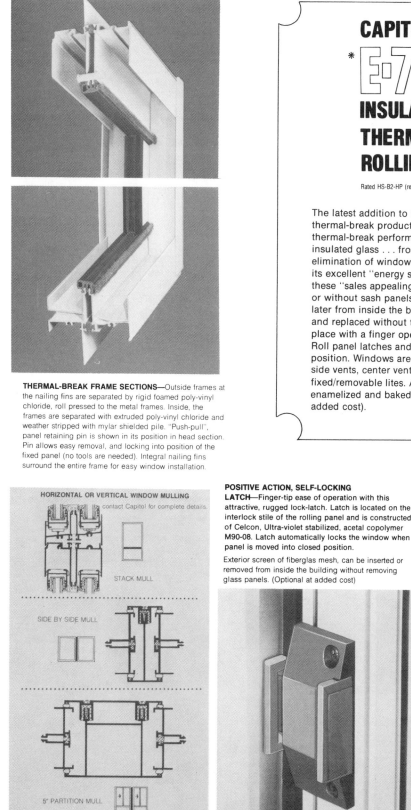

THERMAL-BREAK FRAME SECTIONS—Outside frames at the nailing fins are separated by rigid foamed poly-vinyl chloride, roll pressed to the metal frames. Inside, the frames are separated with extruded poly-vinyl chloride and weather stripped with mylar shielded pile. "Push-pull", panel retaining pin is shown in its position in head section. Pin allows easy removal, and locking into position of the fixed panel (no tools are needed). Integral nailing fins surround the entire frame for easy window installation.

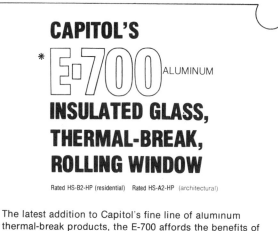

CAPITOL'S
*E-700 ALUMINUM
INSULATED GLASS, THERMAL-BREAK, ROLLING WINDOW

Rated HS-B2-HP (residential)   Rated HS-A2-HP (architectural)

The latest addition to Capitol's fine line of aluminum thermal-break products, the E-700 affords the benefits of thermal-break performance and the convenience of insulated glass . . . frost-free window frames . . . virtual elimination of window glass condensation. In addition to its excellent "energy saving" qualities, the E-700 offers these "sales appealing" features: Window can install with or without sash panels mounted—panels can be installed later from inside the building . . . Panels can be removed and replaced without tools and fixed panel is held in place with a finger operated "push-pull" retaining pin . . . Roll panel latches and automatically locks in closed position. Windows are available in arrangements such as side vents, center vents, sub-, side-, transom lites and fixed/removable lites. All models are available in PPG enamelized and baked-on colors (standard colors at no added cost).

*U.S. Patent No. 4,151,682

POSITIVE ACTION, SELF-LOCKING LATCH—Finger-tip ease of operation with this attractive, rugged lock-latch. Latch is located on the interlock stile of the rolling panel and is constructed of Celcon, Ultra-violet stabilized, acetal copolymer M90-08. Latch automatically locks the window when panel is moved into closed position.

Exterior screen of fiberglas mesh, can be inserted or removed from inside the building without removing glass panels. (Optional at added cost)

INTERLOCKING MEETING STILES—The key to the thermal-break feature, at the panel interlocks, is revealed in this photo. Fixed sash interlock stile is at left and rolling sash interlock stile at the right. The stile interlocks are insulated and parted with rigid Celcon separator to maintain thermal-break performance. Insulated glass is cushioned in flexible PVC and full length flexible vinyl flaps seal panels at each interlock.

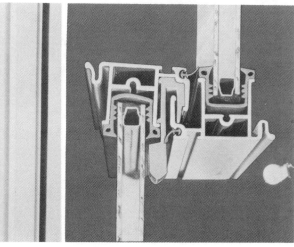

Courtesy of Ethyl Capitol Products Corporation

Figure 26-13 Pages reprinted from typical manufacturer's catalog

Fully Effective Thermal Performance Without Breaking (weakening) the Fixed and Rolling Panels!

This detail shows, that we have, without exception, succeeded at each and every part of this design to completely separate the exterior "cold" window surfaces from the interior "warm" window surfaces - both at frame, sash and glass, by the use of enclosed air spaces (as the most effective insulation) and by applied low conductive plastic insulators. And this has been achieved by only thermo-breaking the frame around the center divider and by shielding the unbroken sashes with the divider either from their exposure to the inside or to the outside of the building.

In other words: The fixed outside (cold) sash has been insulated from the interior, and the rolling (warm) sash has been insulated from the exterior.

And we can confirm our theories with a thermal test report by the AAMA authorized Electrical Test Laboratories in New York.

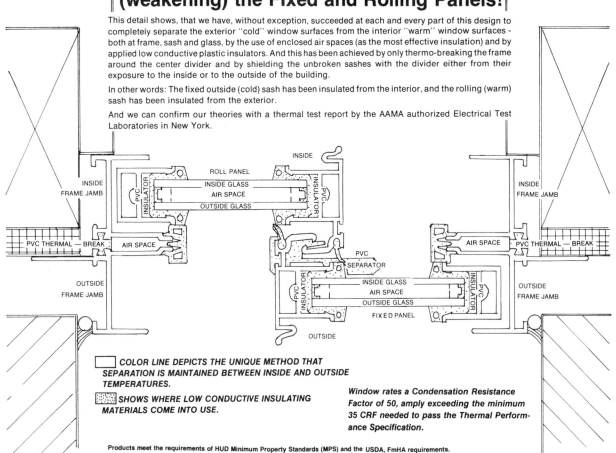

COLOR LINE DEPICTS THE UNIQUE METHOD THAT SEPARATION IS MAINTAINED BETWEEN INSIDE AND OUTSIDE TEMPERATURES.

SHOWS WHERE LOW CONDUCTIVE INSULATING MATERIALS COME INTO USE.

Window rates a Condensation Resistance Factor of 50, amply exceeding the minimum 35 CRF needed to pass the Thermal Performance Specification.

Products meet the requirements of HUD Minimum Property Standards (MPS) and the USDA, FmHA requirements.

SILL AND ROLLING PANEL SECTION—Sloping sill drains water to the outside and is machined to receive one-way weep valves that prevent air infiltration. The thermal-break feature between the inside and outside frame and the weatherstripping are clearly shown here. The roll panel stile shows the Celcon corner section that acts both as a housing for the roller and a guide for the panel as it moves in the sill cavity.

ALUMINUM ROLLERS—Celcon housing is designed to make possible a two position roller height. Roller wheels are of aluminum and ride on a rigid vinyl track for smooth panel operation. Notice that panel frames do not butt at corners. They telescope to achieve a rigid corner and eliminate gaps at the corner joints.

FIXED PANEL REMOVAL—Retaining pins at the frame head and sill lock the fixed window panels in place. Fixed panel cannot be removed when roll panel is in closed position. Retaining pins can be moved without the use of tools to easily remove the fixed panel.

To provide a weather tight seal at the machined areas at the top and bottom of the interlocks, "finger-operated" PVC sealer blocks are used.

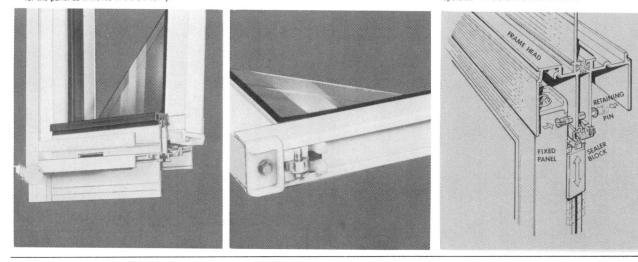

Figure 26-13 (continued)

SPECIFICATIONS

GENERAL: Windows shown are identified as series E-700, rated HS-B2-HP (residential) and series E-705, rated HS-A2-HP (Architectural), as manufactured by Capitol Products Corporation. All horizontally rolling windows and window arrangements (side-vents, center-vents, sub-, side-, transom lites and fixed removable lites) shall be thermally insulated aluminum windows. AAMA requires that the largest size of any particular model we offer for sale be tested. That size has been tested and meets AAMA's specifications. No other size of the model will be tested and no representation as to air infiltration is made except as to the model tested. Windows furnished for job will duplicate the test window in all mechanical parts and details within standard commercial tolerances as required under ANSI/AAMA 302.9-77.

OPERATION: Window frames can be installed without sashes. Fixed and rolling sashes shall be loaded into frames from the building interior and shall be removable without the use of any tools for cleaning, maintenance or re-glazing. The rolling sash shall automatically lock when in closed position. Insect screens shall be installed or removed from the building interior without removing a sash or the use of tools, and shall be reversible for handed rolling sash operation.

MATERIALS: All frame members, rails and stiles shall be fabricated from extruded aluminum alloy 6063-T5 with nominal wall thicknesses of .062″ for the Series E700 windows, and with nominal wall thicknesses of .062″ for all members except the sill, which shall have a nominal wall thickness of .078″, for the Series E705 windows.

Thermal-break frame separators shall be of extruded rigid and rigid-foamed poly-vinyl chloride. Glazing gaskets shall be 70 durometer extruded poly-vinyl chloride. Weatherstripping shall be 50 durometer extruded poly-vinyl chloride.

WEATHERSTRIPPING: Weatherstripping at the frame perimeters and the interlocks shall be extruded 50 durometer poly-vinyl chloride. Adjustable, dual-durometer PVC end seals shall be provided at all interlock ends at the point of contact on the frame separators.

Weather stripping at head and sill and at rolling sash to be mylar shielded pile.
Fasteners shall be stainless steel. Hardware shall be aluminum and Celcon, Ultra-violet stabilized, Acetal Copolymer M90-08. Sash bumpers shall be adhesive backed poly-vinyl chloride.

CONSTRUCTION: Frame members shall be thermally insulated by a crimped insulator and a snap-on type dual-durometer separator which functions as insulator and weatherstrip for the exterior fixed sash and for the rolling sash. Jambs shall be machined at both ends to receive telescopingly the head and sill. The joint shall be made watertight by applying a die-cut gasket and fastening with two screws per joint. Sills shall be machined to receive one-way valves and fixed sash retaining pins. Window thermal design to be such that no thermal break of sash aluminum components is required. All sash stiles shall be tubular aluminum shapes for torsional rigidity and shall be machined at both ends to telescope into top and bottom rails and fastened with one screw per joint together with top rail guides, fixed sash bottom rail setting block and rolling sash roller housing.

Top and bottom rails shall be machined at both ends to receive stiles and shall have weep slots for draining the bottom rail glazing cavity. Interlock stiles shall be designed to provide a pocket for the insect screen frame and shall be machined to receive the spring loaded lock and lock keeper.

HARDWARE: Exterior fixed sashes shall have injection molded rail guides and sash lock keeper. Interior rolling sashes shall have injection molded two-position roller housings, aluminum wheels and injection molded, spring loaded lock at interlock stile. The exterior fixed sash shall be locked in place by a push-pull retaining pin at head and sill. Sills shall have PVC one-way weep valves. Jambs shall have adhesive backed bumpers at mid-height.

FINISH: All aluminum frame and sash members shall have a finish that provides a smooth uniform appearance. Electrostatically applied paint over alodine base provided in standard colors at no added cost. Extruded PVC and exposed injection molded parts shall be of charcoal grey color.

GLASS: Glass shall be ½″ sealed double glazing with standard glass thickness as required by AAMA.

SCREENS: Frames shall be fabricated of a roll formed aluminum rail with corners accurately mitred and fitted to a hairline joint. Screen cloth shall be fiberglas. Frames shall have 2 springloaded plungers on one side rail and a finger pull on the opposite side rail. Screens and finish on screen frames shall be available at additional cost.

INSTALLATION: (by others) Frames should be installed straight plumb and level without springing or twisting, and securely fastened in place in accordance with recommendations or details. Mastic or caulking compounds must be applied, before installation, between fin and adjacent construction to provide weather-tight installation and to maintain the integrity of the energy saving features of the product. Loading of sashes and final adjustments shall be made by the installers to assure proper sash operation and window performance.

PERFORMANCE LEVEL SPECIFICATIONS	Air Infiltration	25mph wind	Water Resistance		Operating Force		Uniform Load Deflection		AAMA requires that the largest size of any particular model we offer for sale be tested. That size has been tested and meets AAMA's specifications. No other size of the model will be tested and no representation as to air infiltration is made except as to the model tested.
	measured	allowed	measured	allowed	measured	allowed	measured	allowed	
	cfm/ft		psf no flow		pounds		inches		
B2	.37	.375	2.86	2.86	10	15	.197	.331	
A2	.37	.375	4.34	3.34	20	20	.200	.351	

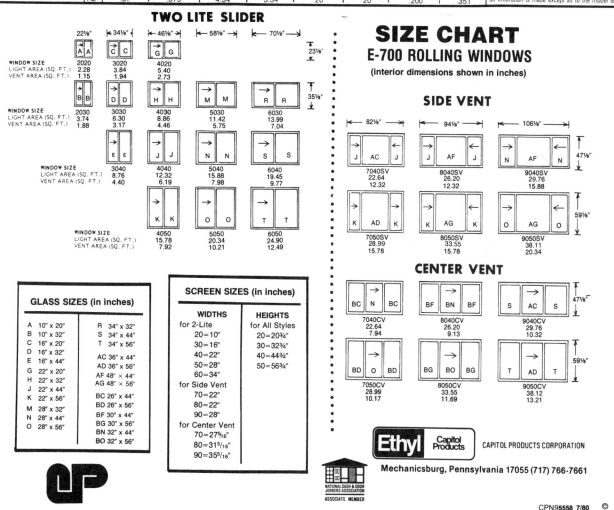

Figure 26-13 (continued)

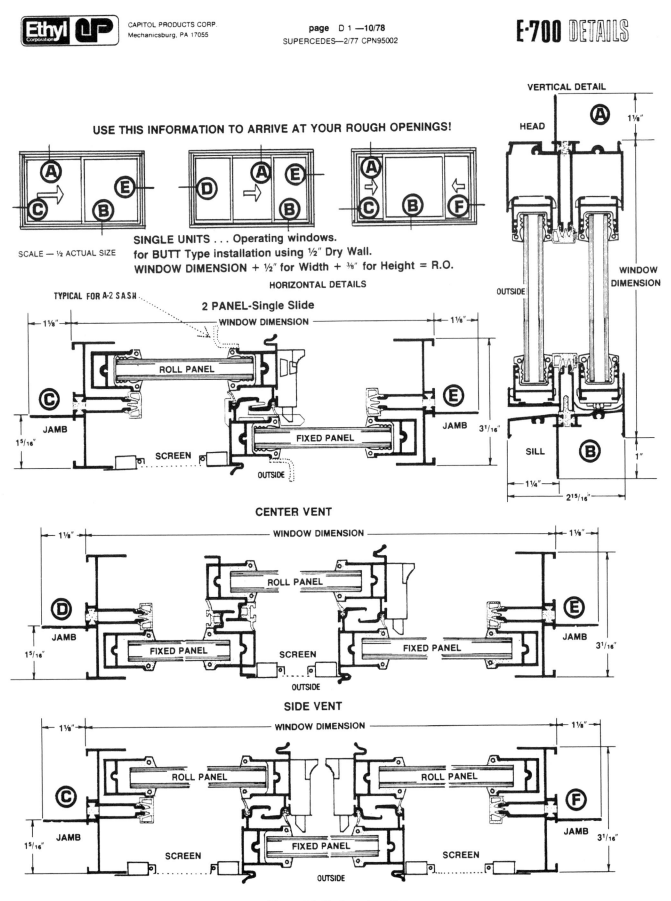

Ethyl ⊔P Corporation

CAPITOL PRODUCTS CORP.
Mechanicsburg, PA 17055

page   D 1 —10/78
SUPERCEDES—2/77 CPN95002

E-700 DETAILS

USE THIS INFORMATION TO ARRIVE AT YOUR ROUGH OPENINGS!

SCALE — ½ ACTUAL SIZE

SINGLE UNITS . . . Operating windows.
for BUTT Type installation using ½″ Dry Wall.
WINDOW DIMENSION + ½″ for Width + ⅜″ for Height = R.O.

VERTICAL DETAIL

HEAD

OUTSIDE

WINDOW DIMENSION

SILL

HORIZONTAL DETAILS

TYPICAL FOR A-2 SASH

2 PANEL-Single Slide

WINDOW DIMENSION

ROLL PANEL

FIXED PANEL

JAMB

JAMB

SCREEN

OUTSIDE

CENTER VENT

WINDOW DIMENSION

ROLL PANEL

JAMB

FIXED PANEL

FIXED PANEL

SCREEN

JAMB

OUTSIDE

SIDE VENT

WINDOW DIMENSION

ROLL PANEL

ROLL PANEL

JAMB

FIXED PANEL

JAMB

SCREEN

SCREEN

OUTSIDE

Figure 26-13 (continued)

Ethyl Corporation **CP** CAPITOL PRODUCTS CORP.
Mechanicsburg, PA 17055

page D 2 —10/78
SUPERCEDES—New CPN95002

E-700 DETAILS

TYPICAL BUTT TYPE INSTALLATION DETAILS

ROUGH OPENING WIDTH = WINDOW DIMENSION + ½"   ROUGH OPENING HEIGHT = WINDOW DIMENSION + ⅜"

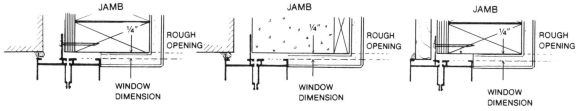

SCALE — 3" = 1'

> **IMPORTANT**—Mastic or caulking compounds must be applied before installation, between window fin and adjacent construction to provide weather-tight installation and maintain the integrity of the products energy saving features.

THE CAPITOL MULLING AND FIXED LITE WINDOW WALL SYSTEM

Fin and finless windows can be combined in fixed lite and operating window configurations. Either Side-by-Side or Stack Mulled. Fixed Lites can be mixed and matched with side vent, center vent and two-lite rolling windows to achieve window wall arrangements.

Windows can be ordered with fins already removed from "mulled side" of window, or fins can easily be removed in the field for mulling.

Fixed Lite System features Removable Inserts. Glass is enclosed in its own sash frame which can be removed from the main window frames.

Figure 26-13 (continued)

✓ CHECK YOUR PROGRESS

Can you perform these tasks?

☐ Identify the following parts of a window: sash, rails, stiles, muntins, glazing, side jambs, head jambs, sill, stop, exterior casing, and interior casing.

☐ Identify each part of a window on a window detail.

☐ Explain the basic features of the following door types: panel, flush, hollow core, molded, and solid core.

☐ Explain what is meant by applied stop and integral stop.

☐ Find the rough opening, unit size, and glass size for windows in a manufacturer's catalog.

ASSIGNMENT

Refer to the Lake House drawings when necessary to complete the assignment.

1. Name the lettered parts (a through f) in Figure 26-14.

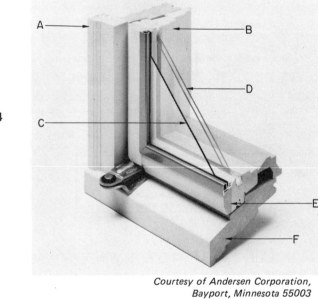

Figure 26-14

Courtesy of Andersen Corporation,
Bayport, Minnesota 55003

2. What is the nominal size of the window in the south end of the Lake House dining room?

3. In the catalog sample shown in Figure 26-13, what is the width and height of the rough opening for a #4030 window?

4. In the catalog sample, what are the rough opening dimensions for a window with a 28-inch by 44-inch glass size?

5. In the catalog sample in Figure 26-13, what is the glass size of the window in bedroom #2 of the Lake House?

6. In the catalog sample, what is the R.O. for the window in bedroom #2 of the Lake House?

7. Is the exterior door in the Lake House kitchen to be prehung or site hung?

8. What type and size are the doors in the Lake House playroom closet?

UNIT 27 Exterior Wall Covering

OBJECTIVES

After completing this unit, you will be able to perform the following tasks:

- Describe the exterior wall covering planned for all parts of a building.
- Explain how flashing, drip caps, and other devices are used to shed water.
- Describe the treatment to be used at corners and edges of the exterior wall covering.

WOOD SIDING

Wood is a popular siding material because it is available in a variety of patterns, it is easy to work with, and it is durable. Wood siding includes horizontal boards, vertical boards, shingles, plywood, and hardboard.

Boards can be cut into a variety of shapes for use as horizontal siding, Figure 27-1. These boards are nailed to the wall surface starting at the bottom and working toward the top. With wood siding, a starting strip of wood furring is nailed to the bottom of the wall. This starting strip holds the bottom edge of the first piece of siding away from the wall. Hardboard siding often has an insert on the back. This insert replaces the starting strip, Figure 27-2. Each board covers the top edge of the one below. The amount of each board left exposed to the weather is called the *exposure* of the siding, Figure 27-3.

Vertical siding patterns can be created by boards, plywood, or hardboard, Figure 27-4. These materials are applied directly to the wall with no special starting strips. However, where horizontal joints are necessary, they should be lapped with

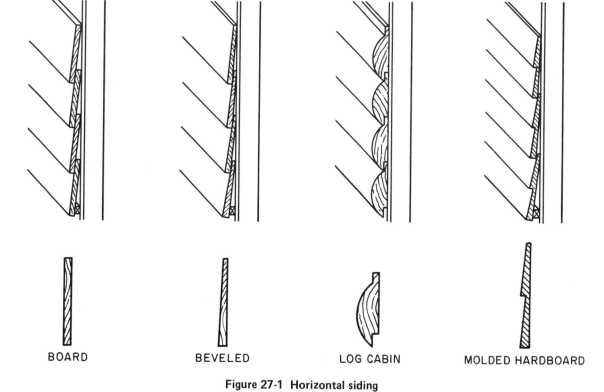

| BOARD | BEVELED | LOG CABIN | MOLDED HARDBOARD |

Figure 27-1 Horizontal siding

173

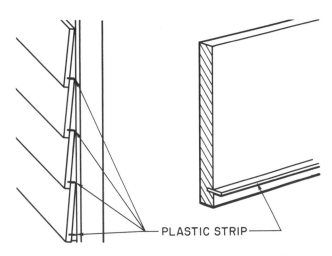

Figure 27-2 Many manufacturers of hardboard siding include a plastic insert which helps align the pieces.

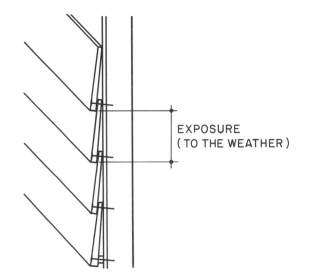

Figure 27-3 The exposure is the amount of the siding exposed to the weather.

rabbet joints or *Z* flashing should be applied. The flashing can be concealed with battens, Figure 27-5.

Shingles take longer to apply, but make an excellent siding material. In place of a starting strip of furring, the bottom *course* (row) of shingles is doubled.

Drip Caps and Flashing

It is important to prevent water from getting behind the siding where it can cause dry rot. Where a horizontal surface meets the siding, water is apt to collect. Flashing is used to prevent this water from running behind the siding. Aluminum is the most common flashing material. The aluminum flashing is nailed to the wall before the siding is applied. The lower edge of the flashing extends over the horizontal surface far enough to prevent the water from running behind the siding. Areas to

be flashed are noted on building elevations. The flashing is shown on the detail drawings or building elevations, Figure 27-6.

The heads of windows and doors may form a small horizontal surface. Wood drip cap molding can be used in these places to shed water, Figure 27-7. Drip caps are shown on details and elevations.

Corner and Edge Treatment

Regardless of the kind of siding used, the edges must be covered to prevent water from soaking into the end grain or running behind the siding. Around windows, vents, doors, and other wall openings; the trim around the opening usually covers the end grain of the siding. In inside corners, a strip of wood can be used to form a corner bead for the siding. Outside corners in plywood, hardboard sheets, or vertical boards are usually

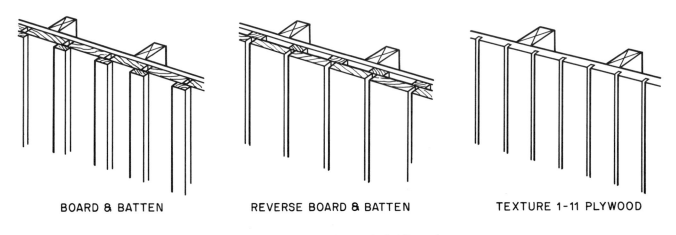

BOARD & BATTEN   REVERSE BOARD & BATTEN   TEXTURE 1-11 PLYWOOD

Figure 27-4 Three popular vertical siding patterns

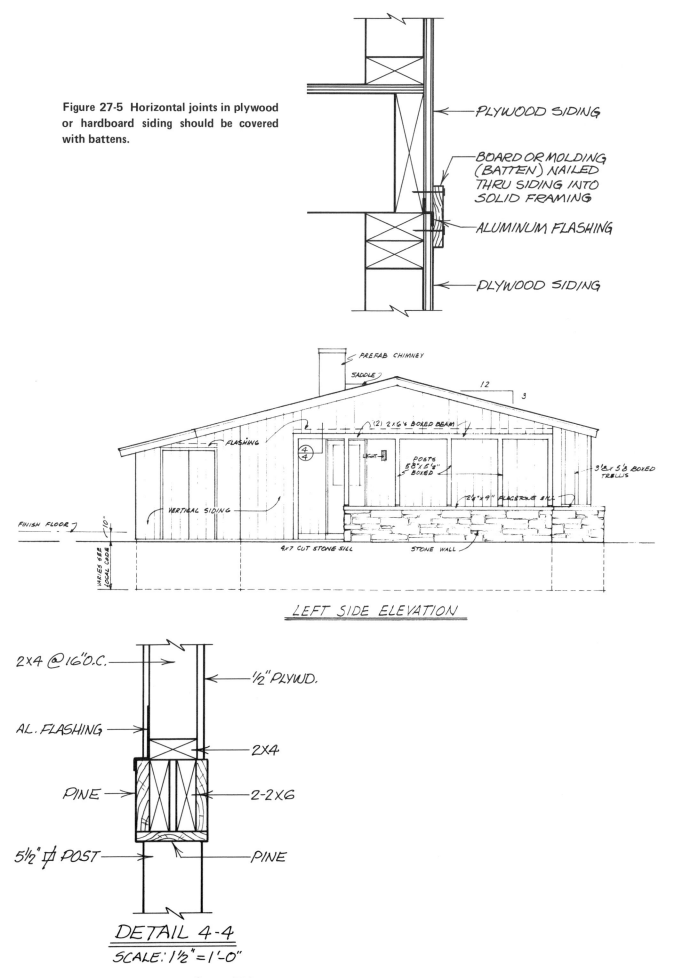

Figure 27-5 Horizontal joints in plywood or hardboard siding should be covered with battens.

PLYWOOD SIDING

BOARD OR MOLDING (BATTEN) NAILED THRU SIDING INTO SOLID FRAMING

ALUMINUM FLASHING

PLYWOOD SIDING

PREFAB CHIMNEY

SADDLE

12
3

(2) 2×6's BOXED BEAM

FLASHING

4
4

LIGHT

POSTS 5⅝"×5½" BOXED

3'8×5'8 BOXED TRELLIS

2'4"×9" FLAGSTONE SILL

VERTICAL SIDING

FINISH FLOOR

10"

VARIES SEE LOCAL CODE

9×7 CUT STONE SILL

STONE WALL

LEFT SIDE ELEVATION

2×4 @ 16"O.C.

½" PLYWD.

AL. FLASHING

2×4

PINE

2-2×6

5½" ⊡ POST

PINE

DETAIL 4-4
SCALE: 1½" = 1'-0"

Figure 27-6 Flashing is shown on elevations and details.

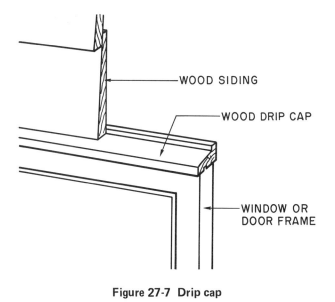

Figure 27-7 Drip cap

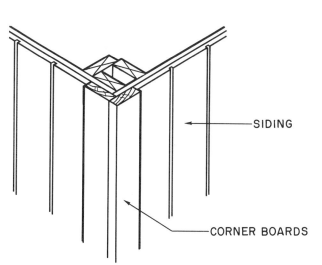

Figure 27-8 Corner boards can be used with most types of siding.

built with corner boards, Figure 27-8. With horizontal siding, outside corners can be built with corner boards or metal corner caps, Figure 27-9. Shingles are trimmed to form their own outside corners, Figure 27-10, or they are butted against corner boards.

The corner treatment to be used can usually be seen on the building elevations. If no corner treatment is shown on the elevations, look for a special detail of a typical corner.

METAL AND PLASTIC SIDING

Several manufacturers produce aluminum, steel, and vinyl siding and trim. The most common type is made to look like horizontal beveled wood siding. A variety of trim pieces is available for any type of application, Figure 27-11.

STUCCO

Stucco is a plaster made with portland cement. The stucco is applied in two or three coats over wire lath. The wall sheathing is covered with waterproof building paper. Next the *lath* (usually wire netting) is stapled to the wall, Figure 27-12. Finally, the stucco is troweled on — a rough *scratch coat* first, Figure 27-13, then a *brown coat* to build up to the approximate thickness, and finally a finish coat. Outside corners and edges are formed with galvanized metal beads nailed to the wall before the scratch coat is applied.

MASONRY VENEER

Masonry veneer is usually either brick or natural stone. It is called veneer because it is a thin

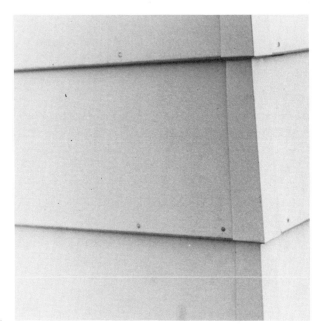

Figure 27-9 Metal corners

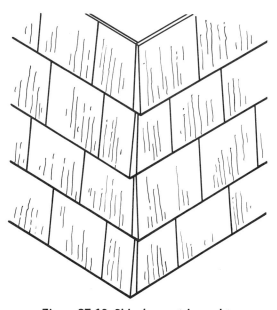

Figure 27-10 Shingles are trimmed to form their own corners.

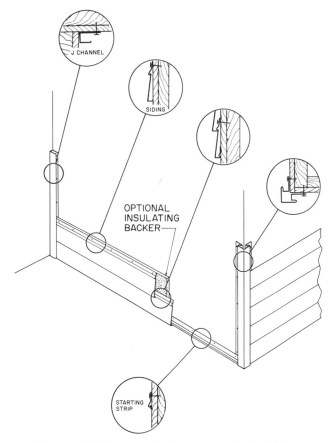

Figure 27-11 Accessories for metal and vinyl siding

Figure 27-12 Wall prepared for stucco

Figure 27-13 Scratch coat applied

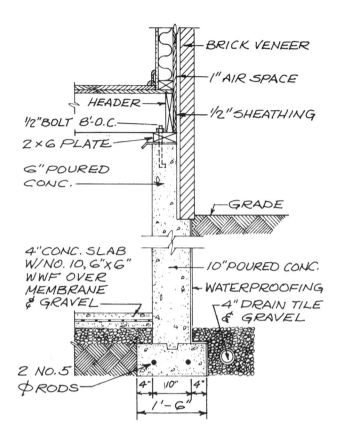

Figure 27-14 Typical foundation section with ledge for masonry veneer

layer of masonry over some kind of structural wall. The structure may be wood framing, metal framing, or concrete blocks. The masonry has to rest on a solid foundation. Normally the building foundation is built with a four-inch ledge to support the masonry veneer. This ledge can be seen in a typical wall section, Figure 27-14.

Interesting patterns can be created by using half bricks and bricks in different positions. Some of the most common *bond patterns*, as they are called, are shown in Figure 27-15. If no pattern is indicated, the bricks are normally laid in running bond.

Above and below windows, above doors, and in other special areas, bricks may be laid in varying positions, Figure 27-16. If bricks are to be laid in any but the stretcher position, they will be shown on the detail drawings, Figure 27-17. The details for openings in masonry walls will also indicate a lintel at the top of the opening. The *lintel* is usually angle iron. It carries the weight of the masonry above the opening.

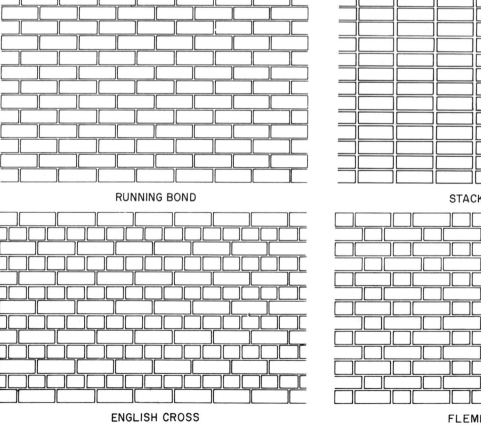

RUNNING BOND

STACKED BOND

ENGLISH CROSS

FLEMISH BOND

Figure 27-15 Frequently used bond patterns

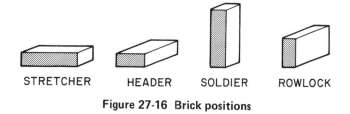

STRETCHER   HEADER   SOLDIER   ROWLOCK

Figure 27-16 Brick positions

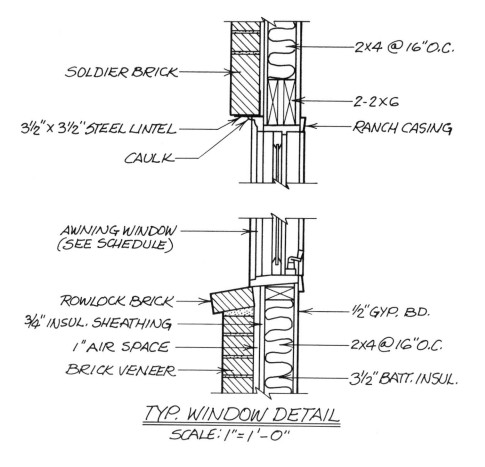

SOLDIER BRICK

3½" × 3½" STEEL LINTEL

CAULK

2X4 @ 16"O.C.

2-2X6

RANCH CASING

AWNING WINDOW
(SEE SCHEDULE)

ROWLOCK BRICK
¾" INSUL. SHEATHING
1" AIR SPACE
BRICK VENEER

½" GYP. BD.

2X4 @ 16"O.C.

3½" BATT. INSUL.

TYP. WINDOW DETAIL
SCALE: 1" = 1'-0"

Figure 27-17 Notice the rowlock and soldier bricks on the window detail

✓ CHECK YOUR PROGRESS

Can you perform these tasks?

☐ Describe the exterior wall covering on all parts of a building.

☐ List the locations of flashing on all parts of a building.

☐ List the dimensions of all drip caps to be installed on a building.

☐ Describe the support for masonry veneer.

☐ List locations where bricks are to be used in stretcher, header, soldier, and rowlock positions.

ASSIGNMENT ──────────────────────────────────────

Refer to the Lake House drawings (in the packet) to complete the assignment.

1. What material is used for the Lake House siding?

2. How are the outside corners of the Lake House siding finished?

3. In some places the Lake House siding extends below the wood framing onto the foundation. How is the siding fastened to the foundation?

4. What prevents water from running under the siding at the heads of the Lake House windows?

5. Detail 1/7 of the Lake House shows aluminum screen nailed behind the top edge of the siding. What is the purpose of the opening covered by this screen?

6. Describe one use of aluminum flashing under the siding on the Lake House.

Miscellaneous Exterior Work
Section 8

 **UNIT 28 Decks**

OBJECTIVES

After completing this unit, you will be able to perform the following tasks:

- Explain how a deck is to be supported.
- Describe how a deck is to be anchored to the house.
- Locate the necessary information to build handrails on decks.

Wood decks are used to extend the living area of a house to the outdoors. A deck may be a single-level platform, or it may be a complex structure with several levels and shapes. However, nearly all wood decks are made of wood planks laid over joists or beams, Figure 28-1. The planks are usually spaced about 1/2 inch apart, so rainwater does not collect on the deck.

The same construction methods are used for decks and porches as for other parts of the house. The parts of deck construction that require special attention or that were not covered earlier in the text are discussed here.

SUPPORT

The deck must be supported by stable earth. The support must also extend below the frostline in cold climates. The most common method of support is by wood posts which rest on concrete footings, Figure 28-2. All wood used in the construction of a deck should be treated with a chem-

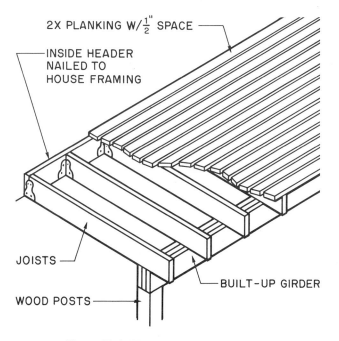

Figure 28-1 Typical deck construction

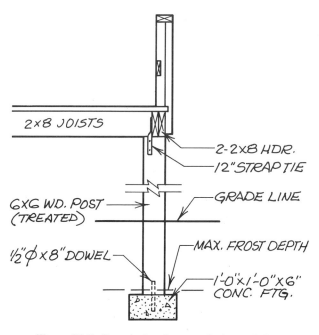

Figure 28-2 Foundation for a typical wood deck

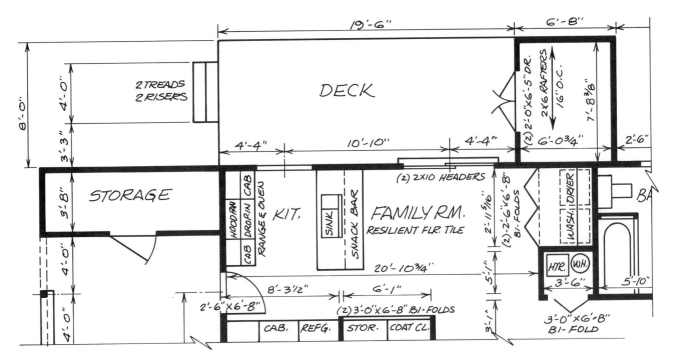

Figure 28-3 Deck shown on floor plan

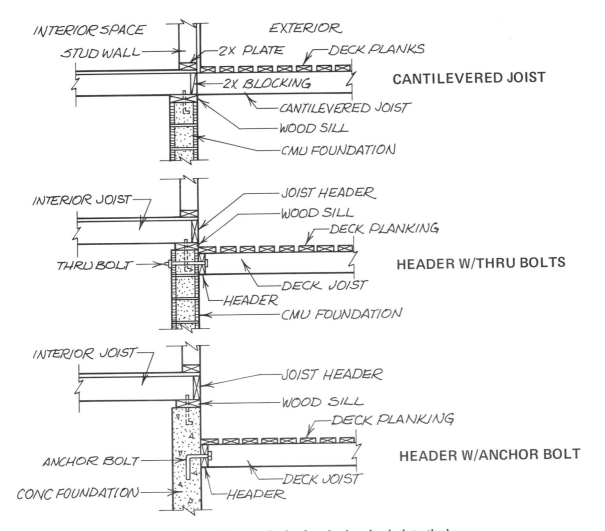

Figure 28-4 Three methods of anchoring the deck to the house

ical to prevent decay. For example, Deck Detail 3/6 for the Lake House includes a note that all wood is to be CCA treated. CCA is the abbreviation for *chromated copper arsenate* — a common wood preservative.

Decks are usually included on the floor plans for the house. The floor plans show the overall dimensions of the decks and the locations of posts, piers, or other support, Figure 28-3. If the decks are complex, they may be shown on a separate plan or detail. On the Lake House, the decks are shown on the floor framing plan.

Typically the posts or piers support a beam or girder, which in turn supports joists. The beam may be solid wood or built-up wood. The joists can be butted against or rested on top of the beam in any of the ways discussed for floor framing.

ANCHORING THE DECK

The frame of the deck is typically fastened to the building. The most common ways of anchoring the deck to the building are shown in Figure 28-4. If the deck is at the same level as the house floor, the deck joists can be cantilevered from the floor joists. In this case, blocking is required between the joists. If through bolts are used, the holes are drilled when the deck is built. If the deck is secured with anchor bolts in a concrete foundation, the anchor bolts must be positioned when the foundation is poured.

The anchor bolts or through bolts hold a joist header. The deck joists either rest on top of the header or butt against it. If the joists butt against the header, they are supported by joist hangers or a ledger strip.

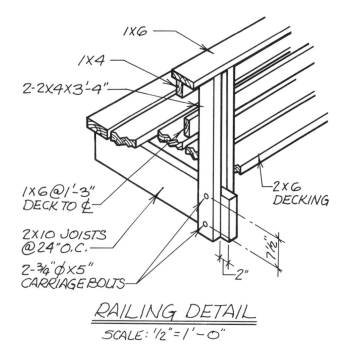

RAILING DETAIL
SCALE: 1/2" = 1'-0"

Figure 28-5 Railing detail

RAILINGS

Most decks have a railing because they are several feet from the ground. Although metal railings are available in ready-to-install form, the architectural style of most wood decks calls for a carpenter-built wood railing. The simplest type of railing is made of uprights and two or three horizontal rails. The uprights are bolted to the deck frame and the rails are bolted, screwed, or nailed to the uprights. The style of the railing and the hardware involved are usually indicated on a detail drawing, Figure 28-5.

✓ **CHECK YOUR PROGRESS**

Can you perform these tasks?

☐ List the dimensions for the size and location of posts or columns to support a deck or porch.

☐ Describe the framing for a deck.

☐ Explain how a deck is to be tied to the building.

☐ Describe the construction of railings for a deck.

☐ List the sizes of all parts of a deck.

ASSIGNMENT

Refer to the Lake House drawings (in the packet) to complete the assignment.

1. What supports the south edge of the decks located outside the Lake House living and dining rooms?

2. How far from the outside of the house foundation is the centerline of these supports?

3. How far apart are these supports?

4. How many anchor bolts are required to fasten both of these decks to the Lake House foundation?

5. What is the purpose of the aluminum flashing shown on Deck Detail 3/6?

6. What material is used for the railings on the Lake House south decks?

7. How many lineal feet of horizontal rails are there on these decks?

8. What is the total rise from the lower deck to the higher deck? Which deck is higher?

9. What supports the west edge of the deck between the Lake House kitchen and the garage?

UNIT 29   Finish Site Work

OBJECTIVES

After completing this unit, you will be able to perform the following tasks:

- Describe retaining walls, planters, and other constructed landscape features shown on a set of drawings.
- Find the dimensions of paved areas.
- Identify new plantings and other finished landscaping shown on a site plan or land-scape plan.

As the exterior of the building is being finished or soon after it is finished; the masons, carpenters, and landscapers begin the finished landscape work. Any constructed features (called *site appurtenances*) are completed first. Then trees and shrubs are planted. Finally, lawns are planted.

RETAINING WALLS

Retaining walls are used where sudden changes in elevation are required, Figure 29-1. The retaining wall retains, or holds back, the earth. Where the height of the retaining wall is several feet, the earth may put considerable stress on the wall. Therefore, it is important to build the wall according to the plans of the designer. In some cases, a section through the wall may be included to show the thickness of the wall, its foundation, and any reinforcing steel, Figure 29-2. For low retaining

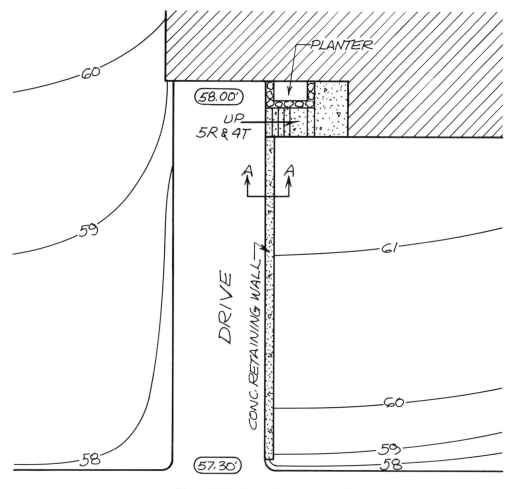

Figure 29-1 Plan of retaining wall

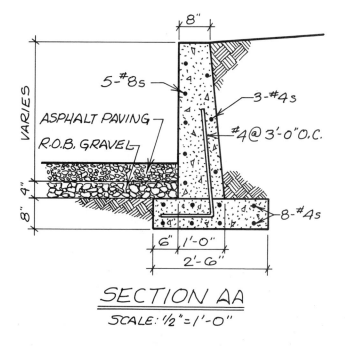

SECTION AA
SCALE: 1/2"=1'-0"

Figure 29-2 Typical retaining wall construction detail

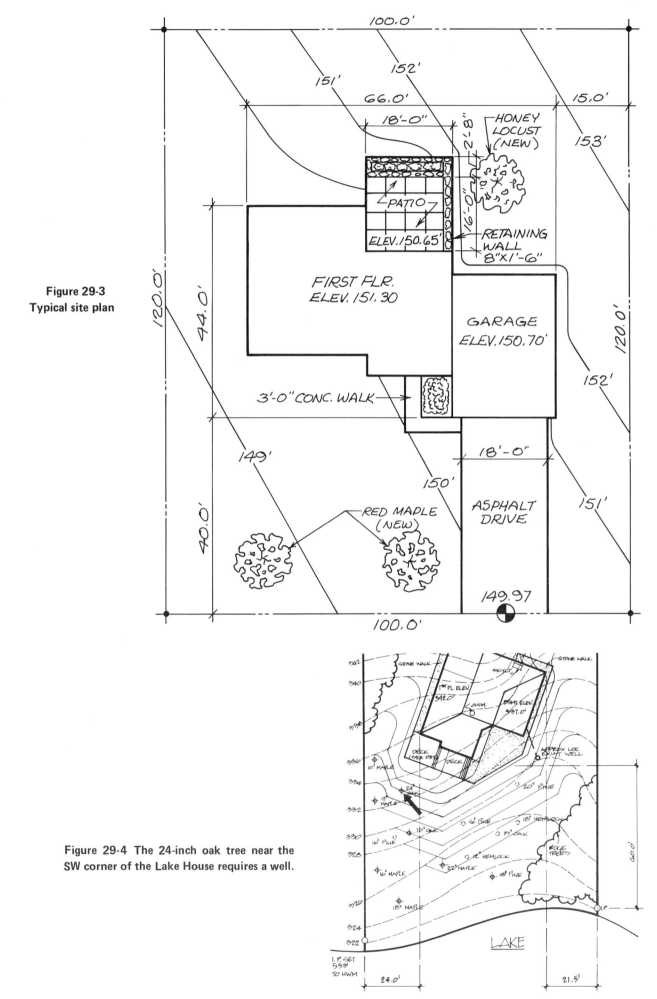

Figure 29-3
Typical site plan

Figure 29-4 The 24-inch oak tree near the SW corner of the Lake House requires a well.

Courtesy of Robert C. Kurzon

walls, the site plan may be the only drawing included, Figure 29-3.

A low retaining wall is sometimes built around the base of a tree when the finished grade is higher than the natural grade. This retaining wall forms a *well* around the tree, allowing the roots of the tree to "breathe." An example of a tree that will require a well can be seen in Figure 29-4, taken from the Lake House site plan. The 24" oak is at an elevation of approximately 333 feet, but the finished grade at this point is 336 feet. Therefore, a well 3 feet deep is required.

PLANTERS

Planters are sometimes included in the construction of retaining walls or attached to the build-

ing. In these cases, the information needed to build the planter is included with the information for the building or retaining wall, Figure 29-5. The planter is built right along with the house or retaining wall. If a planter that is separate from other construction is included, it is usually shown with dimensions on the site plan. A special section may be included with the details and sections to show how the planter is constructed, Figure 29-6.

The planter should be lined with a waterproof membrane, such as polyethylene (common plastic sheeting), or coated with asphalt waterproofing. This keeps the acids and salts in the soil from seeping through the planter and staining it. The planter should also include some way for water to escape. This can be through the bottom or through weep

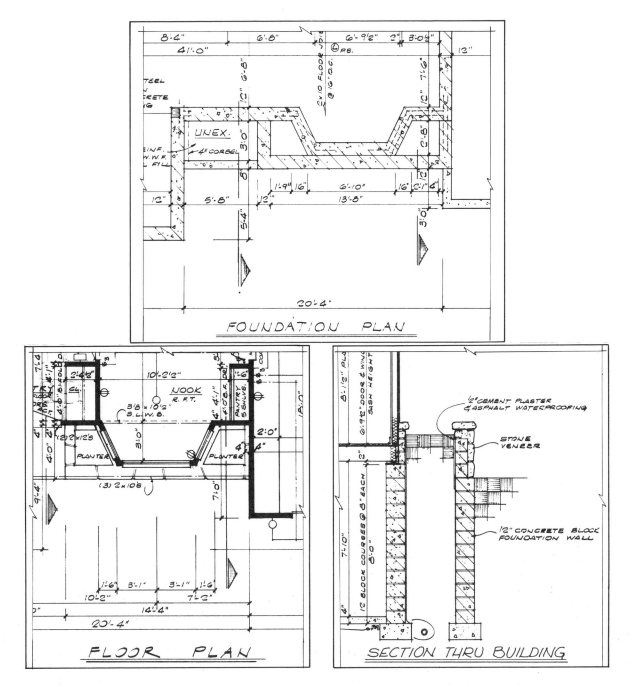

Courtesy of Home Planners, Inc.

Figure 29-5 Because this planter is a part of the foundation, it is included on the normal drawings for the house.

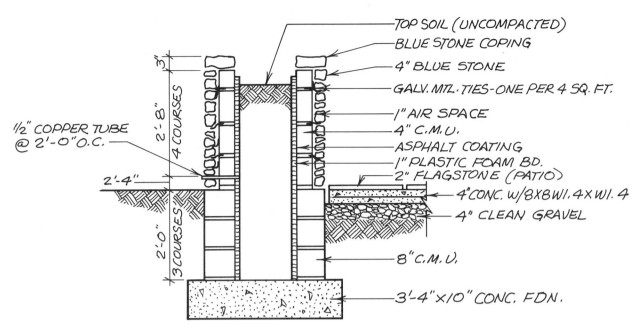

TOP SOIL (UNCOMPACTED)
BLUE STONE COPING
4" BLUE STONE
GALV. MTL. TIES-ONE PER 4 SQ. FT.
1" AIR SPACE
4" C.M.U.
ASPHALT COATING
1" PLASTIC FOAM BD.
2" FLAGSTONE (PATIO)
4" CONC. W/8X8 WI. 4 X WI. 4
4" CLEAN GRAVEL
8" C.M.U.
3'-4" X 10" CONC. FDN.

TOP SOIL (UNCOMPACTED)

1/2" COPPER TUBE @ 2'-0" O.C.

3"
2'-8" 4 COURSES
2'-4"
2'-0" 3 COURSES

Figure 29-6 Section through planter and patio

holes. *Weep holes* are openings just above ground level. In cold climates, the planter may be lined with compressible plastic foam. This allows the earth in the planter to expand as it freezes, without cracking the planter.

PAVED AREAS

Paved areas on housing sites are drives, walks, and patios. Drives and walks are usually described most fully in the specifications for the project.

(Specifications are discussed in Unit 35.) However, the site plan includes dimensions and necessary grading information for paved areas, Figure 29-3. These dimensions are usually quite straight forward and easy to understand.

Patios are similar to drives and walks in that they are flat areas of paving with easy-to-follow dimensions. They may differ from drives and walks by having different paving materials, such as slate, brick, and flagstone, for example. Patios may also be made of a concrete slab with a different surface material.

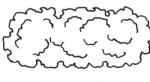

OR

DECIDUOUS (HARDWOOD) TREES

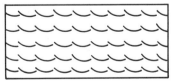

CONIFEROUS (NEEDLE LIKE) TREES

PALM TREES

LOW SHRUBS

SPREADING GROUND COVER

Figure 29-7 Typical symbols for plantings

PLANTINGS

Plantings include three types: grass or lawns, shrubs, and trees. On some projects, such items are planted by the owner. When the builder/contractor does the landscaping, the trees and shrubs are planted first; then, the lawns are planted. Some or all of the trees included in the landscape design may have been left when the site was cleared. Any new trees to be planted are shown with a symbol and an identifying note, as shown in Figure 29-3. There is no widely accepted standard for the sym-

bols used to represent trees and shrubs. Most architects and drafters use symbols that represent deciduous (hardwood) trees, coniferous (evergreen) trees, palms, and low shrubs, Figure 29-7. The trees and shrubs may also be listed in a schedule of plantings, Figure 29-8.

Grass is planted by seeding or sodding. Although a note on the site plan may indicate seeded or sodded areas, more detailed information is usually given on the schedule of plantings or in the specifications.

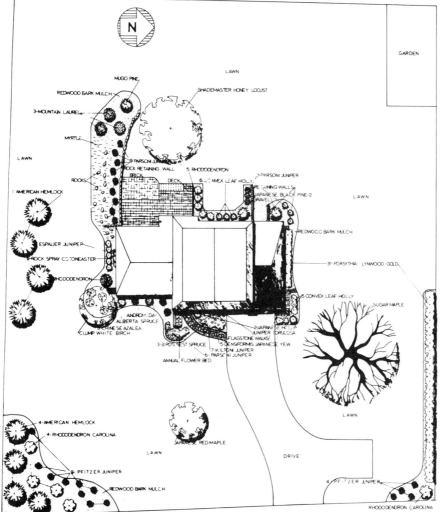

LANDSCAPE PLANTING PLAN FOR:
MR. & MRS. C. BURTON
KENNEDY ROAD
COBLESKILL, N.Y.

LANDSCAPE DESIGN BY DAN PIERRO

PLANT LIST

BOTANICAL NAME	COMMON NAME	NUMBER USED	BOTANICAL NAME	COMMON NAME	NUMBER USED
	~TREES~			~TREES~	
ACER PALMATUM	JAPANESE RED MAPLE	1	JUNIPERUS CHINENSIS TORULOSA	TORULOSA JUNIPER	1
ACER SACCHARUM	SUGAR MAPLE	1	JUNIPERUS CHINENSIS WILTONI	WILTONI JUNIPER	7
BETULA PAPYRIFERA	CLUMP WHITE BIRCH	1	KALMIA LATIFOLIA	MOUNTAIN LAUREL	3
GLEDITSIA TRIACANTHOS INERMIS	SHADEMASTER HONEY LOCUST	1	PICEA ABIES NIDIFORMIS	BIRD'S NEST SPRUCE	3
	~SHRUBS~		PICEA GLAUCA ALBERTIANA	ALBERTA SPRUCE	1
AZALEA MOLLIS	CHINESE AZALEA	5	PIERIS JAPONICA	ANDROMEDA	1
COTONEASTER HORIZONTALIS	ROCK SPRAY COTONEASTER	5	PINUS MUGO MUGHUS	MUGO PINE	1
FORSYTHIA INTERMEDIA	FORSYTHIA LYNWOOD GOLD	31	PINUS THUNBERGI	JAPANESE BLACK PINE	2
ILEX CRENATA	JAPANESE HOLLY	2	RHODODENDRON CAROLINA	RHODODENDRON CAROLINA	11
ILEX CRENATA CONVEXA BULLATA	CONVEX LEAF HOLLY	11	TAXUS CUSPIDATA DENSIFORMIS	DENSIFORMIS JAPANESE YEW	5
JUNIPERUS CHINENSIS ESPALIER	ESPALIER JUNIPER	1	TSUGA CANADENSIS	AMERICAN HEMLOCK	8
JUNIPERUS CHINENSIS PARSONI	PARSONI JUNIPER	18		~GROUND COVERS~	
JUNIPERUS CHINENSIS PFITZERIANA	PFITZER JUNIPER	13	VINCA MINOR	MYRTLE	1200

Figure 29-8 A landscape plan complete with plant list

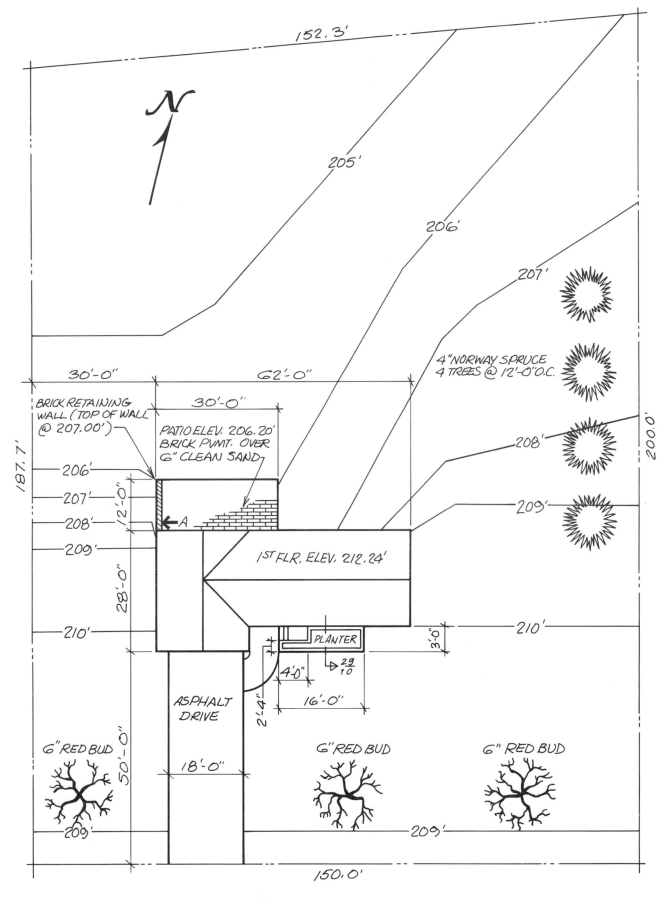

Figure 29-9 Site plan for assignment questions

✓ CHECK YOUR PROGRESS

Can you perform these tasks?
- ☐ List the dimensions of a retaining wall.
- ☐ Describe the footing for a retaining wall.
- ☐ Describe provisions for drainage through a retaining wall.
- ☐ List the dimensions of a planter.
- ☐ Explain how the inside of a planter is to be waterproofed.
- ☐ List the dimensions of all paved areas.
- ☐ Describe the pitch of all paved areas.
- ☐ Name the types of plantings shown on a site or landscape plan.

ASSIGNMENT

Refer to Figures 29-9 and 29-10 to complete the assignment.

1. What is the height of the retaining wall at A?
2. How long is the retaining wall?
3. Of what material is the retaining wall constructed?
4. What is the width and length of the patio?
5. What materials are used in the construction of the patio?
6. Describe the weep holes in the planter.
7. How is the planter treated to prevent acids and salts from staining its surface?
8. How many deciduous trees are to be planted?
9. What is the area of the driveway?
10. Assuming that the driveway is 4 inches thick, how many cubic yards of asphalt does it require? (See Math Review 22.)

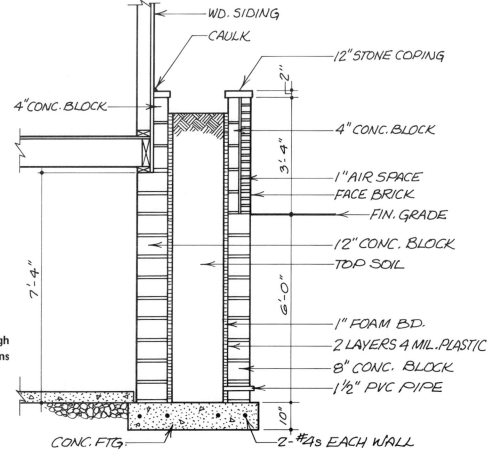

Figure 29-10 Section through planter for assignment questions

Finishing the Interior

Section 9

UNIT 30   Fireplaces

OBJECTIVES

After completing this unit, you will be able to perform the following tasks:

- Describe the foundation, firebox, throat, and chimney of a fireplace using information from a set of construction drawings.

- Explain the finish of the exposed parts of the fireplace, using information from a set of construction drawings.

BASIC CONSTRUCTION AND THEORY OF OPERATION

A fireplace can be divided into four major parts or zones: foundation, firebox, throat area, and chimney, Figure 30-1. Each of these zones has a definite function. To understand the construction details, it is necessary to know how these zones work.

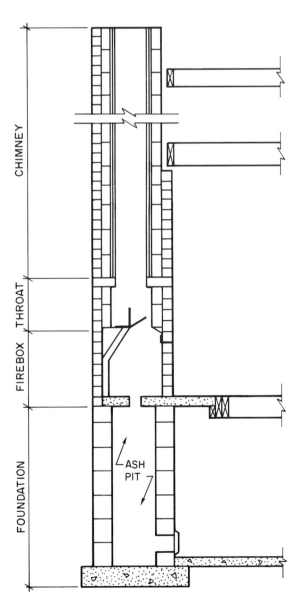

Figure 30-1 Four zones of a fireplace

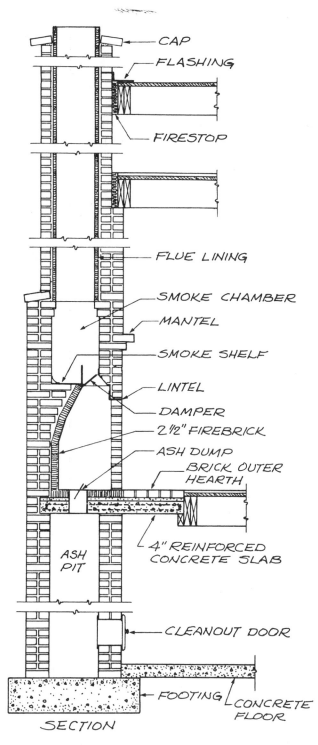

CAP
FLASHING
FIRESTOP
FLUE LINING
SMOKE CHAMBER
MANTEL
SMOKE SHELF
LINTEL
DAMPER
2½" FIREBRICK
ASH DUMP
BRICK OUTER HEARTH
4" REINFORCED CONCRETE SLAB
ASH PIT
CLEANOUT DOOR
FOOTING
CONCRETE FLOOR

SECTION

Courtesy of Richard T. Kreh, Sr.

Figure 30-2 Section view of a masonry fireplace

Foundation

The fireplace foundation serves the same purpose as the foundation of the house — it supports the upper parts and spreads the weight over an area of stable earth. The foundation consists of a footing and walls capable of supporting the necessary weight, Figure 30-2. The fireplace foundation sometimes houses an ash pit. The *ash pit* is a reservoir to hold ashes that are dropped through an *ash dump* (a small door) in the floor of the fireplace, Figure 30-3. When the foundation includes an ash pit, a *cleanout door* is installed near the bottom of the ash pit, Figure 30-4.

Firebox

The firebox is the area where the fire is built. In all-masonry fireplaces, the firebox is constructed of two layers of masonry, Figure 30-2. Each layer is called a *wythe*. The floor of the firebox consists of firebricks laid over a concrete base. The concrete base may extend beyond the face of the firebox to support the hearth. The *hearth*, which may be tile, stone, brick, or slate, forms a noncombustible floor area in front of the fireplace. The outer walls of the firebox are most often common brick. The inner walls are of firebrick to withstand the heat of the fire. The back wall of the firebox slopes in to direct the smoke and gas into the throat area. The masonry over the fireplace opening is supported by a steel lintel. The lintel is long enough so that four inches of it can rest on the masonry at each end.

Throat

The throat of the fireplace is the area where the firebox narrows into the chimney. Modern

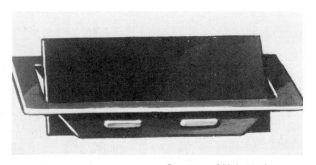

Courtesy of Majestic Company
Figure 30-3 Ash dump

Courtesy of Majestic Company
Figure 30-4 Cleanout door

wood-burning fireplaces are built with a metal damper in the throat, Figure 30-5. The *damper* is a door which can be closed to prevent heat from escaping up the chimney when the fireplace is not in use. The damper is placed on top of the firebox with one-inch clearance on all four sides. This clearance allows the metal damper to expand as it gets hot.

The flat area behind the damper (above the sloped back of the firebox) is called the *smoke shelf*. The smoke shelf is especially important for the proper operation of the fireplace. The cold air coming down the chimney hits the smoke shelf and turns back up with the rising hot gas and smoke from the fire box. This helps carry the smoke and gas up the chimney, Figure 30-6. If the smoke shelf is not built properly and kept clean, the falling cold air can force the smoke and gas back into the firebox.

Courtesy of Majestic Company

Figure 30-5 Damper

Chimney

The chimney carries the smoke and hot gas from the throat to above the house. The top of the chimney must be high enough above the roof, trees, and other nearby obstructions to insure that the air flows evenly across its top. As a general rule the chimney should extend 2 feet through the roof and 2 feet above anything within 10 feet. However, the dimensions on the drawings should always be followed.

To insure fire safety and a smooth inner surface, masonry chimneys are lined with a clay flue. This flue is installed in 2-foot sections as the chimney is built. A 1-inch air space is allowed between the flue lining and the chimney masonry, Figure 30-7.

In recent years insulated, metal chimneys have become quite popular. These chimneys are lightweight; they do not require massive foundation for their support. It also takes less time to install them. Because the outer wall of a metal chimney remains cool, it can be enclosed in wood, Figure 30-8. The chimney sections are slipped together and fastened with sheet metal screws; then, the chimney is framed with wood and covered with the specified siding and trim. A chimney enclosure of this type is called a *chase*.

PREFABRICATED METAL FIREPLACES

Constructing masonry fireplaces is time consuming. Their great weight requires massive foun-

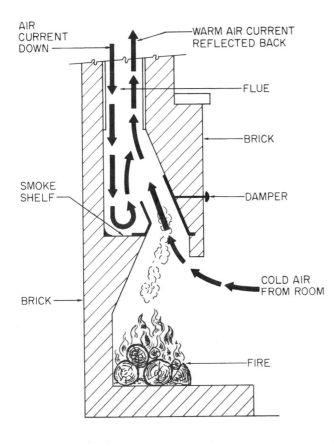

Figure 30-6 The smoke shelf turns the incoming air back up the chimney.

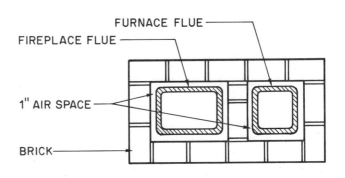

Figure 30-7 Plan view of a two-flue chimney

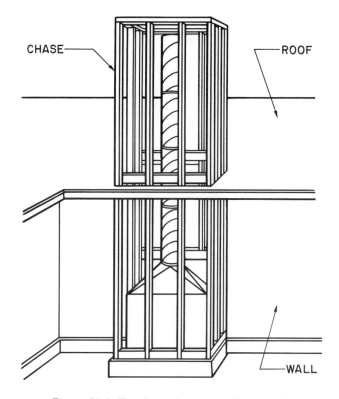

Figure 30-8 The framed enclosure for a metal chimney is called a chase

dations. Engineered, metal fireboxes have been developed which can be installed in very little time and require only modest foundations, Figure 30-9. The prefabricated units are available from several manufacturers, and in a variety of styles. However, they are all similar in that they have double walls and a complete throat with the damper in place. Most also have a firebrick floor.

The double wall improves the heating ability of the metal fireplace. Cool air enters the space between the walls through openings near the bottom. The air absorbs heat from the fire and, because warm air naturally rises, it exits through openings near the top of the unit, Figure 30-10. The outside surfaces of the prefabricated unit are cooled by the circulating air; the unit can be enclosed in wood if recommended by the manufacturer. For a more traditional appearance, the exposed face of the fireplace can be covered with masonry veneer.

Courtesy of Majestic Company

Figure 30-9 Prefabricated metal fireplace

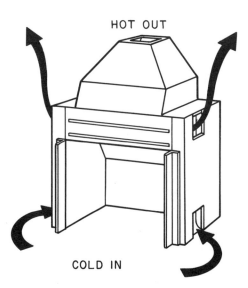

Figure 30-10 In a heat-circulating fireplace, cold air enters the double wall at the bottom and warm air exits at the top.

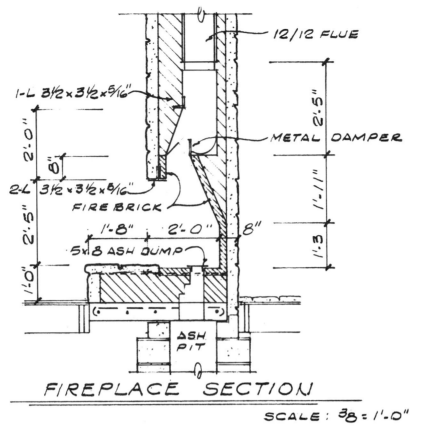

FIREPLACE SECTION

SCALE: 3⁄8 = 1'-0"

Courtesy of Home Planners, Inc.

Figure 30-11 A section view shows the construction of the firebox and throat area.

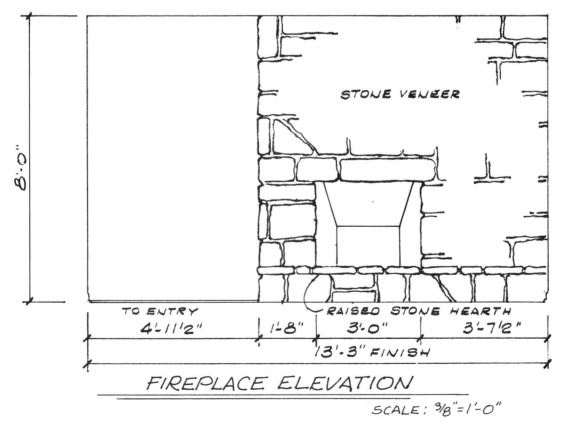

FIREPLACE ELEVATION

SCALE: 3⁄8"=1'-0"

Courtesy of Home Planners, Inc.

Figure 30-12 An elevation view is a good guide to the finished appearance of the fireplace.

FIREPLACE DRAWINGS

Much of the information about a fireplace is shown on the building plans and elevations. The foundation is normally included on the foundation plan for the building. The dimensions and notes show the location of the fireplace foundation, its size, and the size and type of material to be used. The floor plan of the house shows where the fireplace is located and its overall dimensions. The building elevations show the chimney.

More detailed information about the fireplace is shown on the fireplace details, which usually include a cross section, Figure 30-11. The following information is often included on a section view of the fireplace:

- Dimensions of the firebox

- Materials used inside the firebox (firebrick)

- Materials used for the outside of the firebox and chimney

- Ash dump, if any is included

- Lintel over the opening

- Dimensions of the hearth

- Mantle, if any is included

- Dimension from smoke shelf to flue

- Size of the flue

- Materials for the chimney

An elevation of the fireplace shows the exterior finish of the fireplace, Figure 30-12. Only those features that could not be adequately described on the section view are called out on the elevation. However, this view shows the exterior finish — the mantle and the trim — better than the section view does.

✓ CHECK YOUR PROGRESS

Can you perform these tasks?

☐ Identify the firebox, smoke shelf, damper, flue, and hearth on a fireplace section.

☐ List the dimensions of the outside of a fireplace, firebox opening, inside of the firebox, hearth, and height of the chimney for a fireplace.

☐ Describe the construction, including dimensions, for the enclosure of a prefabricated fireplace.

☐ Describe the finish and trim for a fireplace.

 ## ASSIGNMENT

Refer to the Lake House drawings (in the packet) to complete the assignment.

1. Does the Lake House have an all-masonry fireplace or a prefabricated metal fireplace?

2. How wide is the opening of the firebox?

3. How high is the opening of the firebox?

4. What is the opening next to the fireplace?

5. Determine the overall width and length of the fireplace including the hearth.

6. Of what material is the hearth constructed?

7. What is used for a lintel over the firebox opening? (Include dimensions.)

8. Briefly describe the foundation of the fireplace.

9. How far above the highest point on the roof is the top of the chimney?

10. What is the total height from the playroom floor to the top of the chimney?

11. What is the overall height of the brickwork involved in the fireplace construction?

12. The top of the fireplace is covered with plastic laminate on 3/4-inch plywood. How much clearance is there between that plywood and the chimney?

OBJECTIVES

After completing this unit, you will be able to perform the following tasks:

• Identify the parts of stairs.

• Calculate tread size and riser size.

STAIR PARTS

In order to discuss the layout and construction of stairs you need to know the names of the parts of a set of stairs. The main parts of a stair are defined here and also shown in Figure 31-1.

• *Stringers*, sometimes called *strings*, are the main support members. The assembly made up of the stringers and vertical supports is called a stair *carriage.*

• *Treads* are supported by the stringers. The treads are the surfaces one steps on.

• *Risers* are the vertical boards between the treads.

• A *landing* is a platform in the middle of the stairs. Landings are used in stairs that change directions or in very long flights of stairs.

• The *run* of the stairs is the horizontal distance covered by the stairs.

• The *rise* of the stairs is the total vertical dimension of the stairs.

The trim parts and balustrade are shown in Figure 31-2.

• The *nosing* is the portion of the tread that projects beyond the riser.

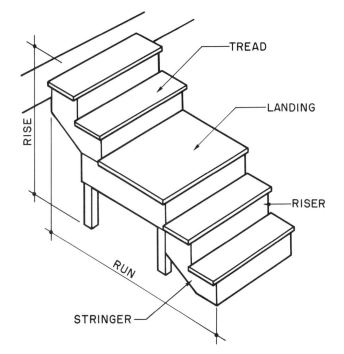

Figure 31-1 Basic stair parts

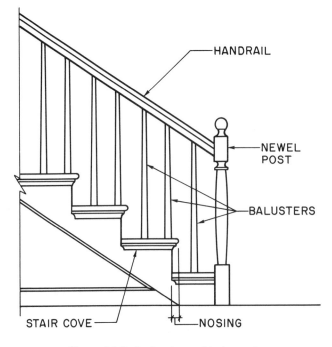

Figure 31-2 Stair trim and balustrade

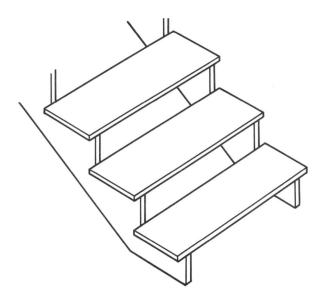

Figure 31-3 Open stringers and open risers

- The underside of the nosing may be trimmed with molding called *stair cove*.

- A *handrail* is usually required on any stairs that are not completely enclosed.

- *Balusters* are the vertical pieces that support the handrail at each step.

- The *newel post* is a heavier vertical support used at the bottom of the stair.

- The balusters, newel post, and handrail together are called a *balustrade*.

TYPES OF STAIRS

Stairs can be built with open stringers or housed stringers. *Open stringers* are cut in a sawtooth pattern to form a surface for fastening the treads, Figure 31-3. *Housed stringers* are routed out, so the treads fit between them, Figure 31-4. Stairs are also called open or closed depending on whether the space between the treads is enclosed with risers. Both the kind of stringers and the risers are shown on a section through the stairs. If the stringers are housed, the top of the far stringer shows on the section view.

Stairs are also named according to their layout as seen in a plan view, Figure 31-5. Straight stairs are the simplest design. They may or may not include a landing. L-shaped and U-shaped stairs are used where space does not permit a straight run. L-shaped and U-shaped stairs have a landing at the change in directions. The carriages for L-shaped and U-shaped stairs include vertical supports under

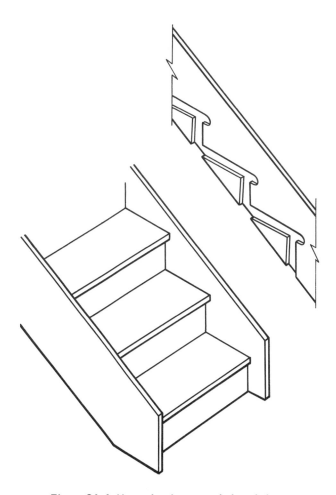

Figure 31-4 Housed stringers and closed risers

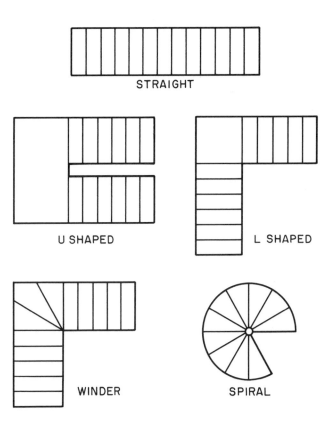

Figure 31-5 Stair layouts

the landing, Figure 31-6. Winder stairs use tapered treads to change directions. Winder stairs are not permitted by some building codes. Spiral stairs require the least floor space. They are made up of winder treads, usually supported by a center column.

Temporary stairs used during construction only are usually of cruder construction. They are often made of 2x10 stringers with wooden cleats to support the treads. Service stairs in unoccupied spaces, such as a cellar, are often made of 2x12 open strings with 2x10 treads. Stairs for exterior decks are also sometimes built with 2x12 open strings, Figure 31-7. However, the carpenter should remember that these stairs are important to the overall appearance of the deck.

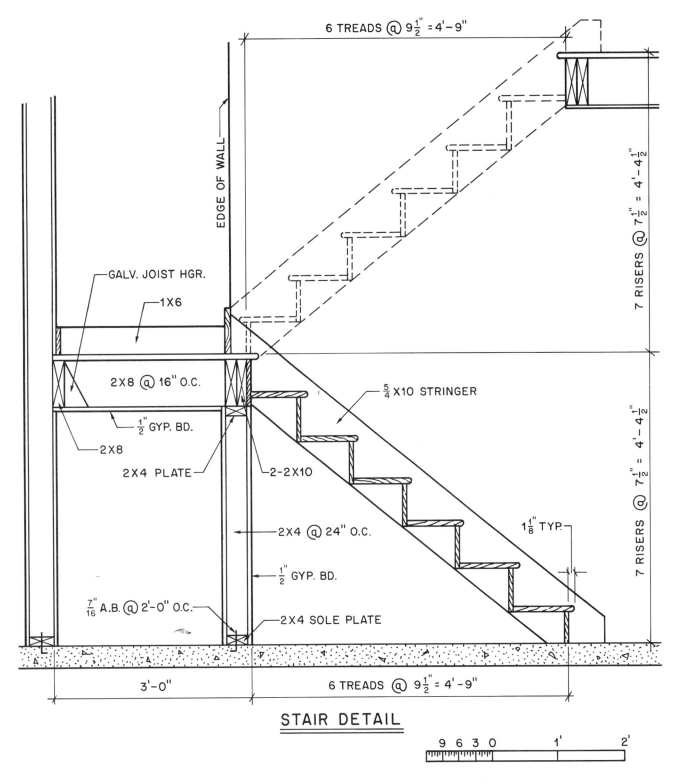

STAIR DETAIL

Figure 31-6 Stair detail with 2x4 studs to support the landing

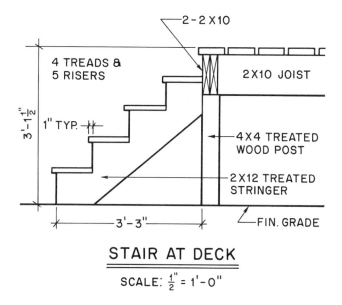

STAIR AT DECK

SCALE: $\frac{1}{2}$" = 1'-0"

Figure 31-7 Typical detail for deck stairs

CALCULATING RISERS AND TREADS

Nearly all stair details include a notation showing the number of treads with their width and the number of risers with their height, Figure 31-6. Occasionally only the total run, total rise, number of treads, and number of risers is given on the drawings, Figure 31-7. Notice that there is one more riser than there are treads. The builder must calculate the size of the treads and risers. The steepness of the stairs depends on the relationship between tread width and riser height. If the treads are wide and the risers low, the stairs are gradual. If the treads are narrow and the risers high, the stairs are steep. To climb stairs with wide treads and high risers requires an uncomfortably long stride. To climb stairs with narrow treads and low risers requires an uncomfortably short stride. To ensure that the relationship between the treads and risers

is comfortable, the following rules should be followed:

1. The sum of one riser and one tread should be between 17 inches and 18 inches.

2. The sum of two risers and one tread should be between 24 inches and 25 inches.

3. Multiplying the width of one tread by the height of one riser should give a product of 70 inches to 75 inches.

To build the stairs in Figure 31-7 within these rules, the risers must be 7 1/2 inches high and the treads must be 9 3/4 inches wide. The height of the risers can be found by dividing the total rise by the number of risers. (3'-1 1/2" or 37 1/2" ÷ 5 = 7 1/2") (See Math Review 10.) The width of the treads is found by dividing the total run by the number of treads. (3'-3" or 39" ÷ 4 = 9 3/4") Now check the tread and riser dimensions against the three rules just listed:

1. 7 1/2 inches (riser) plus 9 3/4 inches (tread) equals 17 1/4 inches.

2. 7 1/2 inches plus 7 1/2 inches (2 risers) plus 9 3/4 inches (1 tread) equals 24 3/4 inches.

3. 9 3/4 inches (tread) multiplied by 7 1/2 inches (riser) equals 73 1/8 inches.

To calculate the sizes of the treads and risers in stairs with landings, treat each part as a separate stair. However, the treads and risers should be the same size in each part.

✓ **CHECK YOUR PROGRESS**

Can you perform these tasks?

☐ Identify the following stair parts on drawings: stringers, treads, risers, landings, nosing, stair cove molding, newel post, handrail, and baluster.

☐ Calculate desired tread width and riser heights from the total run and rise.

☐ List the sizes of all of the parts of a stair.

ASSIGNMENT ————————————————————

Refer to the Lake House drawings (in the packet) to complete the assignment.

Questions 1 through 6 refer to the stairs between the south decks of the Lake House.

1. What is the total run of the stairs between the decks?

2. What is the width of each tread?

3. What is the length of each tread?

4. What is the rise of each step in these stairs?

5. How many stringers are used under these stairs?

6. What size material is used for the stringers?

Questions 7 through 13 refer to the stairs from the kitchen to the bedroom level in the Lake House.

7. How many risers are there and how high is each riser?

8. How many treads are there and how wide (front to back including nosing) is each tread?

9. What is the total rise of the stairs?

10. Is this stair built with open or housed stringers?

11. The railing at this stair extends the length of the kitchen hall. If vertical railing supports are spaced at 16″ O.C., how many vertical supports are used?

12. How long is each vertical railing support?

13. What material is used for the horizontal rails?

UNIT 32 Insulation and Room Finish

OBJECTIVES

After completing this unit, you will be able to perform the following tasks:

• Identify the insulation to be used in walls, floors, and ceilings.

• Identify the wall, ceiling, and floor covering material to be used.

• List all of the kinds of interior molding to be used.

INSULATION

Thermal insulation is any material that is used to resist the flow of heat. In very warm climates, thermal insulation is used to resist the flow of heat from the outside to the inside. In cold climates, thermal insulation is used to resist the flow of heat from the inside to the outside. Insulating material is rated according to its ability to resist the flow of heat. The measure of this resistance is the *R value* of the material. The higher the *R* value, the better the material insulates. Typical *R* values for side-wall or attic insulation range from R-3 to R-38, Figure 32-1.

Thermal insulation is generally available in four forms. *Foamed-in-place* materials are synthetic compounds that are sprayed onto a surface, and then produce an insulating foam by a chemical reaction, Figure 32-2. *Rigid boards* are plastic foams which have been produced in board form at a factory. Another common type of rigid insulation is made of fiberglass which is manufactured in rigid form instead of as flexible blankets, Figure 32-3. Rigid boards are frequently used for foun-

TYPE	TYPICAL THICKNESSES AND R VALUES	COMMENTS
Fiberglass blankets & batts	3/4", R-3 2 1/2", R-8 3 1/2", R-11 6", R-19 9", R-30 12", R-38	Flexible blanket-like material. Available with or without vapor barrier on one side.
Fiberglass blowing wool	R value depends on depth of coverage, but is sligtly less than fiberglass blankets.	Other types of blowing or pouring insulation include cellulose and mineral wool.
Rigid fiberglass board	1", R-4.4	Material is similar to fiberglass blankets, but with rigid binder to create rigid boards. Usually faced with aluminum foil.
Rigid urethane foamed board	1/2", R-3.6 3/4", R-5.4 1", R-7.2 1 1/2", R-10.8 2", R-14.6	Plastic that has been cured in a foamed state to introduce bubbles of air. This creates a rigid board which is usually faced with aluminum foil.
Foamed-in-place urethane	R value depends on thickness depth, but is approximately the same as rigid urethane boards.	Other plastic materials may also be foamed in place.

Figure 32-1 Common types of thermal insulation

Courtesy of The Upjohn Company
Figure 32-3 Foamed-in-place insulation

Courtesy of Owens-Corning Fiberglas Corporation
Figure 32-3 Rigid insulation

dation insulation, under concrete slabs, and for sheathing. *Blanket insulation* is in the form of flexible rolls or batts usually made of fiberglass wool, Figure 32-4. Loose *pouring insulation* can be any of a variety of materials that can be poured into place and has good insulating property. A common pouring insulation is loose fiberglass, Figure 32-5.

The insulation is shown in the building sections. If the insulation is to be installed between studs, joists, or rafters; it will be sized accordingly. For example, if batts are to be used in a 2x4 wall, the insulation will be indicated as 3 1/2 inches thick — the width of a 2x4. Where insulation is used in ventilated spaces, there should be room for the necessary air circulation, Figure 32-6.

Where the insulation includes a vapor barrier, such as Kraft paper, foil, or polyethylene, the vapor barrier is installed on the heated side of the wall. This prevents the moisture in the warm air

from passing through the wall and condensing on the cold side of the wall. Such condensation can reduce the *R* value of the insulation and cause painted surfaces to blister and peel.

WALL AND CEILING COVERING

By far the most widely used wall surface material is gypsum wallboard. The most common thicknesses are 3/8 inch, 1/2 inch, and 5/8 inch. If gypsum board (sometimes called *sheet rock*) is to be used over framing that is spaced more than 16 inches O.C. or over masonry, the designer may call for furring. *Furring* consists of narrow strips of wood, usually spaced at 16 or 12 inches O.C., to which the wall covering is fastened, Figure 32-7. The furring, if any is to be used, and the thickness of the gypsum wallboard are indicated on the wall sections.

Courtesy of Owens-Corning Fiberglas Corporation
Figure 32-4 Blanket insulation

Courtesy of Owens-Corning Fiberglas Corporation
Figure 32-5 Blowing wool

Some walls may be scheduled for gypsum board with another material over it. For example, in bathrooms the walls might be surfaced with plastic-faced hardboard paneling. This material is waterproof and easy to clean. However, special edge molding must be used to prevent water from getting behind the panels, Figure 32-8. Ceramic tile also is used on bathroom walls. Ceramic tile may be installed over a base of water-resistant gypsum board, plaster, or plywood.

Gypsum wallboard is also the most commonly used material for ceilings. Suspended ceilings are the next most common ceiling treatment. Suspended ceilings consist of panels or ceiling tiles supported in a lightweight metal framework. The metal framework is suspended several inches below the ceiling framing on steel wires, Figure 32-9.

Building sections or wall sections usually include a typical wall and ceiling. This is representative of what is planned for most of the walls and ceilings in the house. However, there may be some exceptions, such as the bathrooms, or kitchen where water-resistant wallboard is required. Somewhere within the contract documents, you should find a complete list of all room finishes. This may be on one of the drawings or it may be written into the specifications. Figure 32-10 shows a room finish schedule that might be included on the drawings. It is common for things like finish color to be left for the owner to choose.

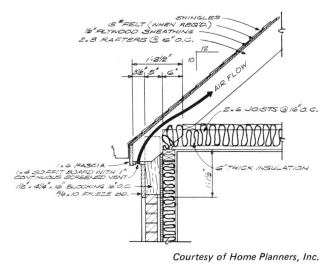

Courtesy of Home Planners, Inc.

Figure 32-6 The insulation must not block the air flow

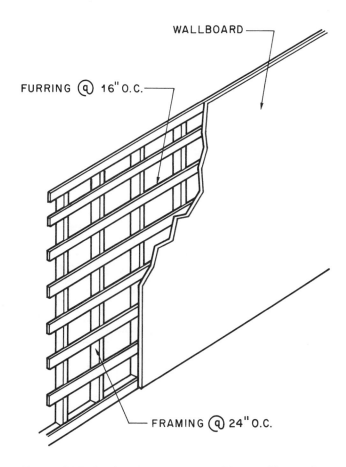

Figure 32-7 Furring is used to provide a nailing surface when framing is spaced too wide or over masonry and concrete.

FINISHED FLOORS

A list of possible floor materials would be very long. However, most of them fall into one of the following categories: wood, carpet, ceramic, masonry, and resilient materials such as vinyl tile. The finished floor covering is easily found in a schedule of room finishes, but the underlayment for each category is different. *Underlayment* is

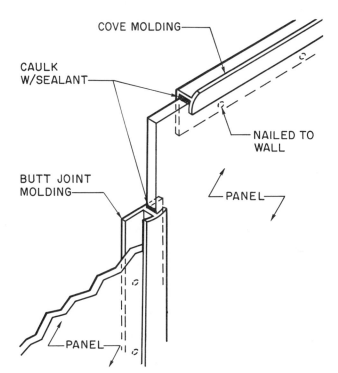

Figure 32-8 The molding on plastic-faced panels to be used in damp areas is caulked with sealant.

Courtesy of Celotex Corporation

Figure 32-9 Suspended ceiling

any material that is used to prepare the subfloor to receive the finished floor.

Architectural drafters seem to differ in how they indicate what underlayment is to be used and how the finished floor is to be installed. Some do not include any underlayment on the drawings, but rely on the builder's knowledge of good construction practices. This is a dangerous practice. If you find this situation, the architect should be asked for a clarification. Sometimes when the drawings do not describe the underlayment, the specifications do include it. Some drafters indicate the underlayment on the floor plans. Other drafters include details and sections of each area with different types of finished floors, so the underlayment can be shown there.

INTERIOR MOLDING

Molding is used to decorate surfaces, protect the edges and corners of surfaces, and to conceal joints or seams between surfaces. Molding may be made of wood, plastic, or metal and is available in many shapes and styles. The shapes of commonly used wood molding have been standardized. Each shape is identified by a number, Figure 32-11.

Most interior molding is shown on detail drawings. The following are some of the most common uses of interior molding:

- Window casing
- Window stool
- Door casing
- Base (bottom of wall)
- Cove (top of wall)
- Chair rail (middle of wall)
- Trim around fireplace mantle
- Trim around built-in cabinets

ROOM	FLOOR	WALLS	CEILING
KITCHEN	QUARRY TILE	GYP. BD. w/WALL PAPER	12"x12" TILE
DINING ROOM	OAK PARQUET	GYP. BD.	GYP. BD.
LIVING ROOM	CARPET/PART. BD.	GYP. BD.	GYP. BD.
FAMILY ROOM	CARPET/PART. BD.	HD. BD. PANEL/GYP. BD.	GYP. BD.
BEDROOM #1	CARPET/PART. BD.	GYP. BD.	GYP. BD.
BEDROOM #2	CARPET/PART. BD.	GYP. BD.	GYP. BD.
BEDROOM #3	CARPET/PART. BD.	GYP. BD.	GYP. BD.
BATH #1	CERAMIC TILE	CERAMIC TILE/GYP. BD.	GYP. BD.
BATH #2	CERAMIC TILE	CERAMIC TILE/GYP. BD.	GYP. BD.
CLOSETS	CARPET/PART BD.	GYP. BD.	GYP. BD.
FOYER	SLATE	GYP. BD. w/WALL PAPER	GYP. BD.

Figure 32-10 Room finish schedule

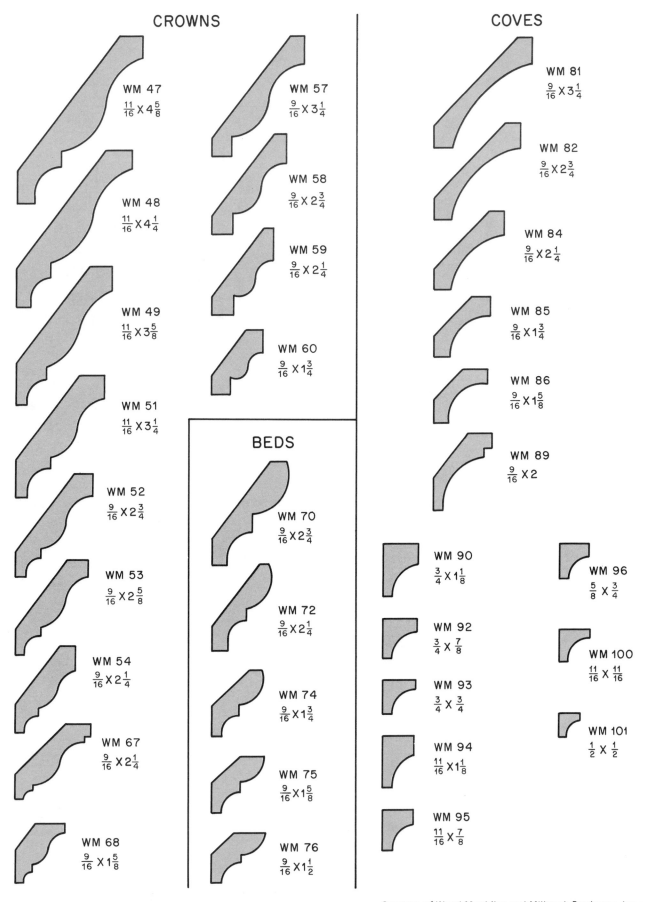

CROWNS

WM 47
$\frac{11}{16} \times 4\frac{5}{8}$

WM 48
$\frac{11}{16} \times 4\frac{1}{4}$

WM 49
$\frac{11}{16} \times 3\frac{5}{8}$

WM 51
$\frac{11}{16} \times 3\frac{1}{4}$

WM 52
$\frac{9}{16} \times 2\frac{3}{4}$

WM 53
$\frac{9}{16} \times 2\frac{5}{8}$

WM 54
$\frac{9}{16} \times 2\frac{1}{4}$

WM 67
$\frac{9}{16} \times 2\frac{1}{4}$

WM 68
$\frac{9}{16} \times 1\frac{5}{8}$

WM 57
$\frac{9}{16} \times 3\frac{1}{4}$

WM 58
$\frac{9}{16} \times 2\frac{3}{4}$

WM 59
$\frac{9}{16} \times 2\frac{1}{4}$

WM 60
$\frac{9}{16} \times 1\frac{3}{4}$

BEDS

WM 70
$\frac{9}{16} \times 2\frac{3}{4}$

WM 72
$\frac{9}{16} \times 2\frac{1}{4}$

WM 74
$\frac{9}{16} \times 1\frac{3}{4}$

WM 75
$\frac{9}{16} \times 1\frac{5}{8}$

WM 76
$\frac{9}{16} \times 1\frac{1}{2}$

COVES

WM 81
$\frac{9}{16} \times 3\frac{1}{4}$

WM 82
$\frac{9}{16} \times 2\frac{3}{4}$

WM 84
$\frac{9}{16} \times 2\frac{1}{4}$

WM 85
$\frac{9}{16} \times 1\frac{3}{4}$

WM 86
$\frac{9}{16} \times 1\frac{5}{8}$

WM 89
$\frac{9}{16} \times 2$

WM 90
$\frac{3}{4} \times 1\frac{1}{8}$

WM 92
$\frac{3}{4} \times \frac{7}{8}$

WM 93
$\frac{3}{4} \times \frac{3}{4}$

WM 94
$\frac{11}{16} \times 1\frac{1}{8}$

WM 95
$\frac{11}{16} \times \frac{7}{8}$

WM 96
$\frac{5}{8} \times \frac{3}{4}$

WM 100
$\frac{11}{16} \times \frac{11}{16}$

WM 101
$\frac{1}{2} \times \frac{1}{2}$

Courtesy of Wood Moulding and Millwork Producers, Inc.
P.O. Box 25278, Portland, Oregon 97225

Figure 32-11 Common shapes of wood molding

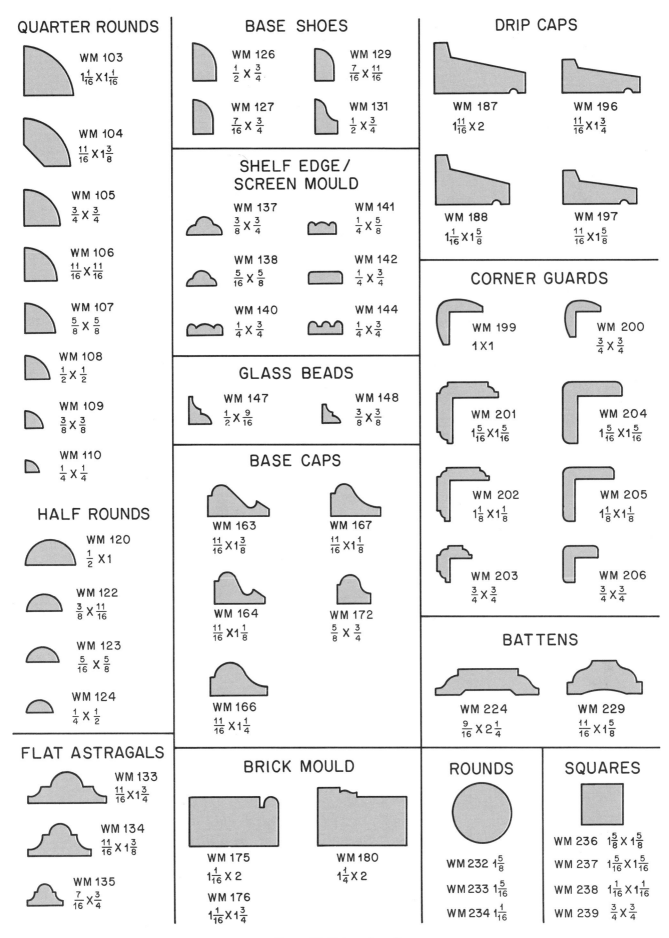

QUARTER ROUNDS

WM 103
$1\frac{1}{16} \times 1\frac{1}{16}$

WM 104
$\frac{11}{16} \times 1\frac{3}{8}$

WM 105
$\frac{3}{4} \times \frac{3}{4}$

WM 106
$\frac{11}{16} \times \frac{11}{16}$

WM 107
$\frac{5}{8} \times \frac{5}{8}$

WM 108
$\frac{1}{2} \times \frac{1}{2}$

WM 109
$\frac{3}{8} \times \frac{3}{8}$

WM 110
$\frac{1}{4} \times \frac{1}{4}$

HALF ROUNDS

WM 120
$\frac{1}{2} \times 1$

WM 122
$\frac{3}{8} \times \frac{11}{16}$

WM 123
$\frac{5}{16} \times \frac{5}{8}$

WM 124
$\frac{1}{4} \times \frac{1}{2}$

FLAT ASTRAGALS

WM 133
$\frac{11}{16} \times 1\frac{3}{4}$

WM 134
$\frac{11}{16} \times 1\frac{3}{8}$

WM 135
$\frac{7}{16} \times \frac{3}{4}$

BASE SHOES

WM 126
$\frac{1}{2} \times \frac{3}{4}$

WM 129
$\frac{7}{16} \times \frac{11}{16}$

WM 127
$\frac{7}{16} \times \frac{3}{4}$

WM 131
$\frac{1}{2} \times \frac{3}{4}$

SHELF EDGE/
SCREEN MOULD

WM 137
$\frac{3}{8} \times \frac{3}{4}$

WM 141
$\frac{1}{4} \times \frac{5}{8}$

WM 138
$\frac{5}{16} \times \frac{5}{8}$

WM 142
$\frac{1}{4} \times \frac{3}{4}$

WM 140
$\frac{1}{4} \times \frac{3}{4}$

WM 144
$\frac{1}{4} \times \frac{3}{4}$

GLASS BEADS

WM 147
$\frac{1}{2} \times \frac{9}{16}$

WM 148
$\frac{3}{8} \times \frac{3}{8}$

BASE CAPS

WM 163
$\frac{11}{16} \times 1\frac{3}{8}$

WM 167
$\frac{11}{16} \times 1\frac{1}{8}$

WM 164
$\frac{11}{16} \times 1\frac{1}{8}$

WM 172
$\frac{5}{8} \times \frac{3}{4}$

WM 166
$\frac{11}{16} \times 1\frac{1}{4}$

BRICK MOULD

WM 175
$1\frac{1}{16} \times 2$

WM 180
$1\frac{1}{4} \times 2$

WM 176
$1\frac{1}{16} \times 1\frac{3}{4}$

DRIP CAPS

WM 187
$1\frac{11}{16} \times 2$

WM 196
$1\frac{11}{16} \times 1\frac{3}{4}$

WM 188
$1\frac{1}{16} \times 1\frac{5}{8}$

WM 197
$\frac{11}{16} \times 1\frac{5}{8}$

CORNER GUARDS

WM 199
1×1

WM 200
$\frac{3}{4} \times \frac{3}{4}$

WM 201
$1\frac{5}{16} \times 1\frac{5}{16}$

WM 204
$1\frac{5}{16} \times 1\frac{5}{16}$

WM 202
$1\frac{1}{8} \times 1\frac{1}{8}$

WM 205
$1\frac{1}{8} \times 1\frac{1}{8}$

WM 203
$\frac{3}{4} \times \frac{3}{4}$

WM 206
$\frac{3}{4} \times \frac{3}{4}$

BATTENS

WM 224
$\frac{9}{16} \times 2\frac{1}{4}$

WM 229
$\frac{11}{16} \times 1\frac{5}{8}$

ROUNDS

WM 232 $1\frac{5}{8}$

WM 233 $1\frac{5}{16}$

WM 234 $1\frac{1}{16}$

SQUARES

WM 236 $1\frac{5}{8} \times 1\frac{5}{8}$

WM 237 $1\frac{5}{16} \times 1\frac{5}{16}$

WM 238 $1\frac{1}{16} \times 1\frac{1}{16}$

WM 239 $\frac{3}{4} \times \frac{3}{4}$

Figure 32-11 (continued)

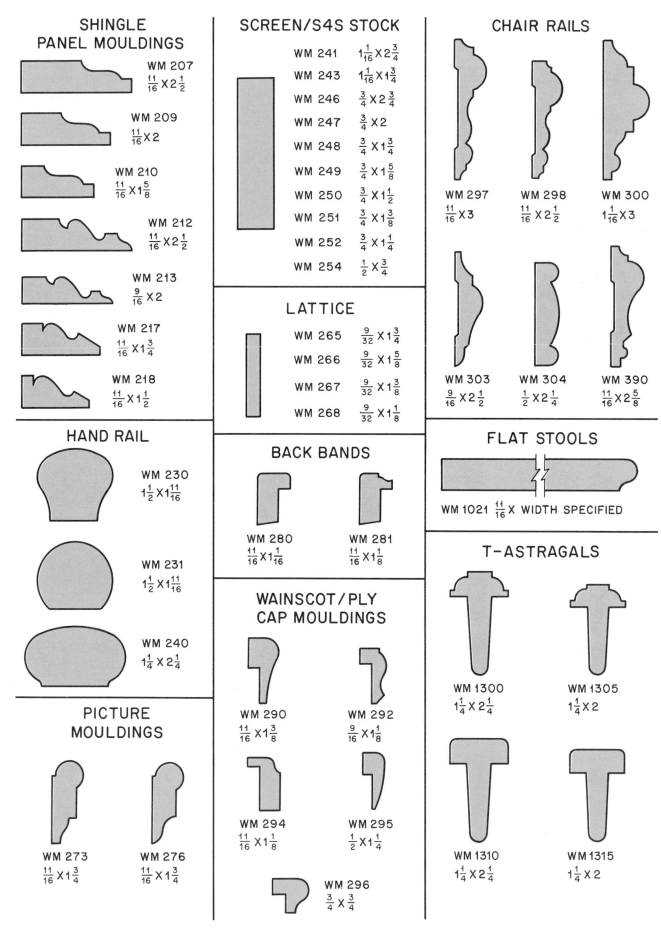

SHINGLE PANEL MOULDINGS

WM 207 $\frac{11}{16} \times 2\frac{1}{2}$

WM 209 $\frac{11}{16} \times 2$

WM 210 $\frac{11}{16} \times 1\frac{5}{8}$

WM 212 $\frac{11}{16} \times 2\frac{1}{2}$

WM 213 $\frac{9}{16} \times 2$

WM 217 $\frac{11}{16} \times 1\frac{3}{4}$

WM 218 $\frac{11}{16} \times 1\frac{1}{2}$

HAND RAIL

WM 230 $1\frac{1}{2} \times 1\frac{11}{16}$

WM 231 $1\frac{1}{2} \times 1\frac{11}{16}$

WM 240 $1\frac{1}{4} \times 2\frac{1}{4}$

PICTURE MOULDINGS

WM 273 $\frac{11}{16} \times 1\frac{3}{4}$

WM 276 $\frac{11}{16} \times 1\frac{3}{4}$

SCREEN/S4S STOCK

WM 241 $1\frac{1}{16} \times 2\frac{3}{4}$

WM 243 $1\frac{1}{16} \times 1\frac{3}{4}$

WM 246 $\frac{3}{4} \times 2\frac{3}{4}$

WM 247 $\frac{3}{4} \times 2$

WM 248 $\frac{3}{4} \times 1\frac{3}{4}$

WM 249 $\frac{3}{4} \times 1\frac{5}{8}$

WM 250 $\frac{3}{4} \times 1\frac{1}{2}$

WM 251 $\frac{3}{4} \times 1\frac{3}{8}$

WM 252 $\frac{3}{4} \times 1\frac{1}{4}$

WM 254 $\frac{1}{2} \times 1\frac{3}{4}$

LATTICE

WM 265 $\frac{9}{32} \times 1\frac{3}{4}$

WM 266 $\frac{9}{32} \times 1\frac{5}{8}$

WM 267 $\frac{9}{32} \times 1\frac{3}{8}$

WM 268 $\frac{9}{32} \times 1\frac{1}{8}$

BACK BANDS

WM 280 $\frac{11}{16} \times 1\frac{1}{16}$

WM 281 $\frac{11}{16} \times 1\frac{1}{8}$

WAINSCOT/PLY CAP MOULDINGS

WM 290 $\frac{11}{16} \times 1\frac{3}{8}$

WM 292 $\frac{9}{16} \times 1\frac{1}{8}$

WM 294 $\frac{11}{16} \times 1\frac{1}{8}$

WM 295 $\frac{1}{2} \times 1\frac{1}{4}$

WM 296 $\frac{3}{4} \times \frac{3}{4}$

CHAIR RAILS

WM 297 $\frac{11}{16} \times 3$

WM 298 $\frac{11}{16} \times 2\frac{1}{2}$

WM 300 $1\frac{1}{16} \times 3$

WM 303 $\frac{9}{16} \times 2\frac{1}{2}$

WM 304 $\frac{1}{2} \times 2\frac{1}{4}$

WM 390 $\frac{11}{16} \times 2\frac{5}{8}$

FLAT STOOLS

WM 1021 $\frac{11}{16} \times$ WIDTH SPECIFIED

T-ASTRAGALS

WM 1300 $1\frac{1}{4} \times 2\frac{1}{4}$

WM 1305 $1\frac{1}{4} \times 2$

WM 1310 $1\frac{1}{4} \times 2\frac{1}{4}$

WM 1315 $1\frac{1}{4} \times 2$

Figure 32-11 (continued)

CASING

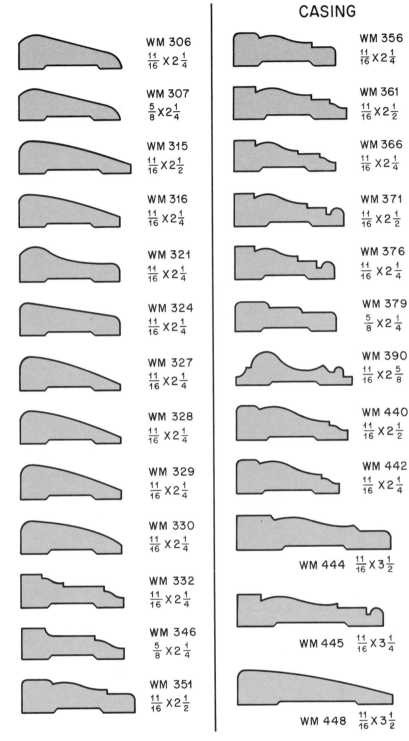

WM 306 $\frac{11}{16}$ X 2$\frac{1}{4}$

WM 307 $\frac{5}{8}$ X 2$\frac{1}{4}$

WM 315 $\frac{11}{16}$ X 2$\frac{1}{2}$

WM 316 $\frac{11}{16}$ X 2$\frac{1}{4}$

WM 321 $\frac{11}{16}$ X 2$\frac{1}{4}$

WM 324 $\frac{11}{16}$ X 2$\frac{1}{4}$

WM 327 $\frac{11}{16}$ X 2$\frac{1}{4}$

WM 328 $\frac{11}{16}$ X 2$\frac{1}{4}$

WM 329 $\frac{11}{16}$ X 2$\frac{1}{4}$

WM 330 $\frac{11}{16}$ X 2$\frac{1}{4}$

WM 332 $\frac{11}{16}$ X 2$\frac{1}{4}$

WM 346 $\frac{5}{8}$ X 2$\frac{1}{4}$

WM 351 $\frac{11}{16}$ X 2$\frac{1}{2}$

WM 356 $\frac{11}{16}$ X 2$\frac{1}{4}$

WM 361 $\frac{11}{16}$ X 2$\frac{1}{2}$

WM 366 $\frac{11}{16}$ X 2$\frac{1}{4}$

WM 371 $\frac{11}{16}$ X 2$\frac{1}{2}$

WM 376 $\frac{11}{16}$ X 2$\frac{1}{4}$

WM 379 $\frac{5}{8}$ X 2$\frac{1}{4}$

WM 390 $\frac{11}{16}$ X 2$\frac{5}{8}$

WM 440 $\frac{11}{16}$ X 2$\frac{1}{2}$

WM 442 $\frac{11}{16}$ X 2$\frac{1}{4}$

WM 444 $\frac{11}{16}$ X 3$\frac{1}{2}$

WM 445 $\frac{11}{16}$ X 3$\frac{1}{4}$

WM 448 $\frac{11}{16}$ X 3$\frac{1}{2}$

WM 683 $\frac{5}{8}$ X 3$\frac{1}{4}$

WM 410 $\frac{11}{16}$ X 4$\frac{1}{4}$

WM 412 $\frac{11}{16}$ X 3$\frac{1}{2}$

WM 413 $\frac{11}{16}$ X 3$\frac{1}{4}$

WM 430 $\frac{9}{16}$ X 4$\frac{1}{4}$

WM 432 $\frac{9}{16}$ X 3$\frac{1}{2}$

WM 433 $\frac{9}{16}$ X 3$\frac{1}{4}$

WM 452 $\frac{11}{16}$ X 2$\frac{1}{2}$

WM 453 $\frac{11}{16}$ X 2$\frac{1}{4}$

WM 472 $\frac{9}{16}$ X 2$\frac{1}{2}$

WM 473 $\frac{9}{16}$ X 2$\frac{1}{4}$

WM 492 $\frac{7}{16}$ X 2$\frac{1}{2}$

WM 493 $\frac{7}{16}$ X 2$\frac{1}{4}$

Figure 32-11 (continued)

BASE MOULDINGS

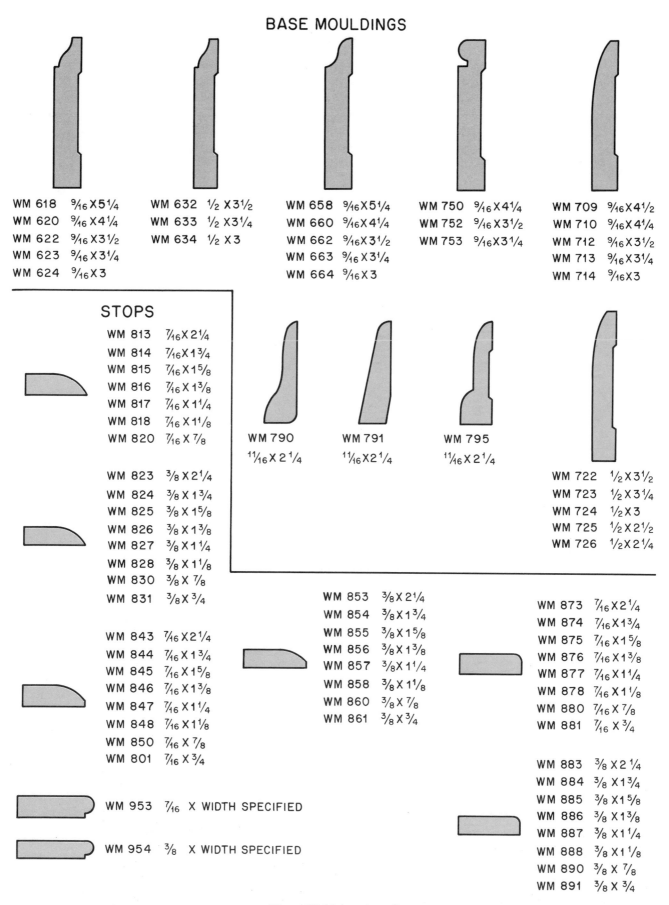

WM 618	$9/16 \times 5 1/4$	WM 632	$1/2 \times 3 1/2$	WM 658	$9/16 \times 5 1/4$	WM 750	$9/16 \times 4 1/4$	WM 709	$9/16 \times 4 1/2$
WM 620	$9/16 \times 4 1/4$	WM 633	$1/2 \times 3 1/4$	WM 660	$9/16 \times 4 1/4$	WM 752	$9/16 \times 3 1/2$	WM 710	$9/16 \times 4 1/4$
WM 622	$9/16 \times 3 1/2$	WM 634	$1/2 \times 3$	WM 662	$9/16 \times 3 1/2$	WM 753	$9/16 \times 3 1/4$	WM 712	$9/16 \times 3 1/2$
WM 623	$9/16 \times 3 1/4$			WM 663	$9/16 \times 3 1/4$			WM 713	$9/16 \times 3 1/4$
WM 624	$9/16 \times 3$			WM 664	$9/16 \times 3$			WM 714	$9/16 \times 3$

STOPS

WM 813 $7/16 \times 2 1/4$
WM 814 $7/16 \times 1 3/4$
WM 815 $7/16 \times 1 5/8$
WM 816 $7/16 \times 1 3/8$
WM 817 $7/16 \times 1 1/4$
WM 818 $7/16 \times 1 1/8$
WM 820 $7/16 \times 7/8$

WM 790 $11/16 \times 2 1/4$
WM 791 $11/16 \times 2 1/4$
WM 795 $11/16 \times 2 1/4$

WM 823 $3/8 \times 2 1/4$
WM 824 $3/8 \times 1 3/4$
WM 825 $3/8 \times 1 5/8$
WM 826 $3/8 \times 1 3/8$
WM 827 $3/8 \times 1 1/4$
WM 828 $3/8 \times 1 1/8$
WM 830 $3/8 \times 7/8$
WM 831 $3/8 \times 3/4$

WM 722 $1/2 \times 3 1/2$
WM 723 $1/2 \times 3 1/4$
WM 724 $1/2 \times 3$
WM 725 $1/2 \times 2 1/2$
WM 726 $1/2 \times 2 1/4$

WM 843 $7/16 \times 2 1/4$
WM 844 $7/16 \times 1 3/4$
WM 845 $7/16 \times 1 5/8$
WM 846 $7/16 \times 1 3/8$
WM 847 $7/16 \times 1 1/4$
WM 848 $7/16 \times 1 1/8$
WM 850 $7/16 \times 7/8$
WM 801 $7/16 \times 3/4$

WM 853 $3/8 \times 2 1/4$
WM 854 $3/8 \times 1 3/4$
WM 855 $3/8 \times 1 5/8$
WM 856 $3/8 \times 1 3/8$
WM 857 $3/8 \times 1 1/4$
WM 858 $3/8 \times 1 1/8$
WM 860 $3/8 \times 7/8$
WM 861 $3/8 \times 3/4$

WM 873 $7/16 \times 2 1/4$
WM 874 $7/16 \times 1 3/4$
WM 875 $7/16 \times 1 5/8$
WM 876 $7/16 \times 1 3/8$
WM 877 $7/16 \times 1 1/4$
WM 878 $7/16 \times 1 1/8$
WM 880 $7/16 \times 7/8$
WM 881 $7/16 \times 3/4$

WM 953 $7/16$ X WIDTH SPECIFIED

WM 954 $3/8$ X WIDTH SPECIFIED

WM 883 $3/8 \times 2 1/4$
WM 884 $3/8 \times 1 3/4$
WM 885 $3/8 \times 1 5/8$
WM 886 $3/8 \times 1 3/8$
WM 887 $3/8 \times 1 1/4$
WM 888 $3/8 \times 1 1/8$
WM 890 $3/8 \times 7/8$
WM 891 $3/8 \times 3/4$

Figure 32-11 (continued)

STOPS

WM 903 $7/16 \times 2\,1/4$	WM 913 $3/8 \times 2\,1/4$	WM 933 $7/16 \times 2\,1/4$	WM 943 $3/8 \times 2\,1/4$
WM 904 $7/16 \times 1\,3/4$	WM 914 $3/8 \times 1\,3/4$	WM 934 $7/16 \times 1\,3/4$	WM 944 $3/8 \times 1\,3/4$
WM 905 $7/16 \times 1\,5/8$	WM 915 $3/8 \times 1\,5/8$	WM 935 $7/16 \times 1\,5/8$	WM 945 $3/8 \times 1\,5/8$
WM 906 $7/16 \times 1\,3/8$	WM 916 $3/8 \times 1\,3/8$	WM 936 $7/16 \times 1\,3/8$	WM 946 $3/8 \times 1\,3/8$
WM 907 $7/16 \times 1\,1/4$	WM 917 $3/8 \times 1\,1/4$	WM 937 $7/16 \times 1\,1/4$	WM 947 $3/8 \times 1\,1/4$
WM 908 $7/16 \times 1\,1/8$	WM 918 $3/8 \times 1\,1/8$	WM 938 $7/16 \times 1\,1/8$	WM 948 $3/8 \times 1\,1/8$
WM 910 $7/16 \times 7/8$	WM 920 $3/8 \times 7/8$	WM 940 $7/16 \times 7/8$	WM 950 $3/8 \times 7/8$
WM 911 $7/16 \times 3/4$	WM 921 $3/8 \times 3/4$	WM 941 $7/16 \times 3/4$	WM 951 $3/8 \times 3/4$

PANEL STRIPS
MULLION CASINGS

WM 955 $3/8 \times 2\,1/4$
WM 956 $3/8 \times 2$
WM 957 $3/8 \times 1\,3/4$

WM 960 $3/8 \times 2\,1/4$
WM 962 $3/8 \times 2$
WM 963 $3/8 \times 1\,3/4$

WM 965 $3/8 \times 2\,1/4$
WM 967 $3/8 \times 2$
WM 968 $3/8 \times 1\,3/4$

WM 970 $3/8 \times 2\,1/4$
WM 972 $3/8 \times 2$
WM 973 $3/8 \times 1\,3/4$

WM 975 $3/8 \times 2\,1/4$
WM 977 $3/8 \times 2$
WM 978 $3/8 \times 1\,3/4$

WM 980 $3/8 \times 2\,1/4$
WM 982 $3/8 \times 2$
WM 983 $3/8 \times 1\,3/4$

WM 985 $3/8 \times 2\,1/4$
WM 987 $3/8 \times 2$
WM 988 $3/8 \times 1\,3/4$

RABBETED STOOLS

SPECIFY WIDTH OF RABBET
AND DEGREE OF BEVEL

WIDTH
BEVEL
$10°$ $14°$

WM 1131 $1\,1/16 \times 3\,5/8$
WM 1133 $1\,1/16 \times 3\,1/4$
WM 1134 $1\,1/16 \times 2\,3/4$

WM 1153 $11/16 \times 3\,1/4$
WM 1154 $11/16 \times 2\,3/4$
WM 1155 $11/16 \times 2\,1/2$
WM 1156 $11/16 \times 2\,1/4$

WM 1161 $1\,1/16 \times 3\,5/8$
WM 1163 $1\,1/16 \times 3\,1/4$
WM 1164 $1\,1/16 \times 2\,3/4$

WM 1193 $11/16 \times 3\,1/4$
WM 1194 $11/16 \times 2\,3/4$
WM 1195 $11/16 \times 2\,1/2$
WM 1196 $11/16 \times 2\,1/4$

Figure 32-11 (continued)

✓ CHECK YOUR PROGRESS

Can you perform these tasks?

☐ Describe the kinds of insulation to be used in walls, around foundations, under slabs, in floors, in attics, and as sheathing.

☐ List the kind and thickness of material used for the interior surface of walls, ceilings, and floors.

☐ Identify areas where furring is to be used and the spacing of the furring.

☐ List all of the types of molding to be used for interior trim.

ASSIGNMENT

Refer to the Lake House drawings (in the packet) to complete the assignment.

1. What size or rating and kind of insulation is to be used in each of the following locations?
   a. Framed exterior walls
   b. Roof
   c. Under heat sink
   d. Masonry walls of playroom

2. What type molding is to be used as casing around interior doors?

3. What type of molding is used at the bottom of interior walls and partitions?

4. What kind of trim is used to cover the lower edges of exposed box beams?

5. Describe the wall finish in the playroom including
   a. what the wall finish material is fastened to
   b. the kind of material used for wall finish

6. What material is used for ceiling finish in the playroom?

7. What material is used for underlayment on typical framed floors?

8. What is the finished floor material at the heat sink?

9. What covers the interior faces of wood box beams?

10. What is the finished wall material in the bedrooms?

OBJECTIVES

After completing this unit, you will be able to perform the following tasks:

- List the sizes and types of cabinets shown on a set of drawings.

- Identify cabinet types and dimensions in manufacturers' literature.

SHOWING CABINETS ON DRAWINGS

Cabinets are typically found in the kitchen, bathroom, and laundry room. Most cabinets are supplied by a manufacturer, so the carpenter's job is only to install them. When custom-built cabinets are to be included, they are assembled in a cabinet shop and delivered to the site.

The layout of the cabinets can be determined by reading the floor plan and cabinet elevations together. The floor plan normally includes reference marks indicating each cabinet elevation, Figure 33-1. The floor plan shows the location of major appliances and may include some overall layout dimensions.

More complete cabinet information is usually shown on the cabinet elevations, Figure 33-2. Cabinet elevations are drawn for each direction from which cabinets can be viewed. These elevations show the types of cabinets, their sizes, and their arrangement. Cabinet types and sizes are recognized by a combination of drawing representation and commonly used letter/number designations. Some drawings rely heavily on standard dimen-

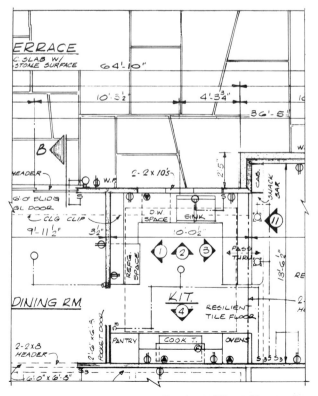

Figure 33-1 Kitchen floor plan with key to cabinet elevations

Courtesy of Home Planners, Inc.

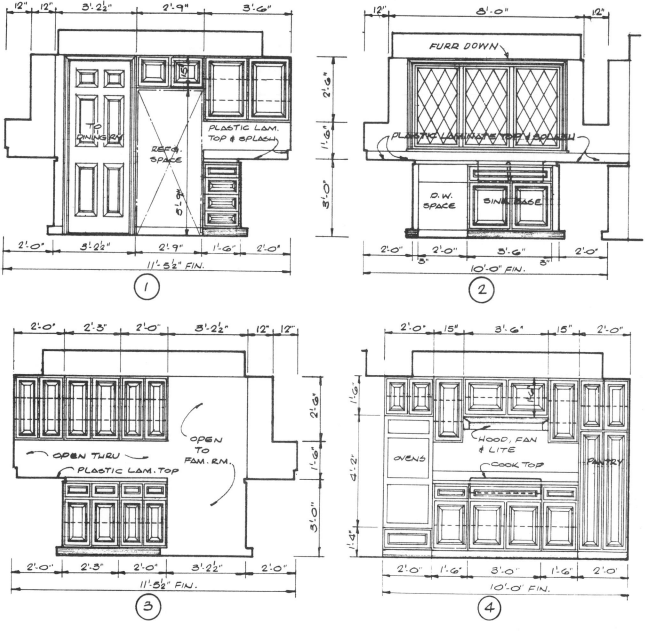

Figure 33-2 Elevations for figure 33-1

Courtesy of Home Planners, Inc.

sioning and pictorial representation. This is true of the case shown in Figure 33-2. Other drawings include a letter/number designation for each cabinet, as shown in Figure 33-3. The letter part of the designation represents the type of cabinet. Architects and drafters vary in how they use letter designations, but they are usually easy to understand after a moments' study. Some typical letter designations are shown in Figure 33-4.

The numbers in the cabinet designation represent the dimensions of the cabinet. Base cabinets are a standard height (usually 3'-0'' including the countertop) and a standard front-to-back depth (usually 2'-0''). Therefore, base cabinet designations only include two digits to represent width in

inches. Wall cabinets are a standard front-to-back depth (usually 1'-0''), but the width and height vary. Therefore, wall cabinet designations include four digits — two for width and two for height. The first and second digits usually indicate height and the third and fourth digits usually indicate width. The following example explains a typical cabinet designation.

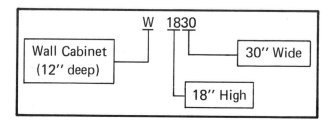

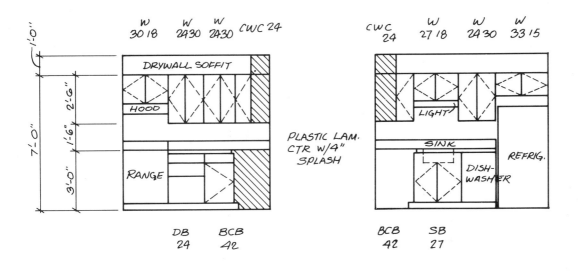

KITCHEN ELEVATIONS

Courtesy of Robert C. Kurzon

Figure 33-3 Letter/number designations for cabinets

In addition to cabinets, most cabinet manufacturers provide a variety of accessories:

- Shelves are finished to match the cabinets, so they can be used in open areas.

- Valances are prefinished decorative pieces to use between wall cabinets (over a sink, for example).

- Filler pieces are prefinished boards used to enclose narrow spaces between cabinets.

- A variety of molding may be used for trim.

READING MANUFACTURERS' LITERATURE

Cabinet manufacturers publish catalogs and specifications describing their cabinets. The numbering systems used in this literature are similar to those used on construction drawings. There may be slight differences between manufacturers,

but these are usually easy to see. Figure 33-5 is reprinted from one manufacturer's literature. A few tips to reading this manufacturer's literature are listed:

- Cabinet heights and standard depths are dimensioned on the left side of the page.

- Referring to these dimensions, you can see that the first and second digits of wall cabinet numbers indicate their height. Therefore, the third and fourth digits indicate width.

- Standard base cabinets with one drawer above and doors below are designated by three digits only. The first digit (4) indicates a standard base cabinet. The second and third digits indicate width.

- Base cabinets with other than one drawer above a door are designated by a letter, but the last two digits still indicate width.

Figure 33-4 Key to typical cabinet designations

LETTER	CABINET TYPE
W	WALL CABINET
WC	CORNER WALL CABINET
B	BASE CABINET
D OR DB	DRAWER BASE CABINET
BC	BASE CORNER CABINET
RC	REVOLVING CORNER CABINET
SF	SINK FRONT (ALSO FOR COOKTOP)
SB	SINK BASE (ALSO FOR COOKTOP)
U	UTILITY OR BROOM CLOSET
OV	OVEN CABINET

✓ CHECK YOUR PROGRESS ——————————————————————

Can you perform these tasks?

☐ Orient cabinet elevations to floor plans.

☐ Describe cabinet types and sizes indicated by typical letter/number designations.

☐ List all of the cabinets shown on the drawings for a house.

ASSIGNMENT ——————————————————————————————

Refer to Lake House drawings (in the packet) to complete the assignment.

Using the catalog pages shown in Figure 33-5 on the following pages, list all of the cabinets for the Lake House kitchen.

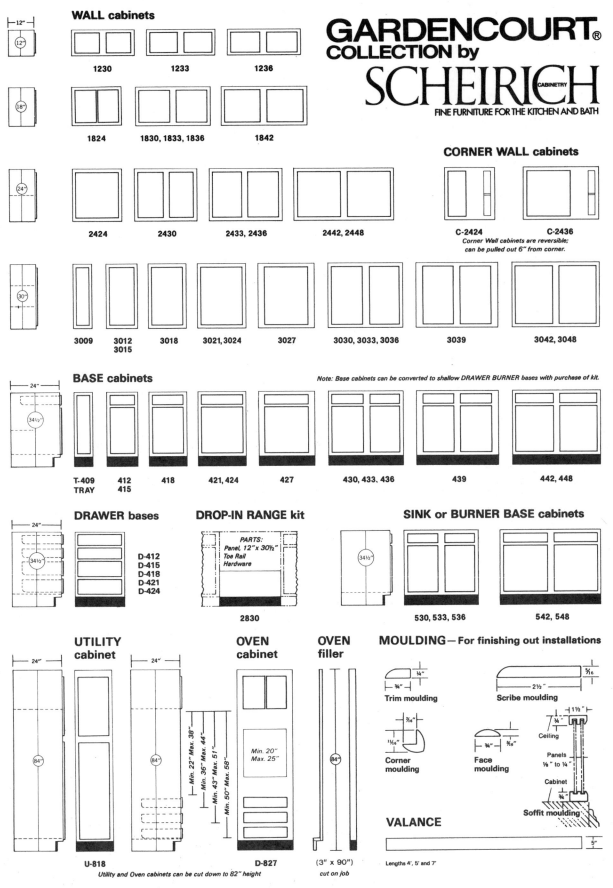

WALL cabinets

GARDENCOURT® COLLECTION by SCHEIRICH CABINETRY
FINE FURNITURE FOR THE KITCHEN AND BATH

1230   1233   1236

1824   1830, 1833, 1836   1842

CORNER WALL cabinets

2424   2430   2433, 2436   2442, 2448   C-2424   C-2436

Corner Wall cabinets are reversible;
can be pulled out 6" from corner.

3009   3012 3015   3018   3021, 3024   3027   3030, 3033, 3036   3039   3042, 3048

BASE cabinets

Note: Base cabinets can be converted to shallow DRAWER BURNER bases with purchase of kit.

T-409 TRAY   412 415   418   421, 424   427   430, 433, 436   439   442, 448

DRAWER bases

D-412
D-415
D-418
D-421
D-424

DROP-IN RANGE kit

PARTS:
Panel, 12" x 30½"
Toe Rail
Hardware

2830

SINK or BURNER BASE cabinets

530, 533, 536   542, 548

UTILITY cabinet

Min. 22" Max. 38"
Min. 36" Max. 44"
Min. 43" Max. 51"
Min. 50" Max. 58"

U-818

Utility and Oven cabinets can be cut down to 82" height

OVEN cabinet

Min. 20"
Max. 25"

D-827

OVEN filler

(3" x 90")
cut on job

MOULDING — For finishing out installations

Trim moulding   Scribe moulding

Corner moulding   Face moulding

Ceiling
Panels ⅛" to ¼"
Cabinet
Soffit moulding

VALANCE

Lengths 4', 5' and 7'

Copyright H. J. Scheirich Co., Louisville, Kentucky 1969 — Revised November 1, 1970; September 1, 1971; January 1, 1973; January 1, 1974

Courtesy of H.J. Scheirich Co.

Figure 33-5 Cabinet catalog

After three years of intensive research and development the H. J. Scheirich Company is proud to present the new Gardencourt Collection. We wished to make the handsomest cabinets possible—and the toughest . . . cabinets so well styled they would fit into any interior, yet so well built they would pass the most rigorous test of all—everyday use in a busy kitchen.

We discarded all the old woodworking concepts and created a wholly new kind of cabinetry of exceptional structural rigidity and dimensional stability. The basic ma-terial for all parts is made of wood fibers bonded solidly together to provide extraordinary strength and resistance to variations in temperature and humidity. Framework, panels, shelves and doors—all are made of the bonded material. Then they are totally sheathed in Vinyl with exterior surfaces in pecan grain finish overlayed with a clear Vinyl for added protection. These are truly "carefree" cabinets. Produced in a modern air-conditioned factory employing highly sophisticated machinery, Gardencourt maintains the well known Scheirich standard of excellence.

All Scheirich cabinets are delivered complete with mounting hardware and detailed instructions for installation. Single door cabinets can hinge left or right. Shipped with doors attached, they can be reversed very simply by moving one hinge screw for each hinge into pre-drilled holes on the opposite inside front of the cabinet frame.

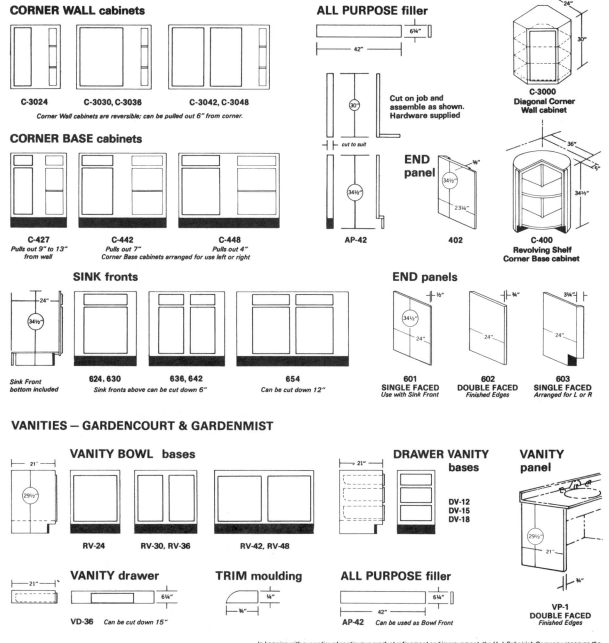

Figure 33-5 Cabinet catalog (continued)

In keeping with our policy of continuous product refinement and improvement, the H. J. Scheirich Company reserves the right to alter specifications without general notice or obligation to make similar changes in units previously shipped.

Contract Documents

Section 10

UNIT 34 *Contracts and General Requirement Specifications*

OBJECTIVES

After completing this unit, you will be able to perform the following tasks:

- List the documents that make up a typical construction agreement.
- Find specific information on the contract documents.

A construction agreement is just that; it is an agreement between parties to perform certain tasks in return for payment. The documents that describe the agreement are the drawings, specifications, bids, and contract. The drawings are explained in the other units of this book. The following paragraphs are a brief explanation of the other contract documents.

SPECIFICATIONS

Working drawings contain as much information as possible about the materials to be used, the position of parts, and the dimensions of all parts of the project. However, it is impossible to include all of the necessary information about the project and the responsibilities of each individual involved. For example, it is not practical to show the grade of lumber to be used for the roof framing, or to indicate who is responsible for cleanup after the construction is finished. *Specifications*, commonly called *specs*, are written documents that give information that is not shown on the drawings. The specs for one-family and two-family houses are usually quite brief, but cover the same basic topics as the specs for bigger projects.

The first section of the specs is usually *General Requirements*, which covers such things as storage, temporary services, responsibility for verifying dimensions, final cleaning, and insurance requirements during construction, Figure 34-1. When the spec form is very brief, the general requirements are not made a part of the specs. In such cases the general information is included in the contract form under the heading *General Conditions*.

Specifications

DIVISION 1: GENERAL REQUIREMENTS

A. ARCHITECT'S SUPERVISION
The architect will have continued supervisory responsibility for this job.

B. TEMPORARY CONVENIENCES
The general contractor shall provide suitable temporary conveniences for the use of all workers on this job. Facilities shall be within a weather tight, painted enclosure complying with legal requirements. The general contractor shall maintain all temporary toilet facilities in a sanitary condition.

C. PUMPING
The general contractor shall keep the excavation and the basement free from water at all times and shall provide, maintain, and operate at his own expense such pumping equipment as shall be necessary.

D. PROTECTION
The general contractor shall protect all existing driveways, parking areas, sidewalks, curbs, and existing paved areas on, or adjacent to, the owner's property.

E. GRADES, LINES, LEVELS, AND SURVEYS
The owner shall establish the lot lines.
The general contractor shall:
1. Establish and maintain bench marks.
2. Verify all grades, lines, levels, and dimensions as shown on the drawings, and report any errors or inconsistencies before commencing work.

3. Lay out the building accurately under the supervision of the architect.

F. FINAL CLEANING
In addition to the general room cleaning, the general contractor shall do the following special cleaning upon completion of the work:
1. Wash and polish all glass and cabinets.
2. Clean and polish all hardware.
3. Remove all marks, stains, fingerprints, and other soil or dirt from walls, woodwork, and floors.

G. GUARANTEES
The general contractor shall guarantee all work performed under the contract against faulty materials or workmanship. The guarantee shall be in writing with duplicate copies delivered to the architect. In case of work performed by subcontractors where guarantees are required, the general contractor shall secure written guarantees from those subcontractors. Copies of these guarantees shall be delivered to the architect upon completion of the work. Guarantees shall be signed by both the subcontractor and the general contractor.

H. FOREMAN
The general contractor shall have a responsible foreman at the building site from the start to the completion of construction. The foreman shall be on duty during all working hours.

I. FIRE INSURANCE
The owner shall effect and maintain builder's risk completed-value insurance on this job.

Figure 34-1 General Requirements for a small project.

The rest of the specs are devoted to describing the technical aspects of the project. Technical specifications are explained later in Unit 35.

BIDS

Many construction contracts are awarded on a bid basis. A *bid* is a contractor's statement that all or certain parts of the project can be completed within a given period of time and for a given price. Notice that the last paragraph of the bid form shown in Figure 34-2 requires the bidder to sign a contract for the work. Therefore, the bid is a binding document and considered a part of the agreement package.

CONTRACT

When the job is awarded, a legal agreement is signed that states the contractor will perform the work and the owner will pay for the work. This agreement is a *contract*. A verbal agreement between two people is a legal contract, but construction contracts are almost always in the form of written documents signed by all parties involved, Figure 34-3. These contracts include provisions for the following:

• **Completion schedule.** It is customary to specify the date by which the project is to be completed. On some large projects, dates for

TO: STUYVESANT PLAZA, INC.
       Executive Park
       Albany, New York 12203

       Att: Mr. D. Lopez

1.   Pursuant to and in compliance with your Invitation to Bid, the Instruction to Bidders and other related documents, the undersigned, _____

_____

having familiarized himself with said documents and local conditions affecting materials, supplies, appliances, devices, equipment, services, and other facilities as required for

1 Executive Park Place

and, in strict accordance with the Bid Requirements, Specifications, Drawings, and addenda thereto, as prepared by KURZON ARCHITECTS and on file at the Offices of the Owner to perform work as follows:

_____

_____

_____

_____

(describe, listing spec. sections and drawings by number)

2.   LUMP SUM BID:

_____Dollars ($)

including any applicable taxes, overhead and profit on all work and all allowances, if any, required by the Contract Document.

3.   CONSTRUCTION PERIOD:

The undersigned proposes to complete the work within the following number of consecutive calendar days from the date _____ days for Lump Sum Bid.

4.   If written notice is mailed or otherwise delivered to the undersigned within sixty (60) days after the opening thereof, or at any time thereafter before this bid is withdrawn, the undersigned agrees to execute and deliver a Contract, in the prescribed form, and furnish any required bonds within ten (10) days after said contract is presented to him for signature.

_____
Bidder

_____   _____
By and Title   Date

_____
Signature

_____
Address

_____
Telephone No.

Courtesy of Robert C. Kurzon

Figure 34-2 Typical bid form.

Building Construction Agreement

This Agreement made and entered into this _____ day of _____, 19 _____.

between _____

_____

called "OWNER" whose address is   _____

and   _____

Called "CONTRACTOR" whose address is _____

_____

It is hereby agreed:

1. DESCRIPTION OF WORK: Contractor shall furnish all labor, materials, and services to construct and complete in good, expeditious,

workmanlike and substantial manner a   _____

hereafter called "project" upon the following described real property: _____

_____

_____

_____

OWNER shall locate and point out the property lines to CONTRACTOR and shall provide boundary stakes by a licensed land surveyor or registered civil engineer if in doubt as to boundaries.

2. PLANS, SPECIFICATIONS AND PERMITS: The project shall be constructed according to the project plans and specifications which have been examined and accepted by the OWNER and which are hereby incorporated by reference and made a part of this Agreement.

Unless otherwise specifically provided in the plans or specifications, CONTRACTOR shall obtain and pay for all required building permits, and OWNER shall pay assessments and charges required by public bodies and utilities for financing or repaying the cost of sewers, storm drains, water service, and other utilities, including sewer and storm drain reimbursement charges, revolving fund charges, hook-up charges and the like.

3. PAYMENT: OWNER shall pay CONTRACTOR the agreed price of $ _____

_____

together with any additional sums as may be provided for herein, in installments as follows: _____

_____

_____

Final payment shall be made thirty (30) days after Notice of Completion has been recorded or upon issuance of a lien free endorsement by a title company to lender.

Upon execution of this Agreement the final payment shall be placed in escrow for CONTRACTOR's benefit. This Agreement constitutes authority for the escrow to pay CONTRACTOR thirty (30) days after recordation of Notice of Completion.

If payments are to be made through a Construction Lender, OWNER represents that the construction loan fund is sufficient to pay the contract price. OWNER will do every thing possible to expedite all payments. OWNER hereby irrevocably authorizes Construction Lender to make all payments directly to CONTRACTOR when due. The name and address of the Construction Lender is   _____

_____ of

_____

If corrective or repair work of a minor nature remains to be accomplished by CONTRACTOR after the project is ready for occupancy CONTRACTOR shall perform such work expeditiously and OWNER shall not withhold any payment pending the completion of such work.

If major items of corrective or repair work remain to be accomplished after the project is ready for occupancy the cost of which aggregates more than one (1%) percent of the gross agreed price as determined by the CONTRACTOR, then OWNER pending completion of such work, may withhold in escrow a sufficient amount to pay for completion of such work, but no more.

4. COMPLETION: CONTRACTOR shall begin work within _____ days after the site is ready for CONTRACTOR and the required

building permit has been issued and the construction loan, if any, has been recorded and shall complete the project within _____ working days, subject to permissable delays as described in Paragraph 5.

5. DELAY: CONTRACTOR shall be excused for any delay in completion of the project caused by acts of God, acts of OWNER or OWNER's agent, employee or independent contractor, stormy weather, labor trouble, acts of public utilities, public bodies or inspectors, extra work, failure of the Owner to make progress and extra work payments promptly, or other contingencies unforeseen by CONTRACTOR and beyond the reasonable control of CONTRACTOR.

6. LABOR AND MATERIALS: CONTRACTOR shall pay all valid charges for labor and material incurred by CONTRACTOR and used in the construction of the project but is excused from this obligation in any period during which OWNER is in arrears in making progress and extra work payments to CONTRACTOR.

7. AGREEMENT, PLANS, AND SPECIFICATIONS: The Agreement, plans and specifications are intended to supplement each other. In case of conflict, however, the specifications shall control the plans, and the provisions of this Agreement shall control both.

8. EXTRA WORK, CHANGES AND INTEREST: (a) Changes shall be made only when a change order is signed by CONTRACTOR and OWNER.

(b) Should OWNER, Construction Lender, or any proper governmental body or building inspector direct any modification or addition to the work covered by this Agreement, the cost shall be added to the agreed price.

(c) Expense incurred because of unusual or unanticipated ground conditions (such as fill, rock, or ground water) shall be paid for as extra work.

(d) Any extra cost incurred by CONTRACTOR at the instance of OWNER shall be paid for as extra work.

(e) Payment for extras and changes shall include the actual cost of labor and materials plus CONTRACTOR's fee and when feasible shall be paid for in advance, otherwise on receipt of statement and there shall be no retention of funds related to extras or changes.

(f) All moneys due CONTRACTOR from OWNER whether for the agreed price or for changes or extras shall bear interest at the rate of ten (10%) percent per annum.

Courtesy of Building Industry Association of Superior California, Inc.

Figure 34-3 Construction contract

9. ALLOWANCES: If the Agreed price includes "allowances," and the cost of performing the work covered by the allowance is either greater or less than the allowance, then the agreed price shall be increased or decreased accordingly. Unless otherwise requested by OWNER in writing, CONTRACTOR shall use his own judgment in accomplishing work covered by an allowance. If OWNER requests that work covered by an allowance be accomplished in such a way that the cost will exceed the allowance, CONTRACTOR shall comply with OWNER's request provided that OWNER pays the additional cost on receipt of statement.

10. NOTICE OF COMPLETION AND OCCUPANCY: OWNER agrees to sign and record a Notice of Completion within five (5) days after the project is complete and ready for occupancy.

11. AGENCY: If OWNER fails to so record Notice of Completion, then OWNER hereby appoints CONTRACTOR as OWNER's agent to sign and record a Notice of Completion on behalf of OWNER. This agency is irrevocable and is an agency coupled with an interest.

12. OCCUPANCY: CONTRACTOR may use such reasonable force or legal means as is necessary to deny occupancy of the project by OWNER or anyone else until CONTRACTOR has received all payments due under this Agreement and until Notice of Completion has been recorded. In the event OWNER occupies the building or any part thereof before CONTRACTOR has received all payment due under this Agreement, such occupancy shall constitute full and unqualified acceptance of all CONTRACTOR's work by OWNER and OWNER agrees that such occupancy shall be a waiver of any and all claims against CONTRACTOR.

13. INSURANCE AND DEPOSITS: OWNER shall procure at his own expense and before the commencement of any work hereunder, fire insurance with coverage for construction, vandalism and malicious mischief; such insurance to be in a sum at least equal to the agreed construction price with loss, if any, payable to any beneficiary under any deed of trust covering the project. Such insurance shall name the CONTRACTOR, all sub-contractors and all suppliers as an additional insured for the protection of OWNER, CONTRACTOR, sub-contractors, and construction lender as their interests may appear.

Should OWNER fail to do so, CONTRACTOR may, but is not required to, procure such insurance as an extra and as agent for and at the expense of OWNER.

If the project is destroyed or damaged by any accident, disaster, or calamity, such as fire, storm, flood, landslide or subsidence, earthquake or by theft or vandalism, any work done by CONTRACTOR in rebuilding or restoring the project shall be paid for by OWNER as extra work.

CONTRACTOR shall carry Workmen's Compensation Insurance for the protection of CONTRACTOR's employees during the progress of the work. OWNER shall obtain and pay for insurance against injury to his own employees and persons under OWNER's direction and persons on the job site at OWNER's invitation.

14. RIGHT TO STOP WORK: If OWNER fails to make any payment whether for the agreed price, changes or extras, CONTRACTOR may stop work until the payments are made and shall not be deemed to have breached this Agreement by reason of such stoppage. OWNER's obligation to make payments is an express condition precedent to CONTRACTOR's obligation to proceed with the work. CONTRACTOR may at his option elect to terminate the Agreement and treat the failure to pay as a breach. Time is of the essence of this Agreement.

15. ARBITRATION: Any controversy or claim arising out of or relating to this Agreement, or the breach thereof, shall be settled by arbitration in accordance with the Construction Industry Arbitration Rules of the American Arbitration Association, and judgment upon the award may be entered in any Court having jurisdiction.

16. WARRANTIES: CONTRACTOR warrants against any defects in workmanship or materials for a period of one (1) year from date of completion. There are no other warranties express or implied.

17. LIMITATIONS: No action of any character arising out of or relating to this Agreement or the performance thereof, shall be commenced by either party against the other more than two (2) years after the completion or cessation of work under this Agreement.

18. ATTORNEYS FEES: In the event of any arbitration or litigation between the parties concerning the work hereunder or any event relating thereto, the party prevailing in such dispute shall be entitled to reasonable attorneys fees.

19. CLEAN-UP: Upon completion of the work, CONTRACTOR will remove debris and surplus material created by his operation from OWNER's property and leave it in a neat and clean condition.

20. ASSIGNMENT: Neither party may assign this Agreement without the written consent of the other party, except for the right of CONTRACTOR to engage such sub-contractors as he deems necessary.

21. NOTICES: Any notice required or permitted under this Agreement may be given by ordinary mail at the addresses contained herein, but such addresses may be changed by written notice by one party to the other from time to time. Notice shall be deemed received one day after deposited in the mail, postage prepaid.

22. ACKNOWLEDGEMENT BY OWNER. OWNER acknowledges he is aware that:

A. NOTICE — "Under the Mechanics' Lien Law, any contractor, subcontractor, laborer, supplier or other person who helps to improve your property but is not paid for his work or supplies, has a right to enforce a claim against your property. This means that, after a court hearing, your property could be sold by a court officer and the proceeds of the sale used to satisfy the indebtedness. This can happen even if you have paid your own contractor in full, if the subcontractor, laborer, or supplier remains unpaid."

B. NOTICE TO OWNER — "Under the Mechanics' Lien Law, any contractor, subcontractor, laborer, materialman or other person who helps to improve your property and is not paid for his labor, services or material, has a right to enforce his claim against your property. "Under the law, you may protect yourself against such claims by filing, before commencing such work or improvement, an original contract for the work of improvement or a modification thereof, in the office of the county recorder of the county where the property is situated and requiring that a contractor's payment bond be recorded in such office. Said bond shall be in an amount not less than fifty percent (50%) of the contract price and shall, in addition to any conditions for the performance of the contract, be conditioned for the payment in full of the claims of all persons furnishing labor, services, equipment or materials for the work described in said contract."

23. _____

_____
The parties have executed this Agreement the day and year first mentioned herein.

CONTRACTORS are required by law to be licensed and regulated by the Contractors' State License Board. Any questions concerning a Contractor may be referred to the registrar of the board whose address is: Contractors' State License Board, 1020 N Street, Sacramento, California, 95814.

_____   _____
OWNER   CONTRACTOR

_____   _____
OWNER   OFFICER

AGREED TO: LICENSE NO. _____

_____
Construction Lender

Figure 34-3 (continued)

WORK ITEMS	PERCENTAGE ALLOTTED FOR WORK ITEM	TOTAL PAYMENT
Plans & Survey	1%	1%
Excavation	2%	3%
Foundation	4%	7%
Joist & Subfloor	4%	11%
Framed	10%	21%
Sheathed	4%	25%
Roofing	3%	28%
Windows	3%	31%
Garage Framing	3%	34%
Siding On	5%	39%
Garage Finished	2%	41%
Plumbing Rough-In	6%	47%
Wiring Rough-In	4%	51%
Chimney & Fireplace	4%	55%
Basement Floor	2%	57%
Furnace Set & Ducts	3%	60%
Insulation	1%	61%
Lathed or Drywall	3%	64%
Plastered or Taped	2%	66%
Interior Decorating	3%	69%
Exterior Painting	3%	72%
Floors Laid	2%	74%
Cabinets	4%	78%
Doors Hung & Trim	2%	80%
Linoleum	2%	82%
Tile & Formica	2%	84%
Water & Sewer	2%	86%
Plumbing Fixtures	3%	89%
Electric Fixtures	1%	90%
Appliances	2%	92%
Floor Finishing/Carpet	4%	96%
Exterior Concrete	2%	98%
Miscellaneous	2%	100%

Figure 34-4 Construction progress payment schedule

completion of the various stages of construction are specified.

- **Schedule of payments.** Usually, the contractor receives a percentage of the contract price for the completion of each stage of construction, Figure 34-4. Payments are typically made three or four times during the stages of construction. Another method is for the contractor to receive partial payment for the work done each month.

- **Responsibilities of all parties involved.** The owner is usually responsible for having the property surveyed. The architect may be responsible for administering the contract. The contractor is responsible for the construction and security of the site during the construction period.

- **Insurance.** Certain kinds of insurance are required during the construction. The contractor is required to have liability insurance. This protects the contractor against being sued for accidents occurring on the site. The owner is required to have property insurance.

- **Workmen's compensation.** This is another form of insurance that provides income for contractors' employees if they are injured at work.

- **Termination of the contract**. The contract describes conditions under which the contract may be ended. Contracts may be terminated if one party fails to comply with the contract, when one of the parties is disabled or dies, and for several other reasons.

There are two kinds of contracts in use for most construction. Each of these offers certain advantages and disadvantages.

Fixed-sum (sometimes called lump-sum) contracts are most often used. With a *fixed-sum contract*, the contractor agrees to complete the project for a certain amount of money. The greatest advantage of this kind of contract is that the owner knows in advance exactly what the cost will be.

However, the contractor does not know what hidden problems may be encountered. Therefore, the contractor's price must be high enough to cover unforeseen circumstances, such as excessive rock in the excavation or sudden increases in the cost of materials.

A *cost-plus contract* is one in which the contractor agrees to complete the work for the actual cost, plus a percentage for overhead and profit. The advantage of this type of contract is that the contractor does not have to allow for unforeseen problems, so the final price is apt to be less. A cost-plus contract is also useful when changes are apt to be made during the course of construction. The main disadvantage of this kind of contract is that the owner does not know exactly what the cost will be until the project is completed.

√ CHECK YOUR PROGRESS

Can you perform these tasks?

- ☐ List several specific kinds of information included in the General Conditions section of the specifications.

- ☐ Explain the bidding process and the importance of the bid form or document.

- ☐ List several specific kinds of information included in the construction agreement or contract form.

 ## ASSIGNMENT

Which of the contract documents gives the following information?

1. What is the depth of the floor joists?
2. On what date must the framing contractor be finished?
3. Who is responsible for washing the windows after the construction is finished?
4. Who is to pay for fire insurance during construction?
5. How soon after the work is completed is the final payment to be made?
6. Who is responsible for providing electricity during construction?
7. How soon after a contractor's bid is accepted must the contractor sign a contract?
8. When is the work to start?

UNIT 35 Technical Specifications

OBJECTIVES

After completing this unit, you will be able to find information in the technical specifications pertaining to any part of a construction project.

Technical specifications include all of the written description of the project that cannot reasonably be shown on the drawings. The amount of detailed information given and the style of the specifications varies between projects. For a simple one-family house, the specifications may be only a Federal Housing Administration (FHA) Description of Materials.

FHA DESCRIPTION OF MATERIALS

Figure 35-1 shows an FHA *Description of Materials* for the Lake House. Notice that the several pages of this form list the grade or quality of materials to be used for each part of the Lake House. However, the *Description of Materials* includes very little information about how the materials are to be installed.

DETAILED SPECIFICATIONS

The specifications for a large industrial project may be several hundred pages long. Specifications of this type are very precise in describing the quality standards for materials, the sizes and grades of materials, how the materials are to be installed, the finishing and cleanup to be done after each step of construction, and any inspection requirements. The specifications may also indicate a particular manufacturer's brand name in some cases. The architect or engineer may allow substitute brand names in place of some of those listed in the specifications. If substitutes are permitted, the specs include a statement such as "or approved equal" after the manufacturer's name. Whenever substitutions are made, they must first be approved by the architect.

CSI Format

In order to make lengthy specifications easier to follow, many specification writers follow the standard format of the Construction Specification Institute (CSI), Figure 35-2. The CSI format is made up of sixteen major divisions. Within each major division, there are several subsections. The major divisions are numbered by thousands. Each subsection is numbered by a more specific number within the thousands grouping of the division.

FHA Form 2005
VA Form 26–1852
Form FmHA 424–2
Rev. 4/77

U. S. DEPARTMENT OF HOUSING AND URBAN DEVELOPMENT
FEDERAL HOUSING ADMINISTRATION
For accurate register of carbon copies, form
may be separated along above fold. Staple
completed sheets together in original order.

Form Approved
OMB No. 63–R0055

☐ Proposed Construction

DESCRIPTION OF MATERIALS

No. _____
(To be inserted by FHA, VA or FmHA)

☐ Under Construction

Property address ___LAKE HOUSE___   City _____   State _____

Mortgagor or Sponsor _____   _____
                         (Name)   (Address)

Contractor or Builder _____   _____
                         (Name)   (Address)

INSTRUCTIONS

1. For additional information on how this form is to be submitted, number of copies, etc., see the instructions applicable to the FHA Application for Mortgage Insurance, VA Request for Determination of Reasonable Value, or FmHA Property Information and Appraisal Report, as the case may be.
2. Describe all materials and equipment to be used, whether or not shown on the drawings, by marking an X in each appropriate check-box and entering the information called for in each space. If space is inadequate, enter "See misc." and describe under item 27 or on an attached sheet. THE USE OF PAINT CONTAINING MORE THAN THE PERCENTAGE OF LEAD BY WEIGHT PERMITTED BY LAW IS PROHIBITED.
3. Work not specifically described or shown will not be considered unless

required, then the minimum acceptable will be assumed. Work exceeding minimum requirements cannot be considered unless specifically described.
4. Include no alternates, "or equal" phrases, or contradictory items. (Consideration of a request for acceptance of substitute materials or equipment is not thereby precluded.)
5. Include signatures required at the end of this form.
6. The construction shall be completed in compliance with the related drawings and specifications, as amended during processing. The specifications include this Description of Materials and the applicable Minimum Property Standards.

1. EXCAVATION:
Bearing soil, type __SANDY LOAM__

2. FOUNDATIONS:
Footings: concrete mix _____; strength psi __2500 @ 28 D.__   Reinforcing __2 - #4__
Foundation wall: material __CONCRETE BLOCK__   Reinforcing __HORIZONTAL @ 24"__
Interior foundation wall: material __CONCRETE BLOCK__   Party foundation wall _____
Columns: material and sizes __TS3 1/2 X 3 1/2 X .250__   Piers: material and reinforcing _____
Girders: material and sizes __SEE DRAWINGS__   Sills: material __2 X 10/CELOTEX SILL SEAL__
Basement entrance areaway _____   Window areaways _____
Waterproofing __2 COATS ASPHALT BELOW GRADE__   Footing drains _____
Termite protection _____ _____
Basementless space: ground cover _____; insulation __R-7.2 FOAM BD__; foundation vents _____
Special foundations _____
Additional information: __3/8"φ X 12" ANCHOR BOLTS @ 3'-0"__

3. CHIMNEYS:
Material __TRIPLE–WALL STEEL__   Prefabricated (make and size) __MAJESTIC 9"__
Flue lining: material __STAINLESS STEEL__   Heater flue size _____   Fireplace flue size _____
Vents (material and size): gas or oil heater _____; water heater _____
Additional information: __INSTALLED IN ACCORDANCE W/MAJESTIC INSTRUCTIONS__

4. FIREPLACES:
Type: ☒ solid fuel; ☐ gas-burning; ☐ circulator (make and size) __MAJESTIC #M28__   Ash dump and clean-out _____
Fireplace: facing __FACE BRICK__; lining _____; hearth __FACE BRICK__; mantel _____
Additional information: _____

5. EXTERIOR WALLS:
Wood frame: wood grade, and species __HEM-FIR #2__   ☒ Corner bracing. Building paper or felt __3/4"CDX PLY. WD.__
Sheathing __R-5.4 FOAM BD.__; thickness __3/4"__; width ___; ☐ solid; ☐ spaced ___" o. c.; ☐ diagonal; ___
Siding __PLYWOOD__; grade __EXT.__; type __T1-11__; size __1/2"__; exposure ___"; fastening __AL. NAILS__
Shingles _____; grade _____; type _____; size _____; exposure ___"; fastening _____
Stucco _____; thickness ___"; Lath _____; weight ___lb.
Masonry veneer _____   Sills _____   Lintels _____   Base flashing _____
Masonry: ☐ solid ☐ faced ☐ stuccoed; total-wall thickness ___"; facing thickness ___"; facing material _____
Backup material _____; thickness ___"; bonding _____
Door sills __OAK__   Window sills _____   Lintels _____   Base flashing _____
Interior surfaces: dampproofing, ___ coats of _____; furring __1 X 3 @ 16" O.C.__
Additional information: _____
Exterior painting: material __STAIN__; number of coats __1__
Gable wall construction: ☒ same as main walls; ☐ other construction _____

6. FLOOR FRAMING:
Joists: wood, grade, and species __DOUG. FIR. #2__; other _____; bridging _____; anchors _____
Concrete slab: ☒ basement floor; ☐ first floor; ☒ ground supported; ☐ self-supporting; mix __2500 PSI @ 28 DAYS__; thickness __4__";
reinforcing __6 X 6 – 10/10 WWM__; insulation _____; membrane __4-MIL POLYETHELENE__
Fill under slab: material __R.O.B. GRAVEL__; thickness __6__". Additional information: __R-14.6 FOAM BD.__
__INSULATION UNDER HEAT SINK__

7. SUBFLOORING: (Describe underflooring for special floors under item 21.)
Material: grade and species __CD GRADE PLYWOOD__; size __1/2"__; type _____
Laid: ☐ first floor; ☐ second floor; ☐ attic _____ sq. ft.; ☐ diagonal; ☐ right angles. Additional information: __ALL FRAMED__
__FLOOR NOT DESCRIBED UNDER ITEM 21 TO HAVE 3/8" PLY. WD. UNDERLAY AT RIGHT ANGLES.__

8. FINISH FLOORING: (Wood only. Describe other finish flooring under item 21.)

LOCATION	ROOMS	GRADE	SPECIES	THICKNESS	WIDTH	BLDG. PAPER	FINISH
First floor ___	LOFT	SELECT	R. OAK	25/32	2 1/2	YES	FILL & 2 COATS POLYURETHANE
Second floor ___							
Attic floor ___	___ sq. ft.						

Additional information: _____

FHA Form 2005
VA Form 26–1852
Form FmHA 424–2

DESCRIPTION OF MATERIALS

Figure 35-1 FHA description of materials for the Lake House

DESCRIPTION OF MATERIALS

9. PARTITION FRAMING:
Studs: wood, grade, and species ___HEM-FIR #2___ size and spacing ___2 X 4 @ 16" O.C.___ Other _____
Additional information: _____

10. CEILING FRAMING:
Joists: wood, grade, and species ___HEM–FIR #2___ Other _____ Bridging _____
Additional information: _____

11. ROOF FRAMING:
Rafters: wood, grade, and species ___HEM-FIR #2___ Roof trusses (see detail): grade and species _____
Additional information: _____

12. ROOFING:
Sheathing: wood, grade, and species ___1/2" CDX PLY. WD.___ ; ☐ solid; ☐ spaced ____" o.c.
Roofing ___COMP. SHINGLES___ ; grade ___#235___ ; size _____; type _____
Underlay ___ASPHALT SATURATED FELT___ ; weight or thickness ___#15___; size _____; fastening _____
Built-up roofing _____ ; number of plies _____; surfacing material _____
Flashing: material ___ALUMINUM___ ; gage or weight ___28 GA.___ ; ☐ gravel stops; ☐ snow guards
Additional information: ___ALUMINUM DRIP EDGE___

13. GUTTERS AND DOWNSPOUTS:
Gutters: material _____ ; gage or weight _____; size _____; shape _____
Downspouts: material _____ ; gage or weight _____; size _____; shape _____; number _____
Downspouts connected to: ☐ Storm sewer; ☐ sanitary sewer; ☐ dry-well. ☐ Splash blocks: material and size _____
Additional information: _____

14. LATH AND PLASTER
Lath ☐ walls, ☐ ceilings: material _____; weight or thickness _____ Plaster: coats ____; finish _____
Dry-wall ☒ walls, ☒ ceilings: material ___GYPSUM___ ; thickness ___1/2"___ ; finish ___PAINTED___ ,
Joint treatment ___TAPE & COMPOUND WITH METAL CORNER BEADS___

15. DECORATING: (Paint, wallpaper, etc.)

Rooms	Wall Finish Material and Application	Ceiling Finish Material and Application
Kitchen	LOW LUSTER ENAMEL - - 2 COATS	ALL CEILINGS - - ONE COAT
Bath	VINYL WALLCOVERING	FLAT LATEX/ONE COAT
Other	FLAT LATEX - - 2 COATS	STIPPLED LATEX

Additional information: _____

16. INTERIOR DOORS AND TRIM:
Doors: type _____ ; material ___WOOD DOORS BIRCH OR PINE___ ; thickness _____
Door trim: type ___RANCH___ ; material ___PINE___ Base: type ___RANCH___ ; material ___PINE___ ; size ___3 1/2"___
Finish: doors ___SEMIGLOSS LATEX ENAMEL - 2 CTS.___ ; trim ___SEMIGLOSS LATEX ENAMEL - - 2 COATS___
Other trim (item, type and location) ___BEAM CASINGS - - PINE W/2 COATS SEMIGLOSS; RAILING & LADDER - - WOOD___
Additional information: ___PARTS, SANDED & 2 COATS POLYURETHANE - - METAL PARTS, 2 COATS RUST-RESISTANT SEMIGLOSS ENAMEL.___

17. WINDOWS:
Windows: type ___SLIDING___ ; make ___CAPITOL___ ; material ___THERM-BRK. AL.___ ; sash thickness _____
Glass: grade ___INSULATING___ ; ☐ sash weights; ☐ balances, type _____; head flashing ___INTEGRAL FLANGE___
Trim: type ___GYP. BD. RETURN___ ; material _____ Paint ___AS WALL___ ; number coats _____
Weatherstripping: type ___FACTORY INSTALLED___ ; material _____ Storm sash, number _____
Screens: ☒ full; ☐ half; type _____ ; number _____; screen cloth material ___ALUMINUM___
Basement windows: type _____; material _____; screens, number _____ ; Storm sash, number _____
Special windows ___SKYLIGHT - - SKYVUE #DV 2852___
Additional information: ___WOOD DRIP CAPS___

18. ENTRANCES AND EXTERIOR DETAIL:
Main entrance door: material ___HOLLOW CORE STEEL___ ; width ___3'-0"___ ; thickness ___1 3/4"___. Frame: material ___WOOD___ , thickness ____"
Other entrance doors: material ___SAME___ ; width _____; thickness ____". Frame: material _____; thickness ____"
Head flashing _____ Weatherstripping: type ___MAGNETIC___ · saddles ___OAK___
Screen doors: thickness ____"; number _____ ; screen cloth material _____ Storm doors: thickness ____"; number _____
Combination storm and screen doors: thickness ____"; number ____; screen cloth material _____
Shutters: ☐ hinged; ☐ fixed. Railings _____ , Attic louvers _____
Exterior millwork: grade and species ___#1 PINE___ Paint ___STAIN___ ; number coats ___1___
Additional information: ___ALL DECK LUMBER TO BE TREATED W/COPPER-BASE PRESERVATIVE___

19. CABINETS AND INTERIOR DETAIL:
Kitchen cabinets, wall units: material ___SCHEIRCH, GARDEN GROVE___ ; lineal feet of shelves _____; shelf width _____
Base units: material ___GARDEN GROVE___ ; counter top ___PLASTIC LAMINATE___ ; edging _____
Back and end splash ___MOLDED___ Finish of cabinets ___FACTORY FINISHED___ ; number coats _____
Medicine cabinets: make ___MIAMI–CAREY___ ; model ___UP-CRP3418/DWN-CRP-306-AL___
Other cabinets and built-in furniture ___BENCH IN L.R. - - AC PLYWOOD, SANDED, 1 COAT FIRZITE___
Additional information: ___2 COATS SEMIGLOSS ENAMEL___

20. STAIRS:

Stair	Treads		Risers		Strings		Handrail		Balusters	
	Material	Thickness	Material	Thickness	Material	Size	Material	Size	Material	Size
Basement	OAK	5/4	PINE	1" NOM.	PINE	5/4				
Main	OAK	5/4	PINE	1" NOM.	PINE	5/4	OAK	5/4" X 3"	STEEL	1"
Attic										

Disappearing: make and model number _____
Additional information: _____

2

Figure 35-1 (continued)

21. SPECIAL FLOORS AND WAINSCOT: *(Describe Carpet as listed in Certified Products Directory)*

	LOCATION	MATERIAL, COLOR, BORDER, SIZES, GAGE, ETC.	THRESHOLD MATERIAL	WALL BASE MATERIAL	UNDERFLOOR MATERIAL
FLOORS	Kitchen	QUARRY TILE (N.I.C.)	OAK	PINE	1" CONC.
	Bath	UNGLAZED CERAMIC TILE	MARBLE	TILE	1" CONC.
	L.R. & D.R.	QUARRY TILE (N.I.C.)		PINE	CONC.

	LOCATION	MATERIAL, COLOR, BORDER, CAP. SIZES, GAGE, ETC.	HEIGHT	HEIGHT OVER TUB	HEIGHT IN SHOWERS (FROM FLOOR)
WAINSCOT	Bath	CERAMIC TILE	4'-0"	4'-0" FROM TUB	TO CLG.

Bathroom accessories: ☒ Recessed; material ___CHINA___ ; number ___5___; ☒ Attached; material ___CHINA___ ; number ___7___
Additional information: _____

22. PLUMBING:

FIXTURE	NUMBER	LOCATION	MAKE	MFR'S FIXTURE IDENTIFICATION NO.	SIZE	COLOR
Sink	1	KITCHEN	MOEN	S3322-4		S.S.
Lavatory	3		UNIVERSAL-RUNDLE	163330	PER DWGS.	BY OWNER
Water closet	3		UNIVERSAL-RUNDLE	4055-15		BY OWNER
Bathtub	1		UNIVERSAL-RUNDLE	118-2	5'-0"	BY OWNER
Shower over tub △	1		MOEN			CROME
Stall shower △	1		KINKEAD INDUSTRIES	MARBLEMOLD	3'-0" X 3'-0"	
Laundry trays						

△☒ Curtain rod   △☒ Door   ☐ Shower pan: material _____SYNTHETIC HARD RUBBER_____
Water supply: ☐ public; ☐ community system; ☒ individual (private) system. ★
Sewage disposal: ☐ public; ☐ community system; ☐ individual (private) system. ★
★Show and describe individual system in complete detail in separate drawings and specifications according to requirements.
House drain (inside): ☐ cast iron; ☐ tile; ☒ other ___PVC___   House sewer (outside): ☐ cast iron; ☐ tile; ☒ other ___PVC___
Water piping: ☐ galvanized steel; ☒ copper tubing; ☐ other _____ Sill cocks, number ___2___
Domestic water heater: type ___ELECTRIC___ ; make and model ___STATE-CENSIBLE___ ; heating capacity _____
_____ gph. 100° rise. Storage tank: material ___GLASS___ ; capacity ___40___ gallons.
Gas service: ☐ utility company; ☐ liq. pet. gas; ☐ other _____ Gas piping: ☐ cooking; ☐ house heating.
Footing drains connected to: ☐ storm sewer; ☐ sanitary sewer; ☐ dry well. Sump pump; make and model _____
_____ ; capacity _____ ; discharges into _____

23. HEATING:
☐ Hot water. ☐ Steam. ☐ Vapor. ☐ One-pipe system. ☐ Two-pipe system.
   ☐ Radiators. ☐ Convectors. ☐ Baseboard radiation. Make and model _____
   Radiant panel: ☐ floor; ☐ wall; ☐ ceiling. Panel coil: material _____
   ☐ Circulator. ☐ Return pump. Make and model _____ ; capacity _____ gpm.
   Boiler: make and model _____ Output _____ Btuh.; net rating _____ Btuh.
Additional information: _____
Warm air: ☐ Gravity. ☐ Forced. Type of system _____
   Duct material: supply _____ ; return _____ Insulation _____ , thickness _____ ☐ Outside air intake.
   Furnace: make and model _____ Input _____ Btuh.; output _____ Btuh.
   Additional information: _____
☐ Space heater; ☐ floor furnace; ☐ wall heater. Input _____ Btuh.; output _____ Btuh.; number units _____
   Make, model _____ Additional information: _____
Controls: make and types _____
Additional information: _____
Fuel: ☐ Coal; ☐ oil; ☐ gas; ☐ liq. pet. gas; ☐ electric; ☐ other _____ ; storage capacity _____
   Additional information: _____
Firing equipment furnished separately: ☐ Gas burner, conversion type. ☐ Stoker: hopper feed ☐; bin feed ☐
   Oil burner: ☐ pressure atomizing; ☐ vaporizing _____
   Make and model _____ Control _____
   Additional information: _____
Electric heating system: type ___BASEBOARD RESISTANCE___ Input _____ watts; @ _____ volts; output _____ Btuh.
   Additional information: ___SEE SCHEDULE___
Ventilating equipment: attic fan, make and model _____ ; capacity _____ cfm.
   kitchen exhaust fan, make and model ___GENERAL ELECTRIC #JV330___
Other heating, ventilating. or cooling equipment _____

24. ELECTRIC WIRING:
Service: ☐ overhead; ☒ underground. Panel: ☐ fuse box; ☒ circuit-breaker; make ___SQUARE D___ AMP's ___200___ No. circuits ___40___
Wiring: ☐ conduit; ☐ armored cable; ☒ nonmetallic cable; ☐ knob and tube; ☐ other _____
Special outlets: ☒ range; ☒ water heater; ☐ other ___CLOTHES DRYER___
☐ Doorbell. ☒ Chimes. Push-button locations ___KITCHEN ENTRANCE___ Additional information: ___ONE TELEPHONE___
___JACK IN EACH LIVING SPACE TO BE COMPATIBLE W/LOCAL TELEPHONE UTILITY___

25. LIGHTING FIXTURES:
Total number of fixtures ___22___ Total allowance for fixtures, typical installation, $ ___$450___
Nontypical installation _____
Additional information: _____

DESCRIPTION OF MATERIALS

Figure 35-1 (continued)

DESCRIPTION OF MATERIALS

26. INSULATION:

LOCATION	THICKNESS	MATERIAL, TYPE, AND METHOD OF INSTALLATION	VAPOR BARRIER
Roof	9"	R-30 FIBERGLASS	4 MIL POLYETHYLENE
Ceiling			
Wall	6"	R-19 FIBERGLASS	4 MIL POLYETHYLENE
Floor			
MASONARY WALLS - - 1', R-7.2 FOAM BD BETWEEN FURRING			

27. MISCELLANEOUS: *(Describe any main dwelling materials, equipment, or construction items not shown elsewhere; or use to provide additional information where the space provided was inadequate. Always reference by item number to correspond to numbering used on this form.)*

    2. HEATSINK – 12" R.Q.B. GRAVEL MINIMUM OVER UNEXCAVATED EARTH. BLOCK UNDER CONCRETE AT HEAT SINK TO BE
        LAID FLAT W/3/8" JOINTS & NO MORTAR.
    19. VANITIES - - SCHEIRICH GARDEN GROVE

HARDWARE: *(make, material, and finish.)*   KWIK-SET: ENTRANCES - - BRONZE
                                                   BATHS - - CHROME
                                                   OTHERS - - BRASS

SPECIAL EQUIPMENT: *(State material or make, model and quantity. Include only equipment and appliances which are acceptable by local law, custom and applicable FHA standards. Do not include items which, by established custom, are supplied by occupant and removed when he vacates premises or chattels prohibited by law from becoming realty.)*
        APPLIANCES NOT INCLUDED ABOVE - - NOT IN CONTRACT

PORCHES:
        DECKS ACCORDING TO DRAWINGS - - ALL LUMBER TO BE CCA TREATED OR EQUAL. FASTENERS TO BE
        GALVANIZED STEEL.

TERRACES:

GARAGES:
        ACCORDING TO PLAN, NO INSULATION, OVERHEAD DOOR - - WOOD FRAME W/HARDBOARD PANELS, ONE SECTION GLAZED.

WALKS AND DRIVEWAYS:
Driveway: width SEE DWG ; base material GRADE ; thickness ____"; surfacing material CLEAN GRAVEL ; thickness 6 "
Front walk: width 3'-0" ; material STONE ; thickness 4 ". Service walk: width ____; material ____; thickness ____"
Steps: material ____; treads ____"; risers ____". Cheek walls ____

OTHER ONSITE IMPROVEMENTS:
(Specify all exterior onsite improvements not described elsewhere, including items such as unusual grading, drainage structures, retaining walls, fence, railings, and accessory structures.)

LANDSCAPING, PLANTING, AND FINISH GRADING:
Topsoil 2 " thick: ☒ front yard; ☒ side yards; ☒ rear yard to PROPERTY LINE feet behind main building.
Lawns *(seeded, sodded, or sprigged)*: ☒ front yard SEED ; ☒ side yards SEED ; ☒ rear yard SEED
Planting: ☐ as specified and shown on drawings; ☐ as follows:
____ Shade trees, deciduous, ____" caliper.   ____ Evergreen trees. ____' to ____', B & B.
____ Low flowering trees, deciduous, ____' to ____'   ____ Evergreen shrubs. ____' to ____', B & B.
____ High-growing shrubs, deciduous, ____' to ____'   ____ Vines, 2-year ____
____ Medium-growing shrubs, deciduous, ____' to ____'
____ Low-growing shrubs, deciduous, ____' to ____'

IDENTIFICATION.—This exhibit shall be identified by the signature of the builder, or sponsor, and/or the proposed mortgagor if the latter is known at the time of application.

Date_____   Signature_____

                                     Signature_____

FHA Form 2005
VA Form 26–1852   4
Form FmHA 424–2

S. GOVERNMENT PRINTING OFFICE : 1977 O - 236-421

Figure 35-1 (continued)

DIVISION 0 – BIDDING AND CONTRACT REQUIREMENTS

00010	PRE-BID INFORMATION
00100	INSTRUCTIONS TO BIDDERS
00200	INFORMATION AVAILABLE TO BIDDERS
00300	BID/TENDER FORMS
00400	SUPPLEMENTS TO BID/TENDER FORMS
00500	AGREEMENT FORMS
00600	BONDS AND CERTIFICATES
00700	GENERAL CONDITIONS OF THE CONTRACT
00800	SUPPLEMENTARY CONDITIONS
00950	DRAWINGS INDEX
00900	ADDENDA AND MODIFICATIONS

SPECIFICATIONS—DIVISIONS 1-16

DIVISION 1 – GENERAL REQUIREMENTS

01010	SUMMARY OF WORK
01020	ALLOWANCES
01030	SPECIAL PROJECT PROCEDURES
01040	COORDINATION
01050	FIELD ENGINEERING
01060	REGULATORY REQUIREMENTS
01070	ABBREVIATIONS AND SYMBOLS
01080	IDENTIFICATION SYSTEMS
01100	ALTERNATES/ALTERNATIVES
01150	MEASUREMENT AND PAYMENT
01200	PROJECT MEETINGS
01300	SUBMITTALS
01400	QUALITY CONTROL
01500	CONSTRUCTION FACILITIES AND TEMPORARY CONTROLS
01600	MATERIAL AND EQUIPMENT
01650	STARTING OF SYSTEMS
01660	TESTING, ADJUSTING, AND BALANCING OF SYSTEMS
01700	CONTRACT CLOSEOUT
01800	MAINTENANCE MATERIALS

DIVISION 2 – SITE WORK

02010	SUBSURFACE INVESTIGATION
02050	DEMOLITION
02100	SITE PREPARATION
02150	UNDERPINNING
02200	EARTHWORK
02300	TUNNELLING
02350	PILES, CAISSONS AND COFFERDAMS
02400	DRAINAGE
02440	SITE IMPROVEMENTS
02480	LANDSCAPING
02500	PAVING AND SURFACING
02580	BRIDGES
02590	PONDS AND RESERVOIRS
02600	PIPED UTILITY MATERIALS AND METHODS
02700	PIPED UTILITIES
02800	POWER AND COMMUNICATION UTILITIES
02850	RAILROAD WORK
02880	MARINE WORK

DIVISION 3 – CONCRETE

03010	CONCRETE MATERIALS
03050	CONCRETING PROCEDURES
03100	CONCRETE FORMWORK
03150	FORMS
03180	FORM TIES AND ACCESSORIES
03200	CONCRETE REINFORCEMENT
03250	CONCRETE ACCESSORIES
03300	CAST-IN-PLACE CONCRETE
03350	SPECIAL CONCRETE FINISHES
03360	SPECIALLY PLACED CONCRETE
03370	CONCRETE CURING
03400	PRECAST CONCRETE
03500	CEMENTITIOUS DECKS
03600	GROUT
03700	CONCRETE RESTORATION AND CLEANING

DIVISION 4 – MASONRY

04050	MASONRY PROCEDURES
04100	MORTAR
04150	MASONRY ACCESSORIES
04200	UNIT MASONRY
04400	STONE
04500	MASONRY RESTORATION AND CLEANING
04550	REFRACTORIES
04600	CORROSION RESISTANT MASONRY

DIVISION 5 – METALS

05010	METAL MATERIALS AND METHODS
05050	METAL FASTENING
05100	STRUCTURAL METAL FRAMING
05200	METAL JOISTS
05300	METAL DECKING
05400	COLD-FORMED METAL FRAMING
05500	METAL FABRICATIONS
05700	ORNAMENTAL METAL
05800	EXPANSION CONTROL
05900	METAL FINISHES

DIVISION 6 – WOOD AND PLASTICS

06050	FASTENERS AND SUPPORTS
06100	ROUGH CARPENTRY
06130	HEAVY TIMBER CONSTRUCTION
06150	WOOD-METAL SYSTEMS
06170	PREFABRICATED STRUCTURAL WOOD
06200	FINISH CARPENTRY
06300	WOOD TREATMENT
06400	ARCHITECTURAL WOODWORK
06500	PREFABRICATED STRUCTURAL PLASTICS
06600	PLASTIC FABRICATIONS

DIVISION 7 – THERMAL AND MOISTURE PROTECTION

07100	WATERPROOFING
07150	DAMPPROOFING
07200	INSULATION
07250	FIREPROOFING
07300	SHINGLES AND ROOFING TILES
07400	PREFORMED ROOFING AND SIDING
07500	MEMBRANE ROOFING
07570	TRAFFIC TOPPING
07600	FLASHING AND SHEET METAL
07800	ROOF ACCESSORIES
07900	SEALANTS

DIVISION 8 – DOORS AND WINDOWS

08100	METAL DOORS AND FRAMES
08200	WOOD AND PLASTIC DOORS
08250	DOOR OPENING ASSEMBLIES
08300	SPECIAL DOORS
08400	ENTRANCES AND STOREFRONTS
08500	METAL WINDOWS
08600	WOOD AND PLASTIC WINDOWS
08650	SPECIAL WINDOWS
08700	HARDWARE
08800	GLAZING
08900	GLAZED CURTAIN WALLS

DIVISION 9 – FINISHES

09100	METAL SUPPORT SYSTEMS
09200	LATH AND PLASTER
09230	AGGREGATE COATINGS
09250	GYPSUM WALLBOARD
09300	TILE
09400	TERRAZZO
09500	ACOUSTICAL TREATMENT
09550	WOOD FLOORING
09600	STONE AND BRICK FLOORING
09650	RESILIENT FLOORING
09680	CARPETING
09700	SPECIAL FLOORING
09760	FLOOR TREATMENT
09800	SPECIAL COATINGS
09900	PAINTING
09950	WALL COVERING

Courtesy of Construction Specification Institute

Figure 35-2 CSI format for specifications

DIVISION 10 – SPECIALITIES

10100	CHALKBOARDS AND TACKBOARDS
10150	COMPARTMENTS AND CUBICLES
10200	LOUVERS AND VENTS
10240	GRILLES AND SCREENS
10250	SERVICE WALL SYSTEMS
10260	WALL AND CORNER GUARDS
10270	ACCESS FLOORING
10280	SPECIALTY MODULES
10290	PEST CONTROL
10300	FIREPLACES AND STOVES
10340	PREFABRICATED STEEPLES, SPIRES, AND CUPOLAS
10350	FLAGPOLES
10400	IDENTIFYING DEVICES
10450	PEDESTRIAN CONTROL DEVICES
10500	LOCKERS
10520	FIRE EXTINGUISHERS, CABINETS, AND ACCESSORIES
10530	PROTECTIVE COVERS
10550	POSTAL SPECIALTIES
10600	PARTITIONS
10650	SCALES
10670	STORAGE SHELVING
10700	EXTERIOR SUN CONTROL DEVICES
10750	TELEPHONE ENCLOSURES
10800	TOILET AND BATH ACCESSORIES
10900	WARDROBE SPECIALTIES

DIVISION 11 – EQUIPMENT

11010	MAINTENANCE EQUIPMENT
11020	SECURITY AND VAULT EQUIPMENT
11030	CHECKROOM EQUIPMENT
11040	ECCLESIASTICAL EQUIPMENT
11050	LIBRARY EQUIPMENT
11060	THEATER AND STAGE EQUIPMENT
11070	MUSICAL EQUIPMENT
11080	REGISTRATION EQUIPMENT
11100	MERCANTILE EQUIPMENT
11110	COMMERCIAL LAUNDRY AND DRY CLEANING EQUIPMENT
11120	VENDING EQUIPMENT
11130	AUDIO-VISUAL EQUIPMENT
11140	SERVICE STATION EQUIPMENT
11150	PARKING EQUIPMENT
11160	LOADING DOCK EQUIPMENT
11170	WASTE HANDLING EQUIPMENT
11190	DETENTION EQUIPMENT
11200	WATER SUPPLY AND TREATMENT EQUIPMENT
11300	FLUID WASTE DISPOSAL AND TREATMENT EQUIPMENT
11400	FOOD SERVICE EQUIPMENT
11450	RESIDENTIAL EQUIPMENT
11460	UNIT KITCHENS
11470	DARKROOM EQUIPMENT
11480	ATHLETIC, RECREATIONAL, AND THERAPEUTIC EQUIPMENT
11500	INDUSTRIAL AND PROCESS EQUIPMENT
11600	LABORATORY EQUIPMENT
11650	PLANETARIUM AND OBSERVATORY EQUIPMENT
11700	MEDICAL EQUIPMENT
11780	MORTUARY EQUIPMENT
11800	TELECOMMUNICATION EQUIPMENT
11850	NAVIGATION EQUIPMENT

DIVISION 12 – FURNISHINGS

12100	ARTWORK
12300	MANUFACTURED CABINETS AND CASEWORK
12500	WINDOW TREATMENT
12550	FABRICS
12600	FURNITURE AND ACCESSORIES
12670	RUGS AND MATS
12700	MULTIPLE SEATING
12800	INTERIOR PLANTS AND PLANTINGS

DIVISION 13 – SPECIAL CONSTRUCTION

13010	AIR SUPPORTED STRUCTURES
13020	INTEGRATED ASSEMBLIES
13030	AUDIOMETRIC ROOMS
13040	CLEAN ROOMS
13050	HYPERBARIC ROOMS
13060	INSULATED ROOMS
13070	INTEGRATED CEILINGS
13080	SOUND, VIBRATION, AND SEISMIC CONTROL
13090	RADIATION PROTECTION
13100	NUCLEAR REACTORS
13110	OBSERVATORIES
13120	PRE-ENGINEERED STRUCTURES
13130	SPECIAL PURPOSE ROOMS AND BUILDINGS
13140	VAULTS
13150	POOLS
13160	ICE RINKS
13170	KENNELS AND ANIMAL SHELTERS
13200	SEISMOGRAPHIC INSTRUMENTATION
13210	STRESS RECORDING INSTRUMENTATION
13220	SOLAR AND WIND INSTRUMENTATION
13410	LIQUID AND GAS STORAGE TANKS
13510	RESTORATION OF UNDERGROUND PIPELINES
13520	FILTER UNDERDRAINS AND MEDIA
13530	DIGESTION TANK COVERS AND APPURTENANCES
13540	OXYGENATION SYSTEMS
13550	THERMAL SLUDGE CONDITIONING SYSTEMS
13560	SITE CONSTRUCTED INCINERATORS
13600	UTILITY CONTROL SYSTEMS
13700	INDUSTRIAL AND PROCESS CONTROL SYSTEMS
13800	OIL AND GAS REFINING INSTALLATIONS AND CONTROL SYSTEMS
13900	TRANSPORTATION INSTRUMENTATION
13940	BUILDING AUTOMATION SYSTEMS
13970	FIRE SUPPRESSION AND SUPERVISORY SYSTEMS
13980	SOLAR ENERGY SYSTEMS
13990	WIND ENERGY SYSTEMS

DIVISION 14 – CONVEYING SYSTEMS

14100	DUMBWAITERS
14200	ELEVATORS
14300	HOISTS AND CRANES
14400	LIFTS
14500	MATERIAL HANDLING SYSTEMS
14600	TURNTABLES
14700	MOVING STAIRS AND WALKS
14800	POWERED SCAFFOLDING
14900	TRANSPORTATION SYSTEMS

DIVISION 15 – MECHANICAL

15050	BASIC MATERIALS AND METHODS
15200	NOISE, VIBRATION, AND SEISMIC CONTROL
15250	INSULATION
15300	SPECIAL PIPING SYSTEMS
15400	PLUMBING SYSTEMS
15450	PLUMBING FIXTURES AND TRIM
15500	FIRE PROTECTION
15600	POWER OR HEAT GENERATION
15650	REFRIGERATION
15700	LIQUID HEAT TRANSFER
15800	AIR DISTRIBUTION
15900	CONTROLS AND INSTRUMENTATION

DIVISION 16 – ELECTRICAL

16050	BASIC MATERIALS AND METHODS
16200	POWER GENERATION
16300	POWER TRANSMISSION
16400	SERVICE AND DISTRIBUTION
16500	LIGHTING
16600	SPECIAL SYSTEMS
16700	COMMUNICATIONS
16850	HEATING AND COOLING
16900	CONTROLS AND INSTRUMENTATION

Figure 35-2 (continued)

Any divisions or subsections that are not needed for a project are skipped over. For example, if none of the specialties of division 10 are to be included, the specifications skip from 09000, *Finishes*, to 11000, *Equipment*. If there are no metal doors or door frames; division 08000, *Doors and Windows*, begins with subsection 08200, *Wood and Plastic Doors*.

Detailed Specifications for Light Construction

The CSI format is an effective system for indexing very large, otherwise cumbersome, specifications. However, several of the divisions and more than half of the subsections are never used in light residential construction. To follow the CSI format on light construction would add to the confusion of the specs because so many numbers would be missing.

A more practical and often-used system is to partially follow the CSI format. The major divisions follow the sequence of the CSI format, but the numbering of subsections is changed or dropped altogether.

The format of the specifications for the Lake House is an adaptation of the CSI format, Figure 35-3. The major divisions follow the CSI numbering. However, subsections are rearranged. Each division includes a brief list of work covered by that division, a description of the materials needed for that work, and subsections for each type of work in the division. To find a particular part of the construction project on the specs, (1) look up the appropriate division on the table of contents, (2) read the *Work Included* at the beginning of the division to be sure you have the right division, (3) find the description of the necessary materials right after *Work Included*, and (4) read the subsection covering the work you are looking for.

√ CHECK YOUR PROGRESS

Can you perform these tasks?

☐ Explain the meaning of all of the information included on an FHA Description of Materials.

☐ Locate information about each part of a building and the work of each trade in technical specifications.

ASSIGNMENT

Questions 1 through 5 refer to Figure 35-1, the FHA Description of Materials, for the Lake House.

1. What size anchor bolts are used at the top of the foundation?
2. Of what material is the fireplace hearth to be constructed?
3. What is the material and weight of the drip edge at the edge of the roof?
4. Of what material are stair risers to be constructed?
5. What is the finish floor material in the kitchen?

Questions 6 through 15 refer to Figure 35-3, which gives detailed specifications for the Lake House.

6. What division includes vapor barriers under concrete slabs?
7. What is the material and thickness of the vapor barrier under the heat sink?
8. What material and rating is the thermal insulation under the heat sink?
9. What is the strength requirement for the concrete in the heat sink?
10. How is the concrete slab in the heat sink to be finished?
11. What make and model number is the skylight?
12. What division of the specs includes installing shower curtain rods?
13. What brand and type of paint is to be used on the kitchen walls?
14. Who is to choose the color of the paint for the kitchen walls?
15. What brand and type of paint is to be used on the trim around the skylight?

SPECIFICATIONS FOR LAKE HOUSE
CONTENTS

(Divisions 12000, 13000, and 14000 are not used.)

GENERAL CONDITIONS

The General Conditions of the Contract For Construction, AIA Document A107, whether or not bound herein, are hereby incorporated into and made a part of this contract and these specifications.

01000 GENERAL REQUIREMENTS

A. ARCHITECT'S SUPERVISION
The architect will have continual supervisory responsibility for this job.

B. TEMPORARY CONVENIENCES
The general contractor shall provide suitable temporary conveniences for the use of all workers on this job. Facilities shall be within a weathertight, painted enclosure complying with legal requirements. The general contractor shall maintain all temporary toilet facilities in a sanitary condition.

C. PUMPING
The general contractor shall keep the excavation and the basement free from water at all times and shall provide, maintain, and operate at his own expense such pumping equipment as shall be necessary.

D. PROTECTION
The general contractor shall protect all existing driveways, parking areas, sidewalks, curbs, and existing paved areas on, or adjacent to, the owner's property.

E. GRADES, LINES, LEVELS, AND SURVEYS
The owner shall establish the lot lines.
The general contractor shall:
1. Establish and maintain bench marks.
2. Verify all grades, lines, levels, and dimensions as shown on the drawings, and report any errors or inconsistencies before commencing to work.
3. Lay out the building accurately under the supervision of the architect.

F. FINAL CLEANING
In addition to the general room cleaning, the general contractor shall do the following special cleaning upon completion of the work:
1. Wash and polish all glass and cabinets.
2. Clean and polish all hardware.
3. Remove all marks, stains, fingerprints, and other soil or dirt from walls, woodwork, and floors.

G. GUARANTEES
The general contractor shall guarantee all work performed under the contract against faulty materials or workmanship. The guarantee shall be in writing with duplicate copies delivered to the architect. In case of work performed by subcontractors where guarantees are required, the general contractor shall secure written guarantees from those subcontractors. Copies of these guarantees shall be delivered to the architect upon completion of the work. Guarantees shall be signed by both the subcontractor and the general contractor.

H. FOREMAN
The general contractor shall have a responsible foreman at the building site from the start to the completion of construction. The foreman shall be on duty during all working hours.

I. FIRE INSURANCE
The owner shall effect and maintain builder's risk completed-value insurance on this job.

02000 SITE WORK

WORK INCLUDED
This work shall include, but shall not be limited by the following:
A. Cleaning site.
B. Excavating, backfilling, grading, and related items.
C. Removal of excess earth.
D. Protection of existing trees to remain on the site.

All excavation and backfill required for heating, plumbing, and electrical work will be done by the respective contractors and are not included under site work.

It is the contractor's responsibility to field inspect existing conditions to determine the scope of work.

02100 CLEARING

A. Clean the area within the limits of the building of all trees, shrubs, or other obstructions as necessary.
B. Within the limits of grading work as shown on the drawings, remove such trees, shrubs, or other obstructions as are indicated on the drawings to be removed, without injury to trunks, interfering branches, and roots of trees to remain. Do cutting and trimming only as directed. Box and protect all trees and shrubs in the construction area to remain; maintain boxing until finished grading is completed.

Figure 35-3 Specifications for the Lake House

C. Remove all debris from the site; do not use it for fill.

02200 EXCAVATION

A. Carefully remove all sod and soil throughout the area of the building and where finish grade levels are changed. Pile on site where directed. This soil is to be used later for finished grading.

B. Do all excavation required for footings, piers, walls, trenches, areas, pits, and foundations. Remove all materials encountered in obtaining indicated lines and grades required.

Beds for all foundations and footings must have solid, level, and undisturbed bed bottoms. No backfill will be allowed and all footings shall rest on unexcavated earth.

C. The contractor shall notify the architect when the excavation is complete so that he may inspect all soil before the concrete is placed.

D. Excavate to elevations and dimensions indicated, leaving sufficient space to permit erection of walls, waterproofing, masonry, and the inspection of foundations. Protect the bottom of the excavation from frost.

02260 BACKFILL

A. All outside walls shall be backfilled to within 6 inches of the finished grade with clean fill. Backfill shall be thoroughly compacted.

B. Unless otherwise directed by the architect, no backfill shall be placed until after the first floor framing is in place. No backfill shall be placed until all walls have developed such strength to resist thrust due to filling operations.

02270 GRADING

A. Do all excavating, filling, and rough grading to bring entire area outside of the building to levels shown on the drawings.

B. Where existing trees are to remain, if the new grade is lower than the natural grade under the trees, a sloping mound shall be left under the base of the tree extending out as far as the branches; if the grade is higher, a well shall be constructed around the base of the tree to provide the roots with air and moisture.

C. After rough grading has been completed and approved, spread topsoil evenly to the previously stripped area. Prepare the topsoil to receive grass seed by removing stone, debris, and unsuitable materials. Hand rake to remove water pockets and irregularities. Seeding will be done by the owner.

D. Furnish and place run of bank gravel as approved under all floor slabs.

03000 CONCRETE

WORK INCLUDED

Provide all materials, labor, equipment, and services necessary to furnish, deliver, and install all work of this Section, as shown on the drawings, as specified herein, and/or as required by job conditions including but not limited to the following:

A. Concrete for all footings and piers
B. Concrete for all slabs on ground
C. Concrete for slab at heat sink
D. Furnishing and installation of all required anchors.
E. Supplying fabrication and placement of all reinforcing bars and mesh and wire reinforcement for concrete where shown, called for, or required with proper supporting devices.
F. Erection of all wood forms required for the concrete work and removal upon completion of the work.
G. The finishing of all concrete work as hereinafter specified.
H. Porous fill below slabs on ground.

03010 MATERIAL

A. *Fine Aggregate*
Fine aggregates for concrete shall consist of natural sand having clean, hard, sharp, uncoated grains free from injurious amounts of dust, lumps, soft or flaky particles, shale, alkali, organic matter, loam or other deleterious substances.

B. *Coarse Aggregate (Stone)*
Coarse aggregates shall consist of crushed stone or gravel having clean, hard, strong, durable, uncoated particles, free from injurious amounts of soft, friable, thin, elongated or laminated pieces, alkali, organic or other deleterious matter.

C. *Water*
All water used in connection with concrete work shall be clean and free from deleterious materials or shall be the water used for drinking daily.

D. *Portland Cement*
Portland cement shall be an approved domestic brand complying with Standard Specifications for Portland Cement, ASTM Designation C-150, Type 1. Only one brand of cement shall be used throughout the course of the work.

E. All concrete is to be machine mixed in an approved mixer with a water metering device. Concrete is to reach a compressive strength of 2500 psi after 28 days.

F. *Reinforcement*
All reinforcing, unless otherwise shown or specified, shall conform to ASTM A-615, Grade 60. Wire mesh reinforcing shall have a minimum ultimate tensile strength of 70,000 psi, and shall conform to ASTM Specifications A-185, latest edition.

03320 INSPECTION & PLACING

A. All reinforcing shall be free of rust, scale, oil, or other coatings that tend to reduce the bond to concrete. All reinforcing is to be tied with 18-gauge wire at intersections and shall be held securely in position during the pouring of concrete.

B. The architect will inspect all footing beds, forms, and reinforcing just prior to placing concrete for footings and slabs.

Figure 35-3 (continued)

C. All concrete shall be placed upon clean surfaces, and properly compacted fill, free from standing water. The concrete shall be compacted and worked into corners and around reinforcing.

D. All concrete to be true and level as indicated on drawings to within ± 1/4 inch in 10 feet.

03330 FINISHING

A. Slabs in occupied spaces shall be troweled smooth and free of trowel marks.

B. Slabs in unoccupied spaces will have wood float finish.

04000 MASONRY

WORK INCLUDED

This work shall include but shall not be limited by the following:

A. Brickwork
B. Concrete blockwork
C. Mortar for brick and blockwork

04010 MATERIALS

A. Delivery and storage:

All materials shall be delivered, stored, and handled so as to prevent the inclusion of foreign materials and the damage of the materials by water or breakage.

Packaged materials shall be delivered and stored in the original packages until they are ready for use.

B. Materials showing evidence of water or other damage shall be rejected.

C. Brick shall be chosen by the owner from approved samples. Brick shall be carefully protected during transportation and shall be unloaded by hand and carefully piled; dumping is not permitted.

D. Concrete block shall be load bearing, hollow, concrete masonry units and shall conform to the standard specifications of ASTM C-145-71.

E. Mortar used for laying brick and concrete block shall consist of one (1) part masonry cement to three (3) parts sand. The mortar ingredients shall comply with the following requirements:

1. Masonry cement: ASTM C-91 T Type 2.
2. Aggregates: ASTM C-144.
3. Water: Clean, fresh, free from acid, alkali, sewage, or organic material.

F. Reinforcing: reinforcing material for masonry walls shall be prefabricated welded steel.

04220 INSTALLATION

A. All work shall be laid true to dimensions, plumb, square, and in bond, and properly anchored. All courses shall be level and joints shall be of uniform width; no joints shall exceed the size specified.

B. Joints shall be finished as follows:

C. All brick shall be laid on full mortar bed with a shoved joint. All joints shall be completely filled with mortar. All horizontal and vertical joints shall be raked 3/8 of an inch deep.

D. All mortar joints for concrete block masonry shall have full mortar coverage on vertical and horizontal face shells.

E. Vertical joints shall be shoved tight. Full mortar bedding shall have ruled joints.

F. Horizontal reinforcement shall be placed in every third bed joint of block work. Reinforcement shall be placed in the first and second bed joints above and below all openings.

G. Concealed work shall have joints cut flush.

H. Fill voids in top course with masonry and set anchor bolts as shown on construction drawings.

I. Protection: cover the wall each night and when the work is discontinued due to weather.

04240 CLEANING AND POINTING

1. Point up all the voids and open joints with mortar. Remove all of the excess mortar and dirty spots from the entire surface.

2. Upon completion, all brickwork shall be thoroughly cleaned with clean water and stiff fiber brushes and then rinsed with clear water. The use of acids or wire brushes is not permitted.

05000 STRUCTURAL STEEL

WORK INCLUDED

This work shall include but shall not be limited by the following:

A. Structural tube columns.
B. Welded flanges and plates.
C. Structural steel rafters and beams.
D. Stanchions at railings: this contractor shall supply fabricated stanchions to be installed by others.
E. Grouting base plates.

05010 MATERIALS

A. All structural steel to conform to ASTM A-36.
B. Welding electrodes to conform to American Welding Society A5.1, E70 series.

05100 FABRICATION & ERECTION

A. Drilling or punching of holes in columns, beams, and rafters shall not be permitted unless approved.

B. Welds shall be by qualified operators and shall achieve complete penetration without voids, cracks, or porosity.

C. Concealed structural steel shall have one coat of approved rust-resistant primer.

06000 WOOD & PLASTICS

WORK INCLUDED

All lumber, plywood, rough hardware, trim, paneling, and finish carpentry joinery and millwork required or implied by drawings and/or specifications. Cabinets and countertops are not included in this section.

06010 MATERIAL

A. Grade or trademark is required on each piece of lumber; only official marks of association under whose rules it is graded will be accepted.

Figure 35-3 (continued)

B. Plywood shall conform to U.S. Product Standards PS-66 and shall be branded or stamped with type and grade.

C. Moisture content shall not exceed 19% for framing lumber, 12% for plywood, 8% for finish millwork.

D. Work that is to be finished or painted shall be free from defects or blemishes on surfaces exposed to view that will show after the finish coat of paint or stain is applied. Defective materials not up to specifications for quality and grade for its intended use, or otherwise not in proper conditions, shall be rejected.

E. Rough lumber shall be dressed four (4) sides, air-dried, well-seasoned, sound, and free from splits, cracks, shakes and wanes, loose or unsound knots, and decay and excessive warp. Species and grades shall be those listed:

Douglas fir or hem-fir for rough carpentry

—Each piece marked as to grade and free from defect

—No. 1 light framing with not more than 25% No. 2 framing allowed for all lumber 2 x 6 or larger

—No. 2 construction grade for studs

Treated lumber: Southern yellow pine, CCA treated per AWPA standards

Finish lumber and millwork: clear white pine or ponderosa pine

F. All nails, spikes, screws, bolts, joist hangers, and timber connectors as indicated, noted or detailed on drawings, and as required to produce a safe, substantial and workmanlike job in all respects

G. Laminated plastic to be Micarta, Formica, or Textolite in decorator colors as selected by the owner

06100   ROUGH CARPENTRY

A. Install all rough wood framing, nailers, edge members, curbs, blocking, grounds, rough sills, backing, furring, and the like as indicated, detailed, noted, or required to properly support, back up, and complete the work this section and of any or all trades under these contracts.

—Securely attach and anchor to adjacent construction as detailed or as approved if detail is not provided.

—Shim to line if so required to provide a uniform base for any other work.

B. Provide double studs adjacent to and headers of size indicated over all openings.

C. Double joists under parallel partitions.

06200   FINISH CARPENTRY

A. Provide all rabbets, splines, ploughs, and other cuts as detailed or required for neat, tight, solid fitting and joining.

B. Finish millwork where indicated to have a clear finish shall be dressed and sanded, free from machine and tool marks, abrasions, raised grain and other defects on surfaces exposed to view. Construction and workmanship of millwork items shall conform to or exceed, the requirements of AWI and good shop practice.

C. Joints shall be tight and so formed as to conceal shrinkage.

D. Interior millwork, running finish, and trim shall be in as long lengths as practicable, shall be spliced

only where necessary, and only when approved by the Architect. All such splices shall be beveled and jointed where solid fastenings can be made.

07000   THERMAL & MOISTURE PROTECTION

WORK INCLUDED

A. Dampproofing basement walls

B. Vapor barriers

C. Thermal insulation

D. Roofing

E. Flashing

F. Caulking and sealants

07010   MATERIALS

A. Dampproofing on basement walls to be Sonneborn Building Products, Semi-mastic Hydrocide 600 or approved equal

B. Vapor barriers under concrete slabs to be 4 mil thick polyethylene

C. Vapor barriers on insulated walls and roofs to be 4 mil thick polyethylene

D. Insulation exposed to earth shall be Dow Styrofoam SM, R-5.4 per inch.

E. All rigid board insulation not exposed to earth shall be Owens/Corning, High-R Sheathing, R-7.2 per inch.

F. Batt insulation shall be Owens/Corning Fiberglass, unfaced, or approved equal.

—3 1/2″ thickness, R-11
—6″ thickness, R-19
—9″ thickness, R-30

G. Roofing underlayment, #15 asphalt-saturated felt

H. Composition roofing, Johns Mannville 12″ x 36″ asphalt shingles, 235 lb per square

I. Metal flashing, 28 gauge aluminum

J. Caulking to be acrylic polymer conforming to F.F. TT-S-00230

07150   DAMPPROOFING

A. Apply two coats of asphalt dampproofing over all masonry wall surfaces to receive earth backfill.

B. Apply polyethylene vapor barrier over gravel fill at all concrete slabs.

C. Vapor barrier sheets to be lapped 6″ minimum.

D. Apply polyethylene vapor to the heated side of insulated frame walls and roofs.

07200   THERMAL INSULATION

A. Pack all voids and cavities in exterior walls and roof. Avoid compressing batt insulation.

B. Cut and fit insulation and vapor barriers as necessary for snug fit.

C. Allow minimum air space of 1/2″ between insulation and roof sheathing.

D. Rigid insulation is to be nailed and glued with Dow Mastic number 11, according to manufacturer's instructions.

07300   ROOFING

A. Apply asphalt-saturated felt underlayment with 6″ lap to all roof surfaces.

B. Install roof shingles according to manufacturer's printed instructions.

C. Roof shingles to be applied with zinc-coated, barbed roofing nails, 4 nails per shingle.

Figure 35-3 (continued)

07600 FLASHING

A. Apply factory-painted aluminum drip edge at all roof edges.
B. Chimney to be flashed at roof with Majestic number 9-6-12 galvanized flashing.
C. Flash all pipes at roof with neoprene flashing of the proper size.
D. Flash at all intersections of roofs and vertical surfaces and as otherwise shown on construction drawings.
E. Flashing to be nailed at 3″ intervals with aluminum roofing nails.

07900 CAULKING

Caulk all windows, doors, and other openings.

08000 DOORS & WINDOWS

WORK INCLUDED
All doors, door frames, windows, skylights, trim for each, and all hardware not included elsewhere.

08010 PRODUCTS

A. Hollow metal doors to be manufactured by Pease, or approved equal. Skin to be 16 gauge steel with phenolic honeycomb core.
B. Wood doors by Iroquois Millwork or approved equal. Birch plywood skin with phenolic impregnated Kraft core.
C. All door frames to be of clear pine in standard patterns as shown on the construction drawings. Side lites to be Iroquois, Weather Guard SL with 5/8″ insulating glass.
D. Windows to be Capitol, Series E-700 aluminum frame, thermal break with factory-applied enamel finish. Windows to include aluminum screens by the same manufacturer.
E. Skylight to be Skyvue, number DV2852.
F. Door trim to be standard WM patterns milled from clear pine.
G. Contractor shall allow $500 for locksets, latches, bi-fold hardware, hinges, weatherstripping, medicine cabinets, closet rods, and shower curtain rods.

08100 DOORS

A. Frames to be plumb and square with accurately fitted joints. Set exposed nails with a nail set.
B. Accurately align doors with frames and adjust hardware as necessary for smooth operation.
C. Install molding as shown on construction drawings with accurately mitered corners.

08500 WINDOWS & SKYLIGHTS

A. Install all windows true and plumb, and according to the manufacturer's recommendations to produce a weathertight installation.
B. Install all hardware and accessories, and check all moving sections for smooth operation.

08900 HARDWARE

A. Install all door and window hardware according to the manufacturer's recommendations and check for smooth operation.
B. Install a closet rod in each closet. Closet rod to be secured through wall finish to blocking installed with rough carpentry.
C. Install shower curtain rods over tub and shower stall. Shower curtain rods to be secured through wall finish to blocking installed with rough carpentry.
D. Install a medicine cabinet (by owner) over each lavatory.

09000 FINISHES

WORK INCLUDED

A. Gypsum wallboard
B. Ceramic tile in baths and toilet rooms
C. Quarry tile floors
D. Painting and varnishing

09250 WALLBOARD

MATERIAL

A. All wallboard material to be the product of one manufacturer; U.S. Gypsum, Flintkote, or approved equal. Drywall in bath, toilet rooms, and tub room to be moisture resistant.

INSTALLATION

A. Gypsum wallboard shall be installed with joints centered over framing or furring.
B. Fasten gypsum wallboard with power-driven drywall screws or ring-shank drywall nails located not over 12 inches O.C. at all edges and in the field.
C. Outside corners are to be protected with metal corner bead.
D. Finish all joints with a minimum of three coats of joint compound and standard gypsum board reinforcing tape in accordance with the manufacturer's printed instructions.
E. Dimples at screwheads or nailheads shall receive three coats of compound.

09300 TILE

MATERIAL

A. Wall tile to be American-Olean Tile Company standard grade bright glazed in color selected. Bathroom accessories to be same manufacturer and color.
B. Floor tile in bath, toilet, and tub rooms to be American-Olean unglazed 1″ x 1″ ceramic mosaic tile in color selected.
C. Quarry tile to be installed in this specification will be 6″ x 6″ shale-and-clay tile provided by owner.
D. Marble thresholds at all doors adjacent to mosaic tile floors shall be Vermont Marble 7/8″ x 3 1/2″.

INSTALLATION

A. Lay out ceramic tile on walls and floors so that no tiles of less than one-half size occur.
B. Cut and fit tile around toilets, tubs, and other abutting devices.
C. Install all floor tile by thin-set method in accordance with TCA recommendations.
D. Install wall tile in mastic cement conforming to the recommendations of the tile manufacturer.
E. Grout all tile work to completely fill joints.
F. Clean all tile surfaces to present a workmanlike job.
G. Install 12 bathroom accessories as follows:
   - 2 soap dishes
   - 3 toilet paper holders
   - 7 towel bars

Figure 35-3 (continued)

09900 PAINTING

MATERIAL

A. Exterior stain: one coat Minwax exterior stain or approved equal, in color by owner
B. Interior walls — flat: two coats Martin Senour, Bright Life alkyd flat or approved equal, in colors by owner
C. Interior walls — semigloss: two coats Martin Senour, Bright Life alkyd semigloss or approved equal, in colors by owner
D. Interior painted woodwork: two coats Martin Senour, Bright Life alkyd semigloss, or approved equal, in colors by owner
E. Metal: two coats DeRusto rust-resistant enamel or approved equal, in colors by owner
F. Polyurethane: two coats United Gilsonite Laboratories, ZAR gloss
G. Primers: all primer to be that recommended by manufacturer of top coat.

APPLICATION

A. Repair all minor defects by patching, puttying, or filling as normally performed by painting contractors.
B. Prime uncoated wood surfaces with tinted primer and touch up previously painted surfaces.
C. Sand all surfaces smooth before each coat of paint to produce a smooth and uniform job at completion.
D. Protect all adjacent surfaces and other work incorporated into project against damage or defacement.
E. Coat all surfaces according to the following painting schedule and as indicated on the drawings. All colors are to be selected by the owner.

EXTERIOR WALLS & TRIM	Stain
METAL RAILINGS	Rust-resistant enamel
PLAYROOM	Walls: Flat / Ceiling: Flat
ALL BATHS	Walls: Semigloss / Ceilings: Semigloss
HALLS & CLOSETS	Walls: Flat / Ceilings: Flat
LIVING ROOM	Walls: Flat / Ceiling: Flat
DINING ROOM	Walls: Flat / Ceiling: Flat
KITCHEN	Walls: Semigloss / Ceiling: Semigloss
BEDROOMS	Walls: Flat / Ceilings: Flat
LOFT	All except floor and ladder: Flat
LOFT FLOOR & LADDER	Polyurethane
INTERIOR DOORS	Polyurethane
INTERIOR STAIRS	Polyurethane
UNSCHEDULED INTERIOR TRIM	Semigloss

10000 FIREPLACE

WORK INCLUDED

Provide and install fireplace, chimney, and accessories as indicated on the construction drawings. Related masonry is not included in this section.

10310 EQUIPMENT

A. Majestic Company fireplace number M28.
B. Majestic Company chimney with 8-inch flue.
C. All chimney accessories required to conform with printed instructions of Majestic Company.

10320 INSTALLATION

Fireplace and chimney are to be installed according to printed instructions of Majestic Company and as indicated on the construction drawings. All equipment is to be installed level and plumb and finished to provide a workmanlike appearance.

11000 EQUIPMENT

WORK INCLUDED

A. Provide and install kitchen cabinets and countertop.
B. Provide and install vanity cabinets. This section does not include lavatories or related plumbing.
C. Provide necessary cutouts for installation of kitchen sink, by plumbing contractor; cooktop, by owner; grill, by owner; oven, by owner.

11910 MATERIAL

A. Cabinets and vanities shall be H.J. Scheirich Company, Garden Grove style.
B. Countertops shall be Formica brand or approved equal with molded backsplash in color by owner.

11920 INSTALLATION

A. Cabinets shall be installed level and true with no less than four screws per base unit and present a workmanlike appearance.
B. Countertop shall be installed with concealed screws at 18-inch intervals front and back.
C. Exposed ends of countertop shall be veneered with matching plastic laminate.
D. Cutouts for other equipment shall be neatly trimmed to proper dimensions according to the equipment manufacturer.

15000 MECHANICAL

This division includes heating, ventilating, air conditioning, and plumbing. It is omitted here because these topics have not been covered earlier in the textbook.

16000 ELECTRICAL

This division covers all aspects of electrical work. It is omitted here because these topics have not been covered earlier in the textbook.

Figure 35-3 (continued)

Part III

TOPICS FOR FURTHER STUDY

Part III provides an opportunity for you to extend your ability to read construction drawings. Once you have mastered the contents of Parts I and II, you should be able to read and thoroughly understand most residential construction drawings. Part III helps you apply the skills developed earlier to other types of residential construction and to the work of the mechanical and electrical trades.

The Town House drawings were selected for reference in Part III for two important reasons. They represent quality construction in a geographic region where many construction practices are different from those in the rest of North America. The construction student of the 1980s will find more work in multi-family housing than perhaps any other type of construction.

The complete set of drawings for the Town House is far too large to include in the packet that accompanies this textbook. Therefore, selected sheets or portions of sheets are used here. To use the available space most efficiently, some of the drawings have been reorganized and combined — with parts of two sheets printed on one. The original sheet and drawing numbers have been retained; therefore, all references printed on the drawings are still applicable.

Multifamily Construction

Section 11

UNIT 36 Orienting the Drawings

OBJECTIVES

After completing this unit, you will be able to perform the following tasks:

- Locate a particular building or plan within a large development.

- Explain the relationships between drawings for construction projects where several plans are to be adjoined in one building.

- Visualize a building design by reading the drawings.

IDENTIFYING BUILDINGS AND PLANS

Multifamily dwellings are often built in large developments — with many similar buildings in a single development. Some developments are completed in phases. One phase is completely constructed and begins earning income for the developer before the next stage is started. Figure 36-1 shows the site plan for Hidden Valley, the development for which the Town House was designed. Hidden Valley is to be developed in four phases. Each phase is outlined with a heavy broken line on the site plan.

The first phase of Hidden Valley includes eleven buildings. Each building is labeled as to building type and the parts that make up the building, Figure 36-2. Building type II is made up of four separate units, each with its own floor plan. Each unit is built like a separate building joined to the next, Figure 36-3. Each plan is identified by a letter or letter and numeral. (The term *plan* is used to refer to a particular arrangement of rooms or design. This should not be confused with the use of *plan* to refer to a type of drawing.)

The plans in building type II are A, B1, B2, and A R. Each plan type is described on separate drawings. Drawings 1 through 5 are for plan A. Drawings 6 through 10 are for plan B.

One technique that is used to create similar, yet different, plans is to reverse them. A reversed plan is created by building the plan as though seen in a mirror. The reversed plan has the same features and the same dimensions, but their arrangement is reversed, Figure 36-4. In building type II of the Hidden Valley development, the south end is a reversal of plan A. This is designated on the site plan and on Figure 36-2 as A R. The letter *R* indicates a reverse plan.

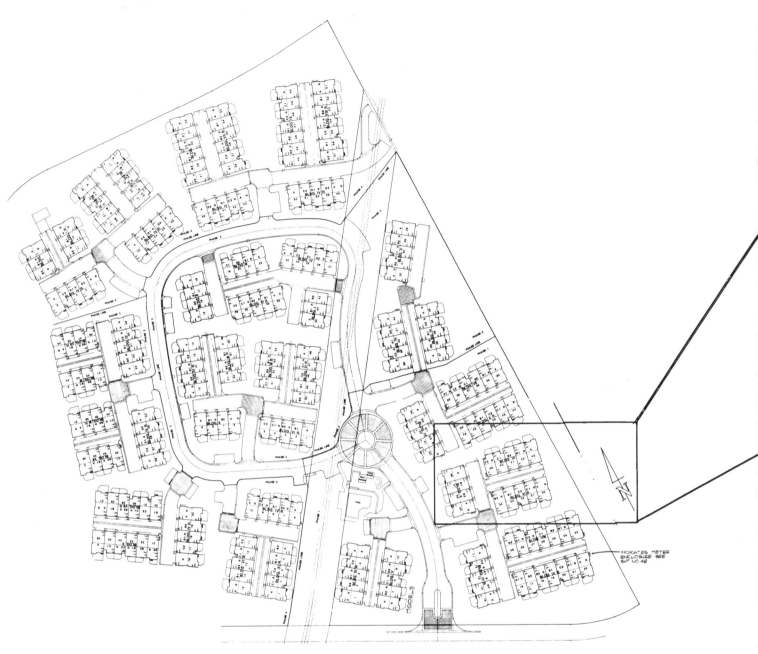

Courtesy of berkus-group architects

Figure 36-1 Site plan

ORGANIZATION OF THE DRAWINGS

The drawings for a large project must be systematically organized, so information can be found easily. Architects follow a similar pattern in organizing their drawings. The general order of drawing sheets is similar to that of specifications: site plans are first; foundation plans and floor plans, next; building elevations, third; then, structural and architectural details; mechanical, next to last; and electrical, last. The cover sheet usually includes an index or table of contents for the drawing set, Figure 36-5.

The index for the Hidden Valley drawing does not indicate mechanical or electrical drawings. In the western part of the United States, it is common practice not to prepare separate drawings for mechanical or electrical work on town houses. The essential information for these trades is included on the drawings for the other trades.

VISUALIZING THE PLAN

The first step in becoming familiar with any plan should be to mentally walk through the plan. This technique was described earlier in Unit 20 for the Lake House. Refer to the Hidden Valley drawings as you mentally walk through Plan A of the Town House.

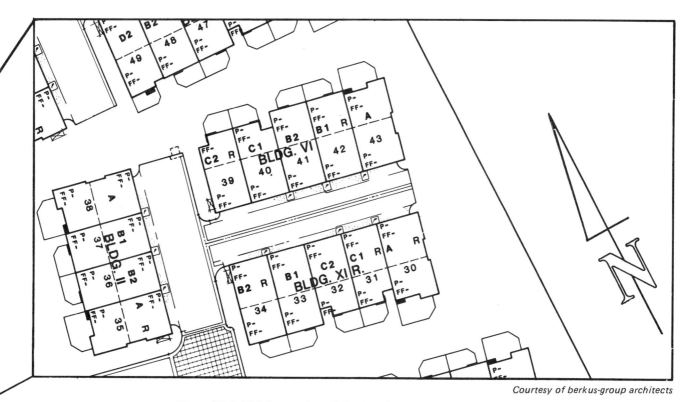

Figure 36-2 This is a section of the site plan at the size it was drawn.

Courtesy of berkus-group architects

Enter through the overhead garage door on the east side of the lowest level. The garage is an open area with stairs leading up to the main floor. Next to the stairs there is an area of dropped ceiling beneath the stairs on the main floor. A note on Drawing 2 refers to Section B.

This note introduces a new consideration for anyone reading construction drawings. Although the architect reviews the drawings carefully, there is always a possibility that errors will appear as one did in this instance. The dropped ceiling beneath the stairs is actually shown on Section E.

The drawings for earlier units of this textbook were reviewed and corrected more thoroughly than can be reasonably expected for actual construction drawings. The drawings for the Town House are printed exactly as they were prepared for the construction job. A few errors may remain on these drawings. Any errors that do remain give you valuable experience in detecting and dealing with error.

At the top of the garage stairs, there is an entry area with an exterior door. A boxed number

A1   B1   B2   A1 REV.

Courtesy of berkus-group architects

Figure 36-3 Building type II includes four housing units.

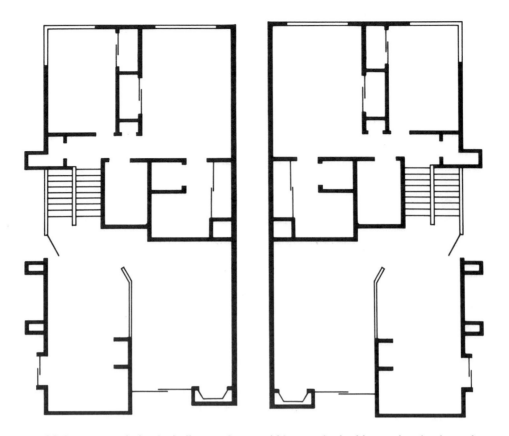

Figure 36-4 A reversed plan is similar to what would be seen by looking at the plan in a mirror.

33 refers to the plan notes printed on Drawing 2. This note indicates a stub wall dividing the entry from the dining room. Boxed note references are used throughout the Town House drawings. Reading each related note as it is encountered will help you to better understand the plan.

As you enter the dining room and kitchen to the west of the entry area, the floor material changes. The kitchen ceiling is dropped to 7'-6" as dimensioned on Section A-A, Drawing 4. The areas with dropped ceiling above are shown by light cross hatching on the floor plan. The kitchen has base cabinets on three walls and a peninsula of cabinets separating the kitchen and dining room. Opposite the pennisula of base cabinets, there is a 4'-8" alcove with cabinets, Figure 36-6.

Visually walk back through the dining room along a 36" high guard rail, plan note 13, to a 2-riser stair down to the living room. At the (front) west end of the large, 18'-4"x13'-4" living room there is a fireplace, Figure 36-7. To the left (north) of the fireplace is a 7'-0"x8'-0" sliding glass door that opens onto a small courtyard. A note on the floor plan indicates a *36 MOJ RADIUS ABOVE*. The legend of symbols and abbreviations on sheet O shows that MOJ means *measure on the job*. Looking at the Front Elevation, Drawing 5, you

can see that this note refers to a window with a 36-inch radius at the top.

From the entry area, another stair leads up to a hall serving the bedrooms and bath. Across the hall, a pair of 2'-6"x6'-8" hollow-core doors open into bedroom #2. By the stairs, a door opens into bathroom #2. This is a small bathroom with a tub, water closet, and lavatory. At the south end of the hall, you enter the master bedroom. There is an 8'-0"x5'-0" aluminum sliding window at the east end of the master bedroom. On the north wall near the hall door, there is a wardrobe closet with a shelf and pole (S&P). At the west end of the room there is a master bath. The master bath has a 42"x60" tub, a vanity (called pullman here), a large wardrobe closet, and a toilet room.

To thoroughly understand the plan, you should read all of the plan notes on Drawing A-2 and locate the features to which they refer. Then locate each of the details, indicated by a circle. The numerals above the horizontal line in these circles indicate the detail being referenced. The letter and numeral below the horizontal line indicate on which drawing the detail appears. As you find these plan notes and details, refer to the framing plans, building sections, and elevations to thoroughly understand each.

table of contents

		COVER SHEET AND INDEX
a		SITE PLAN
b		GENERAL NOTES
c		GENERAL NOTES
1	**PLAN A**	FOUNDATION PLAN
2		FLOOR PLAN
3		FRAMING PLAN
4		SECTIONS
5		ELEVATIONS
6	**PLAN B**	FOUNDATION PLAN
7		FLOOR PLAN
8		FRAMING PLAN
9		SECTIONS
10		ELEVATION (1)
11		ELEVATION (2)
12	**PLAN C**	FOUNDATION PLAN
13		FLOOR PLAN
14		FRAMING PLAN
15		SECTIONS
16		ELEVATIONS (1)
17		ELEVATIONS (2)
18	**PLAN D**	FOUNDATION PLAN
19		FLOOR PLAN
20		FRAMING PLAN
21		SECTIONS
22		ELEVATIONS (1)
23		ELEVATION (2)
24	PLAN A-B	INTERIOR ELEVATIONS -PLAN A&B
25	PLAN C-D.	INTERIOR ELEVATIONS -PLAN C&D
26		BUILDING TYPE I
27		BUILDING TYPE II
28		BUILDING TYPE III
29		BUILDING TYPE IV
30		BUILDING TYPE V
31		BUILDING TYPE VI
32		BUILDING TYPE VII
33		BUILDING TYPE VII
34		BUILDING TYPE VIII
35		BUILDING TYPE VIII
36		BUILDING TYPE IX
37		BUILDING TYPE IX
38		BUILDING TYPE X
39		BUILDING TYPE X
40		BUILDING TYPE XI
41		BUILDING TYPE XII
42		METER ENCLOURE
43		REC. CENTER FOUND FLR. PLAN
44		REC CENTER SECTION ELEV.
D1		FOUNDATION DETAILS
D2		FRAMING DETAILS
D3		FRAMING DETAILS
D4		FRAMING DETAILS

Courtesy of berkus-group architects

Figure 36-5 Table of contents for Hidden Valley construction drawings

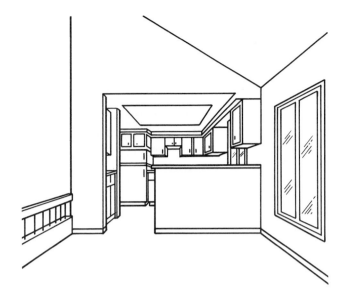

Figure 36-6 Town House kitchen as seen from entry

Figure 36-7 End wall of living room

√ **CHECK YOUR PROGRESS** ——————————————————————

Can you perform these tasks?

☐ Locate a particular building on the site plan for a large development.

☐ Locate a particular plan within a multi-plan building.

☐ Locate the drawings for a particular plan type within a large drawing set consisting of several buildings and plan types.

 ASSIGNMENT ——————————————————————————

Refer to the Town House drawings (including those in the packet and Figures 36-1 and 36-2) to complete the assignment.

1. How many buildings are included in phase one?

2. What plans are included in building type VI?

3. In phase one, building II, on which side (compass direction) is the courtyard?

4. How thick is the concrete slab for the garage floor in plan B?

5. What size are the floor joists under the dining room in plan B?

6. In plan B, what supports the kitchen floor joists under the back wall of the kitchen?

7. On which drawing would you find elevations of the kitchen cabinets for plan A? For plan B?

8. On which drawing would you find details of concrete piers under bearing posts for girders?

9. What is the height of the handrail at the dining room/living room stairs in plan B?

10. For each of the major rooms of plan B listed below, indicate the overall dimensions. Do not include closets, stairs, or minor irregularities. Allow for the thickness of all walls. Walls are dimensioned to the face of the framing.

    a. Living room   d. Deck   f. Bedroom #2
    b. Dining room   e. Library   g. Master bedroom
    c. Kitchen

11. What is the tread width and riser height for the stairs between the living room and dining room in plan B?

12. What important feature of the living room is beside the entry in plan B?

13. How long are the studs in the partition between the master bedroom and bedroom #2 in plan B?

14. How high above finished grade is the top of the privacy fence in front of plan B?

UNIT 37 *Regional Variations*

OBJECTIVES

After completing this unit, you will be able to describe important variations in the construction methods used for the buildings in the drawing packet with this textbook.

The Duplex and the Lake House were designed and built in the Eastern states. The Town House was designed and built in the West. Each of these buildings is quite different from the others in design and construction methods. Some of the differences are the result of differences in the availability of materials; some, the result of different building codes and structural requirements. Other differences are the result of factors such as life-style or heating and air-conditioning requirements. Some of the characteristics of buildings in the same regions and of the same basic types as the Duplex, Lake House, and Town House are presented in the following paragraphs.

THE DUPLEX

The Duplex is a very simple building designed for upstate New York. Some of the important points that the architect considered in designing the Duplex are listed:

- Very cold winters
- Increasing scarcity of wood in this region
- Ready availability in this region of metal building products
- Type of property: low-investment, income-earning property for the owner

Climate Considerations

In this region, temperatures of 20° below zero (Fahrenheit) are common. This drives the frost deep into the earth. Therefore, foundations must be more than three feet deep. The foundation and perimeter of the concrete slab are insulated to prevent excessive heat loss to the frozen earth. Also, all exterior walls and ceilings are heavily insulated. Notice that even the sheathing is rated according to its *R* value.

Use of Metals

In the part of the country where the Duplex was built, metal is used extensively to replace wood as a building material. Metal is sometimes less expensive than wood, it resists the damaging effects of the severe weather, and it can be installed quickly. For these reasons, aluminum siding and aluminum cornices were chosen for the Duplex. Plywood siding is used at the corners. The vertical pattern of the plywood breaks up the long, straight lines of the horizontal aluminum siding. The plywood siding also serves as corner bracing to prevent wracking of the walls. The plywood corners alone would not be sufficient to prevent wracking in some regions. Upstate New York does not have the threat of earthquakes or extreme winds; therefore, the plywood corners are adequate.

Cost Considerations

The Duplex is designed as a low-cost project. The design is simple, and the drawings can be completed in relatively little space. Most homes in the Northeast have basements, but the Duplex is designed without a basement to conserve costs. The rectangular shape of the Duplex also is inexpensive. Generally, the fewer corners a building has and the fewer floor levels it has, the less expensive it is to build.

THE LAKE HOUSE

The Lake House is designed as a year-round vacation home on a lake in Virginia. Although the winters are not as extreme as those in upstate New York, the temperature is frequently well below freezing. The thermal insulation methods used in the Duplex are also used in the Lake House. In addition, the Lake House receives a lot of its heat from the sun.

Cost was less of a restriction in designing the Lake House. Therefore, the Lake House has many corners and many levels. For this reason, more drawings are needed to completely describe the Lake House.

Both the Duplex and the Lake House were designed in the Eastern states. In this part of the country, it is common to prepare separate mechanical and electrical drawings. Therefore, the floor plans for these projects do not include much information about the mechanical and electrical work.

THE TOWN HOUSE

The Town House is quite different from the Duplex or the Lake House because it was designed in southern California and it is part of a large project for high-rent housing. In designing buildings for this region, the architect must consider the threat of earthquakes and high winds. Also, the

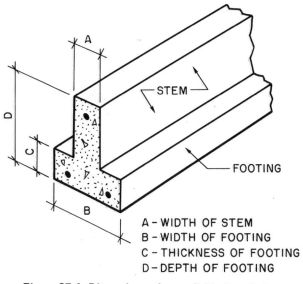

A – WIDTH OF STEM
B – WIDTH OF FOOTING
C – THICKNESS OF FOOTING
D – DEPTH OF FOOTING

Figure 37-1 Dimensions of monolithic foundation

warmer temperatures affect the foundation design and insulation requirements.

Foundations

In the Southwest where frost is not a concern but earthquakes are, foundations are usually shallow and poured in one piece, Figure 37-1. A one-piece foundation is called a *monolithic* foundation. The foundation design varies depending on what it supports and where it is within the building.

For example, find the detail callout for the foundation between the house and garage in plan A (near the middle of Drawing 1). The callout is $\frac{2b}{d1}$. Now find the callout for the detail of the exterior foundation of the garage (in the top left corner of Drawing 1). The callout is $\frac{4}{d1}$. Now find details 2 and 4 on Drawing d-1 and notice the differences. The foundation at the exterior walls has a high ledge to support the garage slab. The foundation between the house and the garage has a lower ledge to receive the thickened edge of the garage slab.

The foundation details include letters in place of actual dimensions in many places. Frequent notes direct you to 10/d4. Sheet d4 is not included in the drawing package with this book, but Detail 10 from that sheet is shown in Figure 37-2. The letter coded dimensions of the foundation details refer to the U.B.C. (Uniform Building Code) Footing Requirements shown at the bottom of 10/d4. The foundation recommendations in the top portion of 10/d4 give other information of a general nature. For example, the interior foundations are to be reinforced with one #4 bar at the top and one at the bottom. This corresponds with the two bold dots shown on interior foundation details, such as 3/d1.

Anchors and Ties

Steel hold-downs, tie straps, and framing anchors are used more extensively in the West to resist the destructive forces of earthquakes. The details for the Town House, especially those on Drawing d-1, show several types of hardware labeled only by the manufacturer's name and number. Simpson, the manufacturer named on the Town House details, is a manufacturer of structural ties and anchors. The ties, anchors, and other such items to be used in the Town House are shown in Figure 37-3.

FOUNDATION RECOMMENDATIONS

FOOTINGS				SLAB IN LIVING AREAS		SLAB IN GARAGE AREA	
EXTERIOR		INTERIOR					
DEPTH	REINFORCEMENT	DEPTH	REINFORCEMENT	REINFORCEMENT	MOISTURE	REINFORCEMENT	MOISTURE
E	F	G	F	N/A	N/A	H	X

LEGEND FOR FOUNDATION RECOMMENDATIONS
PHASE 1

G — 12 INCHES BELOW LOWEST ADJACENT GRADE.

E — 18 INCHES BELOW LOWEST ADJACENT GRADE.

F — ONE NO. 4 REBAR AT TOP AND ONE AT BOTTOM.

H — SLAB SHOULD BE DESIGNED AS FLOATING MEMBER AND SHOULD BE SEPARATED FROM PERIMETER FOOTINGS BY ½ INCH CONSTRUCTION FELT OR EQUIVALENT.

X — NO SPECIAL REQUIREMENT.

NOTE

1.   MINIMUM FOOTING WIDTH SHOULD BE 12 INCHES.

2.   EXTERIOR AND INTERIOR FOOTINGS FOR TWO-STORY STRUCTURES SHOULD HAVE A MINIMUM EMBEDMENT OF 18 INCHES BELOW LOWEST ADJACENT GRADE.

3.   INTERIOR SLABS SHOULD BE STRUCTURALLY TIED TO PERIMETER FOOTINGS IN LIVING AREAS.

4.   PROVIDE FOOTING AT GARAGE OPENING.

U.B.C. FOOTING REQUIREMENTS

(THESE ARE THE MINIMUM REQUIREMENTS. SEE SOILS REPORT FOR MORE STRINGENT CONDITIONS ONLY.)

	A WIDTH OF STEM	B WIDTH OF FTG.	C THICK- NESS OF FTG.	D DEPTH OF FTG.
A — 1 — STORY	6"	12"	6"	12"
B — 2 — STORY	8"	15"	7"	18"
C — 3 — STORY	10"	18"	8"	24"

Courtesy of berkus-group architects

Figure 37-2 Detail 10/D4

CB COLUMN BASES

Model No.	W	L	Material Stirrups	Bolts	Uplift Design Loads
CB44	3 9/16″	3 5/8″	3/16″x2″	(2) 5/8″	5030
CB46	3 9/16″	5 1/2″	3/16″x2″	(2) 5/8″	5030
CB48	3 9/16″	7 1/2″	3/16″x2″	(2) 5/8″	5030
CB5	5 1/4″	Specify	3/16″x3″	(2) 5/8″	5030
CB66	5 1/2″	5 1/2″	3/16″x3″	(2) 5/8″	5030
CB68	5 1/2″	7 1/2″	3/16″x3″	(2) 5/8″	5030
CB610	5 1/2″	9 1/2″	3/16″x3″	(2) 5/8″	5030
CB612	5 1/2″	11 1/2″	3/16″x3″	(2) 5/8″	5030
CB7	6 7/8″	Specify	1/4″x3″	(2) 3/4″	7230
CB88	7 1/2″	7 1/2″	1/4″x3″	(2) 3/4″	7230
CB810	7 1/2″	9 1/2″	1/4″x3″	(2) 3/4″	7230
CB812	7 1/2″	11 1/2″	1/4″x3″	(2) 3/4″	7230
CB9	8 7/8″	Specify	1/4″x3″	(2) 3/4″	7230
CB1010	9 1/2″	9 1/2″	1/4″x3″	(2) 3/4″	7230
CB1012	9 1/2″	11 1/2″	1/4″x3″	(2) 3/4″	7230
CB1212	11 1/2″	11 1/2″	1/4″x3″	(2) 3/4″	7230

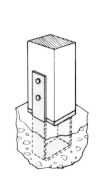

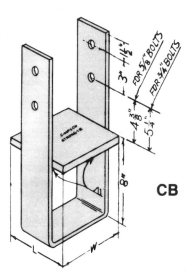

CB

HD HOLDOWNS

| Model No. | Bolt Attachment | | | Average Test Ultimate | Design Load Value* When Installed on Stud Thickness of | | | | |
	Concrete Dia.	Concrete Min.† Embedment	Stud		1 1/2″	2	2 1/2″	3	3 1/2″
HD 2	5/8″	9″	(2)-5/8″MB	13,200	2450	2520	2520	2520	2520
HD 5	3/4″	11″	(2)-3/4″MB	19,000	3375	3610	3610	3610	3610
HD 6	1″	14″	(3)-3/4″MB	18,600	5060	5410	5410	5410	5410
HD 7	1″	14″	(3)-7/8″MB	28,600	6350	6500	6500	6500	6500
HD 7	1 1/8″	15″	(3)-7/8″MB	28,600	6350	7100	7500	7500	7500
HD 2N	5/8″	9″	(2)-5/8″MB	8,800	2450	2520	2520	2520	2520
HD 5N	3/4″	11″	(2)-3/4″MB	11,600	3375	3610	3610	3610	3610
HD 7N	1″	14″	(2)-1″MB	20,300	4800	5640	6480	6500	6500

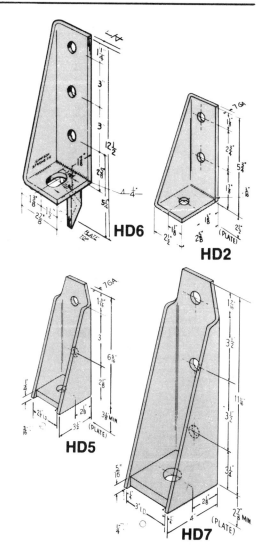

HD6 **HD2**

HD5

HD7

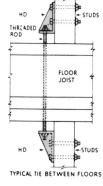

HD
THREADED ROD
STUDS
FLOOR JOIST
HD
STUDS
TYPICAL TIE BETWEEN FLOORS

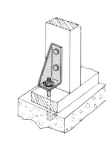

Courtesy of Simpson Company

Figure 37-3 Ties and anchors for Town House

HL ANGLES

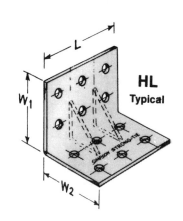

HL
Typical

Model No.	DIMENSIONS			Bolts (total)	Gussets	BOLT LOAD VALUES	
	Mat.	W₁ & W₂	L			Parallel to Grain	Perpend. to Grain
HL33	3/16″	3¼″	2½″	2-⅝″ M.B.	None	1255	725
HL35	3/16″	3¼″	5″	4-⅝″ M.B.	None	2510	1450
HL35G	3/16″	3¼″	5″	4-⅝″ M.B.	One	2510	1450
HL37	3/16″	3¼″	7½″	6-⅝″ M.B.	None	3765	2175
HL37G	3/16″	3¼″	7½″	6-⅝″ M.B.	Two	3765	2175
HL53	3/16″	5¾″	2½″	4-⅝″ M.B.	None	2510	1450
HL55	3/16″	5¾″	5″	8-⅝″ M.B.	None	5025	2900
HL55G	3/16″	5¾″	5″	8-⅝″ M.B.	One	5025	2900
HL57	3/16″	5¾″	7½″	12-⅝″ M.B.	None	7535	4250
HL57G	3/16″	5¾″	7½″	12-⅝″ M.B.	Two	7535	4250
HL43	¼″	4¼″	3″	2-¾″ M.B.	None	1805	970
HL46	¼″	4¼″	6″	4-¾″ M.B.	None	3610	1940
HL46G	¼″	4¼″	6″	4-¾″ M.B.	One	3610	1940
HL49	¼″	4¼″	9″	6-¾″ M.B.	None	5435	2910
HL49G	¼″	4¼″	9″	6-¾″ M.B.	Two	5435	2910
HL73	¼″	7¼″	3″	4-¾″ M.B.	None	3610	1940
HL76	¼″	7¼″	6″	8-¾″ M.B.	None	7225	3880
HL76G	¼″	7¼″	6″	8-¾″ M.B.	One	7225	3880
HL79	¼″	7¼″	9″	12-¾″ M.B.	None	10875	5820
HL79G	¼″	7¼″	9″	12-¾″ M.B.	Two	10875	5820

STC/DTC ROOF TRUSS CLIPS

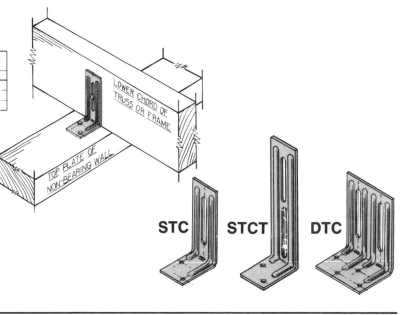

Model No.	DIMENSIONS		MATERIAL Ga.	NAILING	
	Plate Base	Vertical Leg		Base	Slot
STC	1¼″ x 1¾″	1¼″ x 2¾″	18 Ga. Galv.	2-8d	1-8d
STCT	1¼″ x 1¾″	1¼″ x 4¼″	18 Ga. Galv.	2-8d	1-8d
DTC	2½″ x 1¾″	2½″ x 2¾″	18 Ga. Galv.	4-8d	2-8d

STC **STCT** **DTC**

PA PURLIN ANCHORS

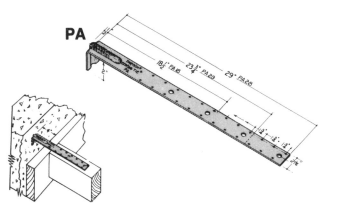

PA

Model No.	Length	Connectors To Purlins	DESIGN LOADS	
			Normal	Max.
PA18	18½″	(12)-16d	1600	2130
PA23	23¾″	(18)-16d	2410	3200
PA28	29″	(24)-16d	3140	4170

Figure 37-3 (continued)

A35 FRAMING ANCHORS

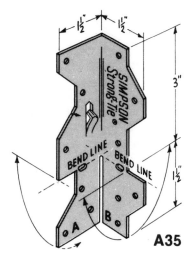

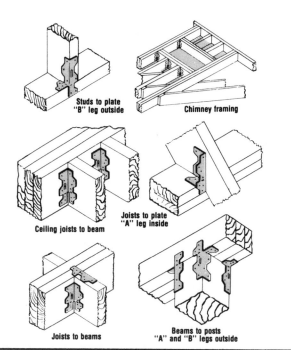

Studs to plate "B" leg outside

Chimney framing

Ceiling joists to beam

Joists to plate "A" leg inside

Joists to beams

Beams to posts "A" and "B" legs outside

ST STRAP TIES

| Model No. | DIMENSIONS | | | FASTENERS | | DESIGN LOADS | | |
| | Material | Width | Length | Nails | Bolts | Nails | BOLTS | |
							Single Shear	Double Shear
ST292	20 ga. galv.	2¹/₁₆"	9⁵/₁₆"	12-16d	—	805	—	—
ST2122	20 ga. galv.	2¹/₁₆"	12¹³/₁₆"	16-16d	—	1170	—	—
ST2115	20 ga. galv.	¾"	16⁵/₁₆"	10-16d	—	670	—	—
ST2215	20 ga. galv.	2¹/₁₆"	16⁵/₁₆"	20-16d	—	1340	—	—
ST6215	16 ga. galv.	2¹/₁₆"	16⁵/₁₆"	20-16d	—	1340	—	—
ST6224	16 ga. galv.	2¹/₁₆"	23⁵/₁₆"	28-16d	—	1875	—	—
ST6236	16 ga. galv.	2¹/₁₆"	33¹³/₁₆"	40-16d	—	2580	—	—
ST9	16 ga. galv.	1¼"	9"	8-16d	—	535	—	—
ST12	16 ga. galv.	1¼"	11⅝"	10-16d	—	670	—	—
ST18	16 ga. galv.	1¼"	17¾"	14-16d	—	935	—	—
ST22	16 ga. galv.	1¼"	21⅝"	18-16d	—	1205	—	—

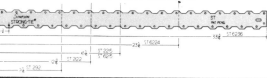

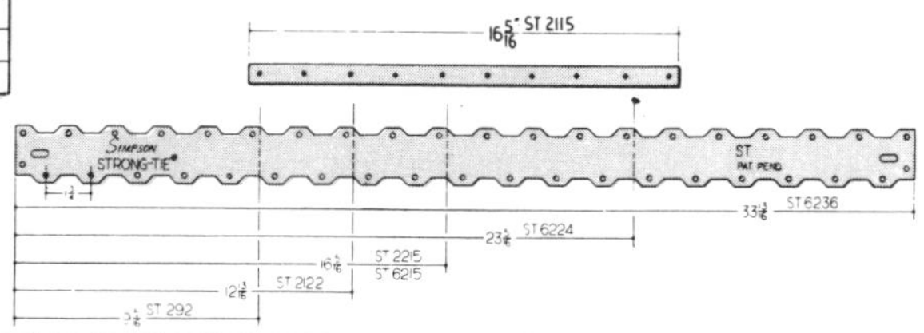

ST

SA STRAP CONNECTORS

| Model No. | Strap Section | DIMENSIONS ALTERNATE CONNECTIONS | | | | DESIGN LOADS* | |
		L₁	L₂	Bolts ea. side	Nails ea. side	Bolts only	Nails only
SA34	7 ga x 2¹/₁₆"	34"	9"	2-¾"	—	2860	—
SA45	7 ga x 2¹/₁₆"	45"	19½"	2-¾"	—	2860	—
SA36	12 ga x 2¹/₁₆"	36"	9"	2-½"	11-16d	1630	1475
SA47	12 ga x 2¹/₁₆"	47"	19½"	2-½"	11-16d	1630	1475
SAL36¹	12 ga x 2¹/₁₆"	36"	9"	2-½"	11-16d	3260	2950
SAL47¹	12 ga x 2¹/₁₆"	47"	19½"	2-½"	11-16d	3260	2950

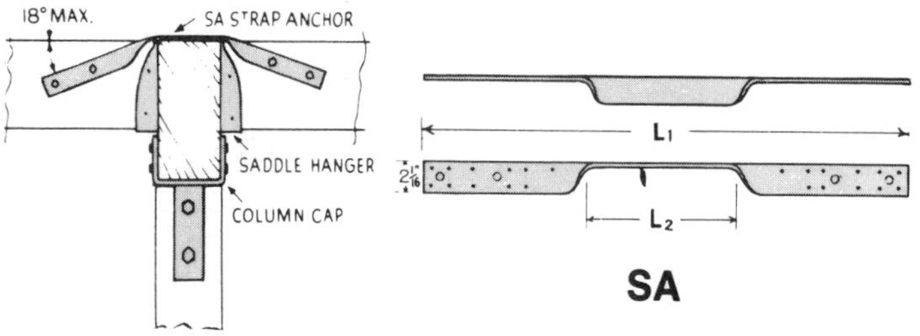

18° MAX.   SA STRAP ANCHOR

SADDLE HANGER

COLUMN CAP

L₁

L₂

SA

WB WALL BRACING

Type	Material	Size
WB106	16-ga. (galv.)	1¼" x 9'5⅝" long
WB126	16-ga. (galv.)	1¼" x 11'4⅜" long

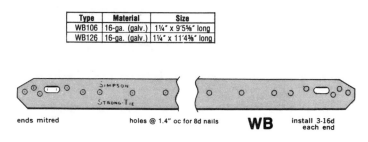

ends mitred

holes @ 1.4" oc for 8d nails

WB   install 3-16d each end

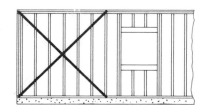

Figure 37-3 (continued)

HUTF JOIST HANGARS

Model No.	Joist Size	DIMENSIONS				TF	Nail Schedule		Aver. Ult.	I.C.B.O. Loads*		
		A	B	H	W		Header	Joist		Uplift	Norm	Max
HU26TF	2x6	1¼"	2"	5⅜"	1 9/16"	2"	(10)-16d	(4)-10d†	4,800	420	1620	2030
HU28TF	2x8	1¼"	2"	7¼"	1 9/16"	2½"	(10)-16d	(4)-10d†	6,000	420	1620	2030
HU210TF	2x10	1¼"	2"	9¼"	1 9/16"	2½"	(12)-16d	(4)-10d†	7,200	420	1620	2030
HU212TF	2x12	1¼"	2"	11⅛"	1 9/16"	2½"	(14)-16d	(6)-10d†	8,400	630	1860	2320
HU214TF	2x14	1¼"	2½"	13⅛"	1 9/16"	2½"	(16)-16d	(6)-10d†	9,600	630	1860	2330
HU216TF	2x16	1¼"	2½"	15⅛"	1 9/16"	2½"	(18)-16d	(8)-10d†	10,800	840	2070	2580
HU34TF	3x4	1¼"	2"	3½"	2 9/16"	2½"	(8)-16d	(2)-10d	8,270	210	2160	2600
HU36TF	3x6	1¼"	2"	5⅜"	2 9/16"	2½"	(10)-16d	(4)-10d	9,830	420	2390	2990
HU36TF	3x8	1¼"	2"	7¼"	2 9/16"	2½"	(12)-16d	(4)-10d	11,390	420	2390	2990
HU310TF	3x10	1¼"	2"	9¼"	2 9/16"	2½"	(14)-16d	(6)-10d	12,950	630	2630	3320
HU312TF	3x12	1¼"	2½"	11"	2 9/16"	2½"	(16)-16d	(6)-10d	13,760	630	2630	3320
HU314TF	3x14	1¼"	2½"	13"	2 9/16"	2½"	(18)-16d	(8)-10d	14,580	840	3350	4180
HU316TF	3x16	1¼"	2½"	15"	2 9/16"	2½"	(20)-16d	(8)-10d	15,400	840	3350	4180
HU44TF	4x4	1¼"	2"	3½"	3 9/16"	2½"	(8)-16d	(2)-10d	8,270	210	2600	2600
HU46TF	4x6	1¼"	2"	5⅜"	3 9/16"	2½"	(10)-16d	(4)-10d	9,830	420	3160	3210
HU48TF	4x8	1¼"	2"	7¼"	3 9/16"	2½"	(12)-16d	(4)-10d	11,390	420	3160	3600
HU410TF	4x10	1¼"	2"	9¼"	3 9/16"	2½"	(14)-16d	(6)-10d	12,950	630	3400	4130
HU412TF	4x12	1¼"	2½"	11"	3 9/16"	2½"	(16)-16d	(6)-10d	13,760	630	4070	4400
HU414TF	4x14	1¼"	2½"	13"	3 9/16"	2½"	(18)-16d	(8)-10d	14,580	840	4310	4710
HU416TF	4x16	1¼"	2½"	15"	3 9/16"	2½"	(20)-16d	(8)-10d	15,400	840	4310	4710
HU66TF	6x6	1¼"	2"	5⅜"	5½"	2½"	(10)-16d	(4)-16d	9,830	210	3210	3210
HU68TF	6x8	1¼"	2"	7¼"	5½"	2½"	(12)-16d	(4)-16d	11,390	420	3600	3600
HU610TF	6x10	1¼"	2"	9¼"	5½"	2½"	(14)-16d	(6)-16d	12,950	630	4130	4130
HU612TF	6x12	1¼"	2½"	11⅛"	5½"	2½"	(16)-16d	(6)-16d	13,760	630	4400	4400
HU614TF	6x14	1¼"	2½"	13⅛"	5½"	2½"	(18)-16d	(8)-16d	14,580	840	4710	4710
HU616TF	6x16	1¼"	2½"	15⅛"	5½"	2½"	(20)-16d	(8)-16d	15,400	840	4710	4710
HU24-2TF	(2)2x4	1¼"	2"	3½"	3⅛"	2½"	(8)-16d	(2)-10d	8,270	210	2540	2600
HU26-2TF	(2)2x6	1¼"	2"	5⅜"	3⅛"	2½"	(10)-16d	(4)-10d	9,830	420	2780	3210
HU28-2TF	(2)2x8	1¼"	2"	7¼"	3⅛"	2½"	(12)-16d	(4)-10d	11,390	420	2780	3470
HU210-2TF	(2)2x10	1¼"	2"	9¼"	3⅛"	2½"	(14)-16d	(6)-10d	12,950	630	3010	3770
HU212-2TF	(2)2x12	1¼"	2½"	11⅛"	3⅛"	2½"	(16)-16d	(6)-10d	13,760	630	3590	4400
HU214-2TF	(2)2x14	1¼"	2½"	13⅛"	3⅛"	2½"	(18)-16d	(8)-10d	14,580	840	3830	4710
HU216-2TF	(2)2x16	1¼"	2½"	15⅛"	3⅛"	2½"	(20)-16d	(8)-10d	15,400	840	3830	4710
HU210-3TF	(3)2x10	1¼"	2"	9¼"	4 11/16"	2½"	(14)-16d	(6)-16d	12,950	630	4130	4130
HU212-3TF	(3)2x12	1¼"	2½"	11⅛"	4 11/16"	2½"	(16)-16d	(6)-16d	13,760	630	4400	4400
HU214-3TF	(3)2x14	1¼"	2½"	13⅛"	4 11/16"	2½"	(18)-16d	(8-)16d	14,580	840	4710	4710
HU216-3TF	(3)2x16	1¼"	2½"	15⅛"	4 11/16"	2½"	(20)-16d	(8)-16d	15,400	840	4710	4710

HUTF

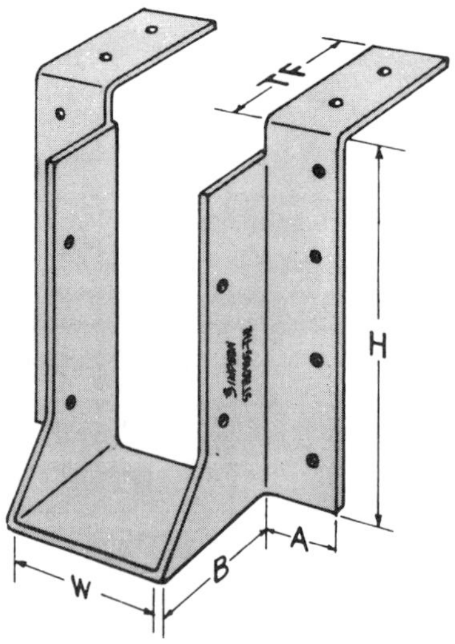

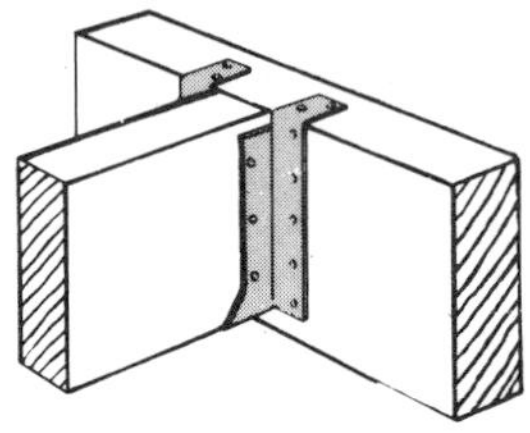

HU JOIST HANGARS

Model No.	Joist Size	DIMENSIONS				Nail Schedule		Aver. Ult.	I.C.B.O. Loads*		
		A	B	H	W	Header	Joist		Uplift	Normal	Max
HU26	2x4 2x6	1"	2"	3 1/16"	1 9/16"	4-16d	2-10d†	2,600	110	535	670
HU28	2x8	1"	2"	5¼"	1 9/16"	6-16d	4-10d†	3,700	420	805	1010
HU210	2x10	1"	2"	7⅛"	1 9/16"	8-16d	4-10d†	4,900	420	1070	1345
HU212	2x12	1"	2"	9"	1 9/16"	10-16d	6-10d†	6,200	630	1340	1680
HU214	2x14	1"	2½"	10⅛"	1 9/16"	12-16d	6-10d†	8,500	630	1610	2015
HU34	3x4	1¼"	2"	3⅜"	2 9/16"	4-16d	2-10d	2,600	210	535	670
HU36	3x6	1¼"	2"	5⅜"	2 9/16"	8-16d	4-10d	5,020	420	1070	1345
HU38	3x8	1¼"	2"	7⅛"	2 9/16"	10-16d	4-10d	7,430	420	1340	1680
HU310	3x10	1¼"	2"	8⅞"	2 9/16"	14-16d	6-10d	9,850	630	1875	2350
HU312	3x12	1¼"	2½"	10⅝"	2 9/16"	16-16d	6-10d	11,700	630	2145	2690
HU314	3x14	1¼"	2½"	12⅜"	2 9/16"	18-16d	8-10d	13,560	840	2410	3010
HU316	3x16	1¼"	2½"	14⅛"	2 9/16"	20-16d	8-10d	15,420	840	2680	3360
HU44	4x4	1¼"	2"	2⅞"	3 9/16"	4-16d	2-10d	2,600	210	535	670
HU46	4x6	1¼"	2"	4⅞"	3 9/16"	8-16d	4-10d	5,020	420	1070	1345
HU48	4x8	1¼"	2"	6⅝"	3 9/16"	10-16d	4-10d	7,430	420	1340	1680
HU410	4x10	1¼"	2"	8⅜"	3 9/16"	14-16d	6-10d	9,850	630	1875	2350
HU412	4x12	1¼"	2½"	10⅛"	3 9/16"	16-16d	6-10d	11,700	630	2145	2690
HU414	4x14	1¼"	2½"	11⅞"	3 9/16"	18-16d	8-10d	13,560	840	2410	3010
HU416	4x16	1¼"	2½"	13⅝"	3 9/16"	20-16d	8-10d	15,420	840	2680	3360
HU66	6x6	1¼"	2"	5"	5½"	8-16d	4-16d	5,020	420	1070	1345
HU68	6x8	1¼"	2"	6⅝"	5½"	10-16d	4-16d	7,430	420	1340	1680
HU610	6x10	1¼"	2"	8⅜"	5½"	14-16d	6-16d	9,850	630	1875	2350
HU612	6x12	1¼"	2½"	10⅛"	5½"	16-16d	6-16d	11,700	630	2145	2690
HU614	6x14	1¼"	2½"	11⅞"	5½"	18-16d	8-16d	13,560	840	2410	3010
HU616	6x16	1¼"	2½"	13⅝"	5½"	20-16d	8-16d	15,420	840	2680	3360
HU24-2	2x4	1¼"	2"	3 1/16"	3⅛"	4-16d	2-10d	2,600	210	535	670
HU26-2	2x6	1¼"	2"	5 1/16"	3⅛"	8-16d	4-10d	5,020	420	1070	1345
HU28-2	2x8	1¼"	2"	6 13/16"	3⅛"	10-16d	4-10d	7,430	420	1340	1680
HU210-2	2x10	1¼"	2"	8 9/16"	3⅛"	14-16d	6-10d	9,850	630	1875	2350
HU212-2	2x12	1¼"	2½"	10 5/16"	3⅛"	16-16d	6-10d	11,700	630	2145	2690
HU214-2	2x14	1¼"	2½"	12 1/16"	3⅛"	18-16d	8-10d	13,530	840	2410	3010

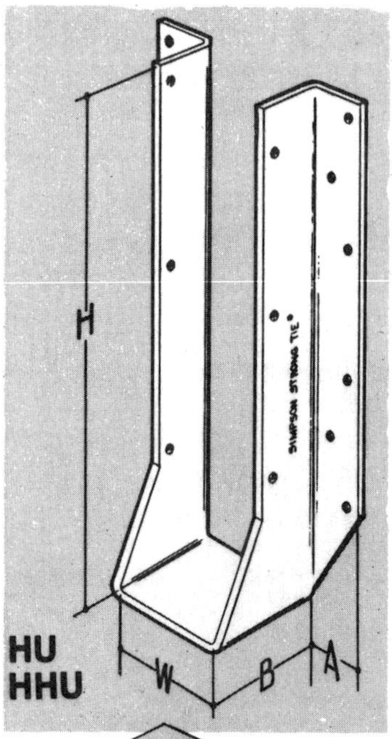

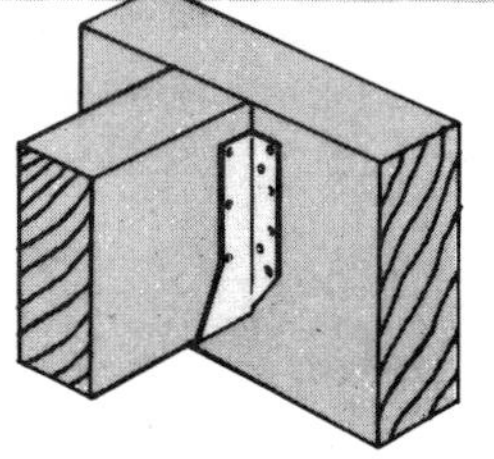

Figure 37-3 (continued)

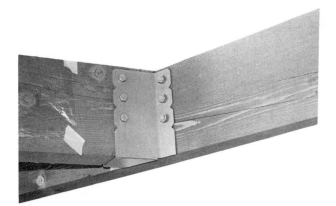

Figure 37-4 Large wooden beams are common on the west coast.

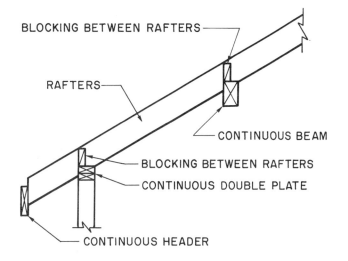

BLOCKING BETWEEN RAFTERS

RAFTERS

CONTINUOUS BEAM

BLOCKING BETWEEN RAFTERS

CONTINUOUS DOUBLE PLATE

CONTINUOUS HEADER

Figure 37-5 Continuous framing is shown by crossing diagonals.
Blocking is shown by a single diagonal line.

The basic methods and materials used for framing on the West coast also differ slightly from those used in other parts of North America. The West has a better supply of tall, straight trees for saw timber; therefore, larger wood beams and posts are common, Figure 37-4. For example, headers over framed wall openings in the Eastern states are usually built of two or more 2xs. The garage door headers for the Town House are 8x16. (See Drawing 3, Garage Floor Framing Plan.)

Another important difference in construction methods is the use of blocking between framing members. Because of the need for extra rigidity in earthquake and high-wind regions, more blocking is required by the building codes in this region. Blocking is placed between rafters and joists at all bearing walls, *purlins* (intermediate supports), and beams. Blocking is shown in section views with a single diagonal line to differentiate it from continuous framing lumber, Figure 37-5.

✓ CHECK YOUR PROGRESS

Can you perform these tasks?

☐ Given drawings for buildings in different parts of the country, list the differences that result from their different locations.

☐ Explain the purpose of each anchor, hold-down, and tie strap shown on a set of drawings.

☐ List the dimensions of the stem and the footing for monolithic foundations.

☐ Describe the reinforcement in a monolithic foundation.

ASSIGNMENT

Refer to the Town House drawings (in the packet) to complete the assignment.

1. How deep is the footing under the garage door in plan B?
2. What reinforcing steel is to be used in the footing at the front of the building in plan B?
3. How deep is the footing under the front entrance in plan B?
4. What size is the beam over the opening between the kitchen and dining room in plan B?
5. How is the kitchen-dining room partition tied to the beam above the opening between these rooms?
6. What is the size and spacing of anchor bolts in the front wall of the house in plan B?
7. How are the joists above the garage in plan B tied to the frame wall under the dining room?
8. What is the size and spacing of the anchor bolts in the privacy walls of the courtyards?
9. What size and spacing is the framing of the pot shelf at the front corner of the kitchen in plan A?
10. How are nonbearing partitions tied to roof trusses above in most places?

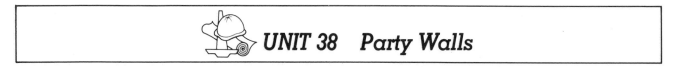

UNIT 38   Party Walls

OBJECTIVES

After completing this unit, you will be able to perform the following tasks:

• Identify and explain fire-resistant construction.
• Identify and explain construction for acoustical insulation.

Multifamily buildings are constructed the same as single-family buildings in most respects. The most important differences between these two classes of buildings is the construction of party walls in multi-family buildings. A *party wall* is a wall that is shared by two separate living units, Figure 38-1. In addition to the usual requirements of a wall, a party wall provides more fire resistance and privacy.

The Town House party walls have stricter fire-resisting requirements in some places than in others. In the Town House drawings, the architect refers to party walls where the fire resistance factor is lower. Where special fire-code requirements must be met, the architect refers to area-separation walls. The terms *party wall* and *area-separation wall* are often used interchangeably. Other architects may use them with reverse meanings.

FIRE-RATED CONSTRUCTION

Fire-rated construction serves two purposes in the event of a fire. It slows the spread of fire, and it maintains structural support longer than non-fire-rated construction.

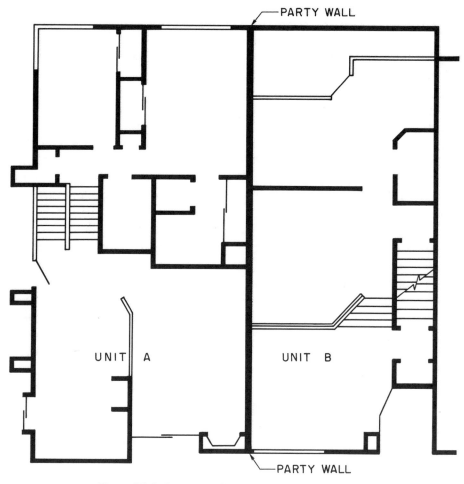

Figure 38-1 A party wall separates two or more units.

Structural Support

In residential buildings, the structural members in fire-rated construction are usually the same as in non-fire-rated construction. These members are protected from fire damage by the nonstructural materials that are used to resist fire spread. Because they are protected from fire damage, these commonly used materials are able to provide structural support longer than unprotected members. Therefore, the important elements of fire-rated residential construction are those that slow the spread of fire.

Slowing Fire Spread

The most obvious way in which a wall, floor, ceiling, or roof can slow the spread of fire is by having a fire-resistant surface. Plaster and gypsum wallboard are fire-resistant materials. They do not burn, and they do not easily transmit the heat of fire to the framing members on the other side. Of course, the thicker the wallboard or plaster, the better it resists the flow of heat. For this reason party walls are often required to have double thicknesses of gypsum wallboard on each side, Figure 38-2.

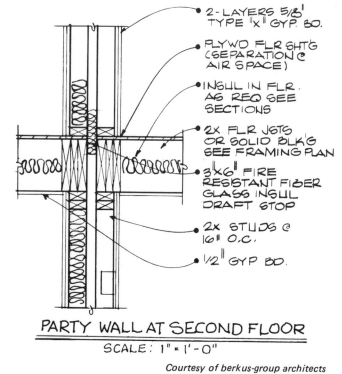

PARTY WALL AT SECOND FLOOR

SCALE: 1" = 1'-0"

Courtesy of berkus-group architects

Figure 38-2 Party walls often have double layers of gypsum wallboard on one or both surfaces.

Notice that the wallboard in Figure 38-2 is indicated as type *x*. This is a special fire-rated wallboard. Although all gypsum plaster is noncombustible, standard wallboard breaks down and crumbles in the high heat of a fire. Fire-rated wallboard holds up much longer in a fire.

Fire-rated construction is frequently used to separate garages and mechanical rooms from living spaces. Party walls are also fire rated to prevent a fire from spreading between housing units. To completely separate the housing units, a fire-rated party wall should extend all the way from the foundation through the roof, Figure 38-3.

Most building codes allow an alternative to extending the party wall through the roof. The fire-rated construction may end at the bottom of the roof as long as the roof is of fire-rated construction, Figure 38-4. This usually means that the

Figure 38-3 Fire walls can be extended through the roof for maximum protection.

roofing material resists fire for as long as the party wall. For example, if the party wall is required to be a one-hour code wall (it resists fire for one hour), the roof must be covered with only material that also resists fire for one hour.

Fire-rated walls must also prevent fire from spreading vertically inside the wall. If left open, the spaces between the studs in a frame wall act like chimneys, allowing flame to spread very quickly from one level to another. To prevent vertical flame spread, stud spaces are not permitted to be more than one story high. The spaces between the studs are closed off with *firestops* at each level, Figure 38-5. Instead of wood firestops, the wall cavity can be blocked off with fire-resistant fiberglass insulation, Figure 38-2.

Openings are usually avoided in fire-rated walls. Where it is necessary to include a door, it is made of fire-resistant material. The fire rating of the door must comply with the building code. Fire-rated doors are often allowed to have slightly lower rating than the walls in which they are installed. Doors in fire walls are equipped with a self-closing mechanism.

SOUND INSULATION

To provide privacy between the housing units, party walls should not allow the sound from one unit to be heard in the next unit. The measurement of the capability of a building element to reduce

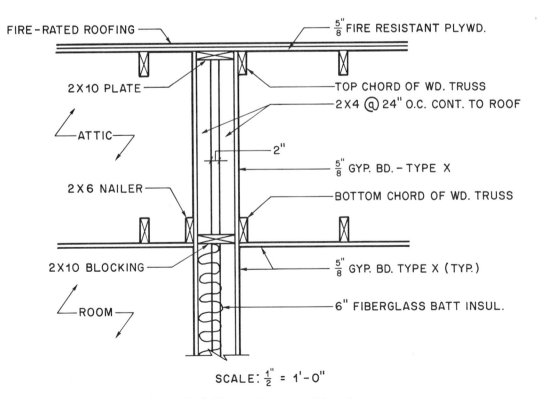

SCALE: $\frac{1}{2}$" = 1'-0"

Figure 38-4 Fire-rated party wall in attic space

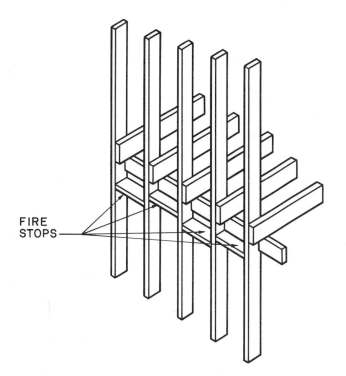

Figure 38-5 Fire stops are installed between studs to prevent vertical drafts inside the wall.

the passage of sound is its *sound transmission classification (STC)*, Figure 38-6.

Sound is transmitted by vibrations in any material: solid, liquid, and gas. To slow the passage of sound, a party wall must reduce the flow of vibrations. The materials used in the construction of most walls vibrate relatively well. Also, they transmit these vibrations to the air inside the wall. The air carries the sound to the other side where it is transmitted to the air on the opposite side of the wall.

The sound transmission classification of a wall can be improved greatly by not allowing the studs to contact both surfaces. This may be accomplished in one of two ways. One method is to attach clips made for sound insulation to the studs; then, fasten the wallboard to these clips. The clips absorb the vibrations. Using the clips and sound-deadening fiberboard results in an STC rating of 52.

An STC of 45 is achieved without clips by using 2x4 studs and 2x6 plates. The studs are staggered on opposite sides of the wall so that no studs contact both surfaces, Figure 38-7. The STC can be increased to 49 by including fiberglass insulation.

STC RATING	EFFECTIVENESS
25	Normal speech can be understood quite easily.
35	Loud speech can be heard, but not understood.
45	Must strain to hear loud speech.
48	Some loud speech can barely be heard.
50	Loud speech cannot be heard.

Figure 38-6 Sound transmission classes

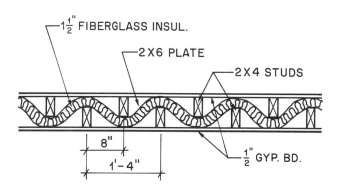

Figure 38-7 Typical sound-insulated wall

✓ CHECK YOUR PROGRESS

Can you perform these tasks?

☐ Identify party walls in a multifamily building.

☐ Identify fire-rated construction on drawings.

☐ List the locations of firestops inside framed walls.

☐ Describe construction to reduce sound transmission.

ASSIGNMENT

Refer to Town House drawings (in the packet) to complete the assignment.

1. List the type, thickness, and number of layers of wallboard at each of the following locations in the Town House:
   a. Area separation wall in plan B dining room
   b. Party wall in plan B kitchen
   c. Party wall in plan B garage
   d. Area separation wall in plan B bedroom #2
   e. Area separation above ceiling in plan B master bedroom
   f. Party wall in plan A living room
   g. Garage ceiling under nook in plan B

2. What is used to stop the vertical spread of fires inside the party walls in the Town House?

3. How is fire prevented from spreading over the top of area separation walls where they meet the Town House roof?

4. What is the total thickness of the party wall at the library in plan B?

5. What is done to stop the transmission of sound through the air space in the party wall of the library in plan B?

6. What is the STC rating of the party wall of the library in plan B?

Mechanical and Electrical
Section 12

| UNIT 39 Plumbing |

OBJECTIVES

After completing this unit, you will be able to perform the following tasks:

- Explain the basic principles of plumbing design.
- Identify the plumbing symbols used on drawings.

All houses include water supply and waste plumbing, and many also have gas plumbing. The water supply plumbing provides fresh hot and cold water to all points of use throughout the house. After the water has been supplied and used, the waste plumbing carries the water and any waste material to the municipal sewer or the septic tank. Waste plumbing is sometimes called *DWV* (drainage, waste, and vent). The gas plumbing supplies fuel gas to the water heater, furnace, range, and other gas-fired fixtures.

PLUMBING MATERIALS

The materials most often used for plumbing are copper, plastic, cast iron, and black iron. A brief description of each is given in the paragraphs that follow.

Copper is frequently used for plumbing because it resists corrosion. However, it is relatively expensive. Copper pipes and fittings may be threaded or smooth for soldered joints.

Plastics for use in plumbing materials are lightweight, noncorrosive, and easily joined. However, most plastic plumbing materials cannot be used around heat above 200°. Also, plastics are not suitable for some applications where high strength is required, although they are sometimes used for general supply and DWV plumbing.

Cast iron is used extensively for DWV plumbing because of its strength and resistance to corrosion. However, it is seldom used for supply plumbing in residential construction.

Black iron is used extensively for gas piping. Black iron pipes and fittings are threaded, so they can be screwed together. Brass fittings are frequently used to join black iron pipe.

Figure 39-1 A coupling is used to permanently join lengths of pipe.

Figure 39-2 A union allows the piping to be disconnected easily.

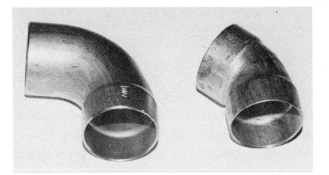

Figure 39-3 90° elbow and 45° elbow

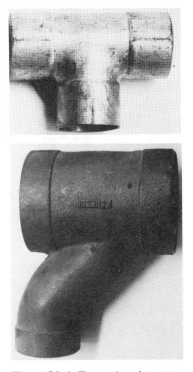

Figure 39-4 Tee and sanitary tee

Fittings

A wide assortment of fittings is used for joining pipe, making turns at various angles, controling the flow of water, and gaining access to the system for service, for example. Most fittings are made of the materials of which pipe is made. Plumbers must be familiar with all types of fittings so they can install their work according to the specifications of the designer.

Couplings, Figure 39-1, are used to join two pipes in a straight line. Couplings are generally used only where a single length of pipe is not long enough.

Unions, Figure 39-2, allow piping to be disconnected easily. A union consists of two parts, with one part being attached to each pipe. Then the two parts of the union are screwed together. When it becomes necessary to disconnect the pipe, the two halves of the union are unscrewed.

Elbows, Figure 39-3, are used to make changes in direction of the piping. Elbows turn either 90° or 45°.

Tees and wyes, Figure 39-4, have three openings to allow a second line to join the first from the side. Tees have a 90° side outlet. Wyes have a 45° side outlet.

Cleanouts, Figure 39-5, allow access to sewage plumbing for cleaning. A cleanout consists of a threaded opening and a matching plug. When cleaning is necessary, the plug is removed and a *snake* or *auger* is run through the line. Cleanouts are installed in each straight run of DWV.

Valves, Figure 39-6, are used to stop, start, or regulate the flow of water. The faucets on a sink or lavatory are a type of valve.

DESIGN OF SUPPLY PLUMBING

In most communities, water is distributed through a system of water mains under or near the

Figure 39-5 A cleanout allows access to the system

Figure 39-6 Each branch of piping should include a shutoff valve.

street. When a new house is constructed, the municipal water department *taps* (makes an opening in) this main. The supply plumbing from the municipal tap to the house is installed by plumbers who work for the plumbing contractor.

The main supply line entering the house must be larger in diameter than the individual branches running from the main to each point of use. There are two basic reasons for this. First, water develops friction as it flows through pipes, and the greater size reduces this friction in the long supply line. Second, when more than one fixture is used at a time, the main supply must provide adequate flow for both. Generally, the main supply line for a one- or two-family house is 3/4-inch or 1-inch pipe.

At the point where the main supply enters the building, a water meter is installed. The water meter measures the amount of water used. The municipal water department relies on this meter to de-termine the proper water bill for the building. The main water shutoff valve is located near the water meter.

From the main shutoff, the supply continues to the water heater. Somewhere between the water heater and the meter, a tee is installed to supply cold water to the house. From the main supply, branches are run to each fixture or point of use.

When a valve is suddenly closed at a fixture, the water tends to slam into the closed valve. This causes a sudden pressure buildup in the pipes and may cause the pipes to *hammer* (a sudden shock in the supply piping). To prevent this, an air chamber is installed at a high point in the system. An *air chamber* consists of a short vertical section of pipe that is filled with trapped air, Figure 39-7. When a valve is suddenly closed, the air chamber acts as a shock absorber. Although water cannot be compressed, air can be. When the pressure tends

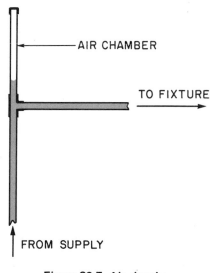

AIR CHAMBER

TO FIXTURE

FROM SUPPLY

Figure 39-7 Air chamber

to build up suddenly, the air in this chamber compresses and cushions the resulting shock. Figure 39-8 shows a typical water supply system.

DESIGN OF WASTE PLUMBING

The main purpose of DWV, as stated earlier, is to remove water after it has been used and to carry away solid waste. To accomplish this, a branch line runs from each fixture to the main building sewer. The main building sewer carries the *effluent* (fouled water and solid waste) to the municipal sewer or septic system.

Traps

The sewer contains foul-smelling, germ-ladened gases which must be prevented from enter-

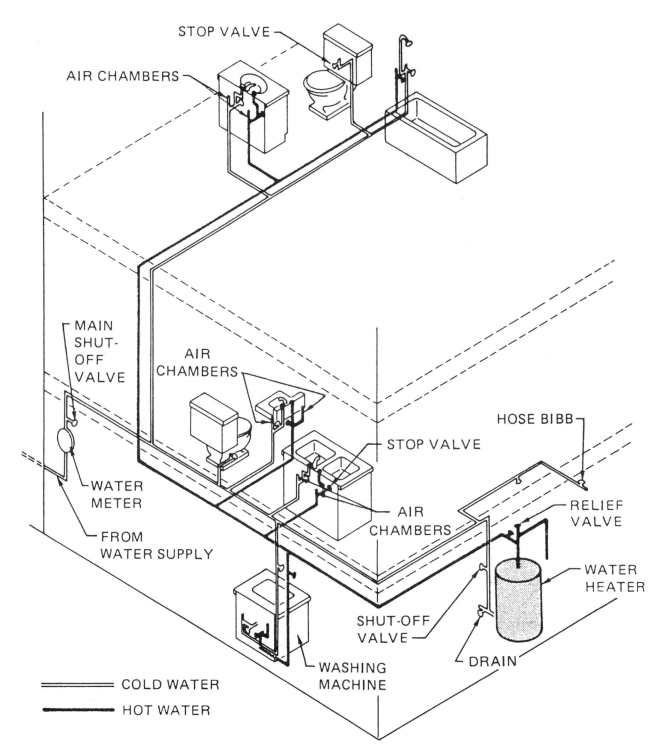

Figure 39-8 Hot- and cold-water piping

ing the house. If waste water simply emptied into the sewer from the pipe, this sewer gas would be free to enter the building. To prevent this from happening, a trap is installed at each fixture. A *trap* is a fitting that naturally fills with water to prevent sewer gas from entering the building, Fig-

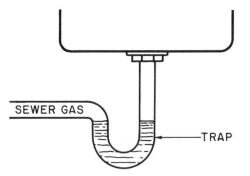

Figure 39-9 A trap fills with water to prevent sewer gas from entering the building.

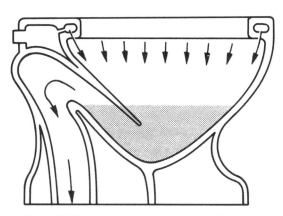

Figure 39-10 A water closet has a built-in trap.

ure 39-9. Not all traps are easily seen. Some fixtures, such as *water closets* (toilets), have built-in traps, Figure 39-10.

Vents

As the water rushes through a trap, it is possible for a siphoning action to be started. (The air pressure entering the fixture drain is higher than that on the other side of the trap. This forces the water out of the trap.)

To prevent DWV traps from siphoning, a vent is installed near the outlet side of the trap. The vent is an opening that allows air pressure to enter the system and break the suction at the trap, Figure 39-11. Because the vent allows sewer gas to pass freely, it must be vented to the outside of the building. Usually all of the fixtures are vented into one main vertical pipe, through the roof, Figure 39-12.

PLUMBING PLANS

For residential construction, the architect does not usually include a plumbing plan with the set of working drawings. The floor plan shows all of the plumbing fixtures by standard symbols. These symbols are easily recognized, because they resemble the actual fixture. The dimensions of the fixtures are provided by the manufacturer on rough-in sheets, Figure 39-13.

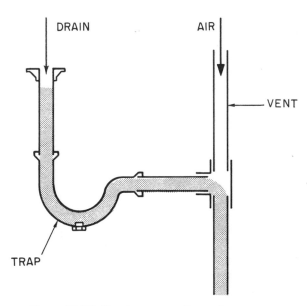

Figure 39-11 Venting a trap allows air to enter
the system and prevents siphoning.

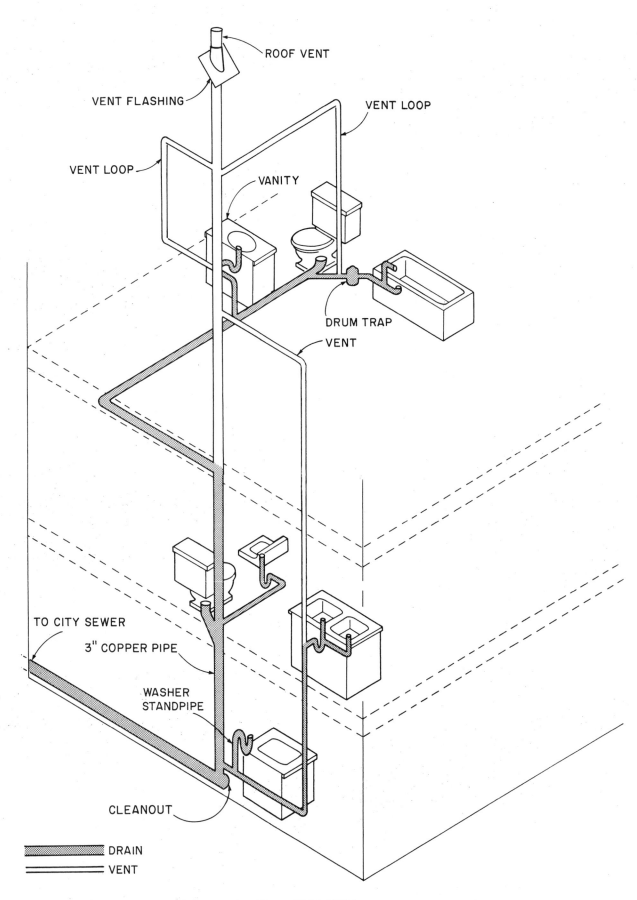

Figure 39-12 DWV system

ROUGHING-IN DIMENSIONS

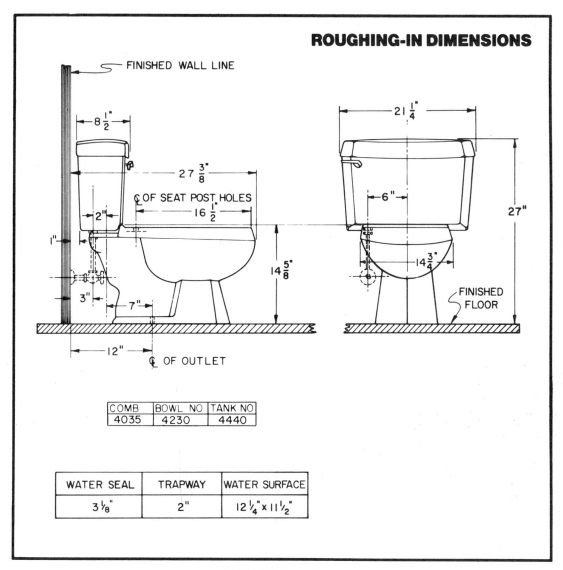

COMB	BOWL NO	TANK NO
4035	4230	4440

WATER SEAL	TRAPWAY	WATER SURFACE
$3\frac{1}{8}$"	2"	$12\frac{1}{4}$" x $11\frac{1}{2}$"

SPECIFICATIONS

FIXTURE SPECIFICATION

☐ **U/R 4035 NEW VENUS** — vitreous china floor-mounted siphon jet close-coupled water-saver combination for 12″ rough with regular-rim bowl, requires only 3½ gallons per flush — tank furnished with Fluidmaster 400A ballcock and adjustable tilt flush valve — (2) bolt caps and U/R 8085 lift-off seat included.

COLOR SPECIFICATION

☐ Acid Resisting White

☐ Acid Resisting Color _____
(show U/R color desired)

Universal-Rundle
Bath Products
® 303 NORTH STREET • NEW CASTLE, PA 16103

Figure 39-13 Typical manufacturer's rough-in sheet

If the building and the plumbing are fairly simple, plumbers may prepare estimates and bids, and complete the work from the symbols on the floor plan only. For more complex houses, the plumbing contractor usually draws a plumbing isometric, Figure 39-14, or a special plumbing plan. The drawing set with this textbook includes a plumbing plan and details for plan A of the Town House. This sheet includes more details than would normally be found on a plumbing plan for a single-family housing unit. The extra detail is included here to help you understand the plumbing plan.

Plumbing plans show each kind of piping by a different symbol. Common plumbing symbols are shown in the Appendix. It will help you understand the plumbing plan if you trace each kind of piping from its source to each fixture. For example, trace the gas piping for the Town House. The gas lines can be recognized by the letter G in the piping symbol. The gas supply is shown as a broken line until it is inside the garage. Broken lines are used to indicate that the pipe is underground or concealed by construction. Although it is not noted on this plan, the plumbing contractor should know that the building code requires the gas line to be run in a sleeve where it passes through the foundation and the concrete slab, Figure 39-15. Just inside the garage wall, the broken line changes to a solid line. At this point, a symbol indicates that the solid line (exposed piping) turns down or away. Here, at this point, the gas piping runs above the concrete slab and along the garage wall. A call-out on this line indicates that the diameter of the pipe is 3/4-inch. At the back of the garage, the gas line has a *T*. Both of the outlets of this *T* are 1/2-inch in diameter. One side of the *T* supplies the forced air unit (F.A.U.). The other side of the *T* continues around behind the F.A.U. to another *T*,

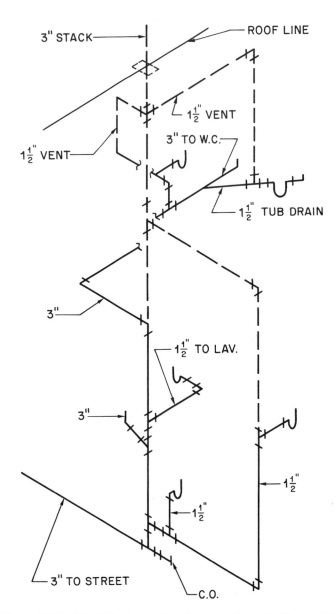

Figure 39-14 Single-line isometric of system shown in Figure 39-12

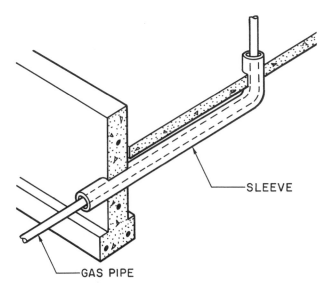

SLEEVE

GAS PIPE

Figure 39-15 Sleeve for running gas piping under and through concrete

and then to the water heater. The side outlet of the second *T* supplies a log lighter in the fireplace. This branch is drawn on the first floor plumbing plan. Notice that the log-lighter branch is reduced further to 1/4-inch diameter.

You should trace each type of plumbing in a similar manner to be sure you understand it. Refer to the details on the drawing for clarification of the complex areas. As you trace each line, look for the following:

- kind of plumbing (hot water, cold water, waste)
- diameter
- fittings
- exposed or concealed
- where line passes through building surfaces

√ CHECK YOUR PROGRESS

Can you perform these tasks?

☐ List the plumbing fittings shown on a plumbing plan.

☐ List the material and size of all piping and fittings for waste plumbing.

☐ List the material and size of all piping and fittings for supply plumbing.

☐ Use manufacturers' literature to list dimensions for the location of plumbing to fixtures.

ASSIGNMENT

Refer to the Town House drawings (in the packet) to complete the assignment.

1. What size pipe supplies the washing machine?

2. What size is the cold water supply to the water heater?

3. What size is the cold water branch to the lavatory in bathroom #2?

4. At what point does the 3/4-inch cold water branch to the kitchen reduce to 1/2-inch for the hose bibb?

5. Does the cold water supply turn up or down as it leaves the bathroom area to supply the kitchen area?

6. List each of the fittings that water will pass through after it drains out of the master bathroom lavatory.

7. List each of the fittings that water will pass through to flow from the main shutoff at the building line to the shutoff on the supply side of the water heater.

8. What size is the waste piping from the water closet in the master bath?

9. What size is the waste piping from the kitchen sink?

10. What size is the waste piping from the washing machine?

UNIT 40   Heating and Air Conditioning

OBJECTIVES

After completing this unit, you will be able to perform the following tasks:

- Locate and identify the heating and air-conditioning equipment shown on the mechanical plan for a house.

- Explain how heat or conditioned air is distributed throughout a building.

FORCED AIR SYSTEMS

One of the most common systems for climate control circulates the air from the living spaces through or around heating or cooling devices. A fan forces the air into large sheet metal or plastic pipes called *ducts*. These ducts connect to openings, called *diffusers*, in the room. The air enters the room and either heats it or cools it as needed.

Air then flows from the room through another opening into the *return duct*. The return duct directs the air from the room over a heating or cooling device, depending on which is needed. If cool air is required, the return air passes over the surface of a cooling coil. If warm air is required, the return air is either passed over the surface of a *combustion chamber* (the part of a furnace where fuel is burned) or a heating coil. Finally, the conditioned air is picked up again by the fan and the air cycle is repeated, Figure 40-1.

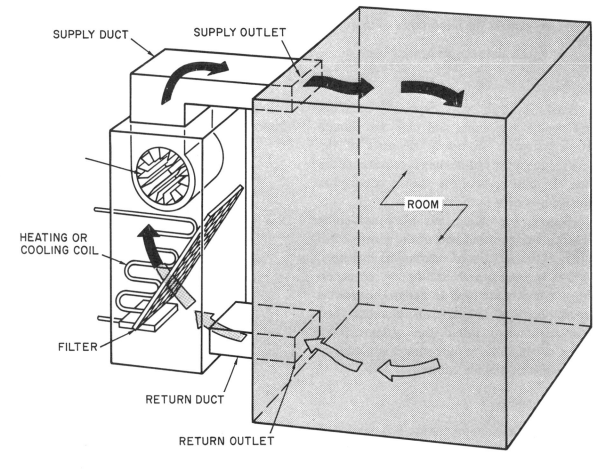

Figure 40-1   The air cycle in a forced-air system

Furnace

If the air cycle just described is used for heating, the heat is generated in a furnace. Furnaces for residential heating produce heat by burning fuel oil or natural gas, or from electric heating coils. If the heat comes from burning fuel oil or natural gas, the *combustion* (burning) takes place inside a combustion chamber. The air to be heated does not enter the combustion chamber, but absorbs heat from the outer surface of the chamber. The gases given off by the combustion are vented through a chimney. In an electric furnace, the air to be heated is passed directly over the heating coils. This type of furnace does not require a chimney.

Refrigeration Cycle

If the air from the room is to be cooled, it is passed over a cooling coil. The most common type of residential cooling system is based on the following two principles:

- As liquid changes to vapor, it absorbs large amounts of heat.
- The boiling point of a liquid can be changed by changing the pressure applied to the liquid. This is the same as saying that the temperature of a liquid can be raised by increasing its pressure and lowered by reducing its pressure.

The principal parts of a refrigeration system are the cooling coil (*evaporator*), *compressor* (an air pump), the *condenser*, and the *expansion valve*, Figure 40-2.

Keep in mind that common refrigerants can boil (change to a vapor) at very low temperatures — some as low as 21°F below zero. Also remember that a liquid boils at a higher temperature when it is under pressure.

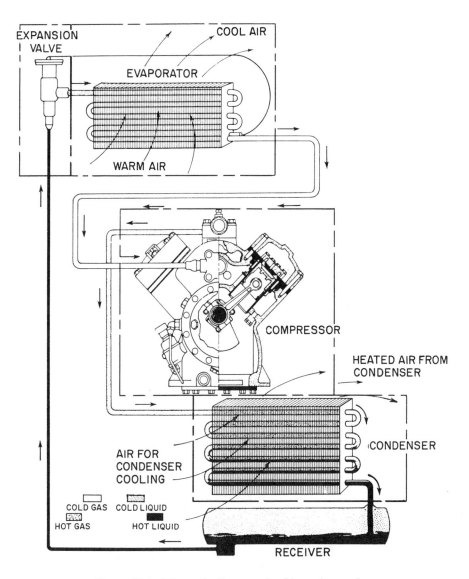

Figure 40-2 Schematic diagram of refrigeration cycle

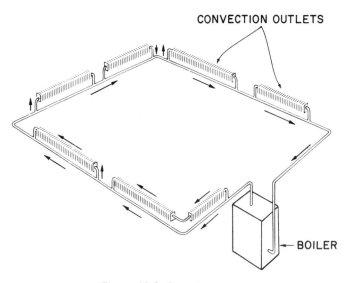

Figure 40-3 One-pipe system

The warm air from the ducts is passed over the evaporator. As the cold refrigerant liquid moves through the evaporator coil, it picks up heat from the warm air. As the liquid picks up heat, it changes to a vapor.

The heated refrigerant vapor is then drawn into the compressor where it is put under high pressure. This causes the temperature of the vapor to rise even more.

Next, the high-temperature, high-pressure vapor passes to the condenser where the heat is removed. In residential systems this is done by blowing air over the coils of the condenser. As the condenser removes heat, the vapor changes to a liquid. It is still under high pressure, however.

From the condenser, the refrigerant flows to the expansion valve. As the liquid refrigerant passes through the valve, the pressure is reduced. This lowers the temperature of the liquid still further, so that it is ready to pick up more heat.

The cold, low-pressure liquid then moves to the evaporator. The pressure in the evaporator is low enough to allow the refrigerant to boil again and absorb more heat from the air passing over the coil of the evaporator.

HOT-WATER SYSTEM

Many buildings are heated by hot-water systems. In a hot-water system, the water is heated in an oil or gas-fired boiler, then circulated through pipes to radiators or convectors in the rooms. The boiler is supplied with water from the fresh water supply for the house. The water is circulated around the combustion chamber where it absorbs heat.

In some systems, one pipe leaves the boiler and runs through the building and back to the boiler. In this type, called a *one-pipe system*, the heated water leaves the supply, is circulated through the

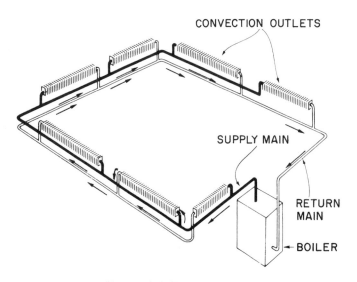

Figure 40-4 Two-pipe system

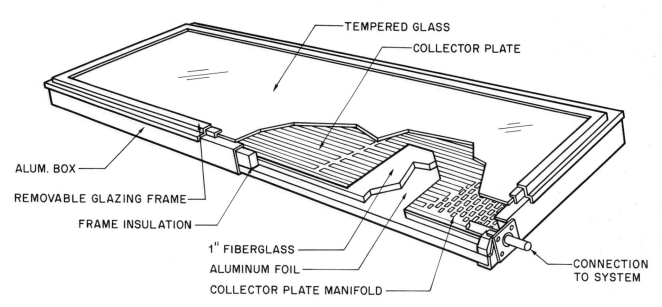

TEMPERED GLASS

COLLECTOR PLATE

ALUM. BOX

REMOVABLE GLAZING FRAME

FRAME INSULATION

1" FIBERGLASS

ALUMINUM FOIL

COLLECTOR PLATE MANIFOLD

CONNECTION TO SYSTEM

Figure 40-5 Flat-plate solar collector

outlet, and is returned to the same pipe, Figure 40-3. Another type, the *two-pipe system*, uses two pipes running throughout the building. One pipe supplies heated water to all of the outlets. The other is a return pipe which carries the water back to the boiler for reheating, Figure 40-4.

Hot-water systems use a pump, called a *circulator*, to move the water through the system. The water is kept at a temperature of 150° – 180°F in the boiler. When heat is needed, the thermostat starts the circulator.

Solar Collectors

Solar energy can also be used as a source of heat for a hot-water system. All that must be done is to concentrate the heat of the sun on pipes carrying water to the heating system.

A flat-plate solar collector is a box that absorbs the heat of the sun and transfers it to water or antifreeze. The top of the box is made of glass or specially formulated plastic. As the rays of the sun pass through the glass top, they strike the col-

lector plate. The collector plate is made up of a network of tubes that carry the water or antifreeze. Below the collector plate, the box is filled with insulation. At each end of the collector plate a larger tube, called a *manifold*, connects the tubes to the system, Figure 40-5.

In operation, solar flat-plate collectors are placed where the rays of the sun strike their surface in the winter, Figure 40-6. The energy from the sun heats the antifreeze solution in the pipes of the collector. This warmed liquid is pumped to a large tank near the regular boiler. The water entering the boiler is circulated through a coil in the tank. This preheats the water going into the boiler, Figure 40-7. The use of solar collectors can reduce the fuel consumption of a boiler by as much as 80 percent.

ELECTRIC RESISTANCE HEAT

There are a number of heating system designs that rely on electric heating elements located in each room. Some such systems have electric heating

Figure 40-6 The collectors are mounted where they will get direct sunlight.

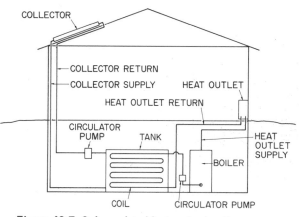

COLLECTOR

COLLECTOR RETURN

COLLECTOR SUPPLY

HEAT OUTLET

HEAT OUTLET RETURN

CIRCULATOR PUMP

TANK

HEAT OUTLET SUPPLY

BOILER

COIL

CIRCULATOR PUMP

Figure 40-7 Solar-assisted hot-water heating system

elements embedded in the floor or ceiling. In these systems, the surface of the room is heated. Another kind of electric heat has heating outlets similar to those used for hot-water heat.

HEATING AND AIR-CONDITIONING EQUIPMENT ON DRAWINGS

As with plumbing, architects do not usually prepare heating and air-conditioning drawings for residential construction. The architect specifies the type of heating and air conditioning to be used. The HVAC (heating, ventilating, and air-conditioning) contractor prepares required drawings as needed for personal use only.

The mechanical drawing for the Town House includes some limited HVAC information. This is enough information so the other trades will know what they may encounter. This mechanical drawing shows the following:

- Location of the air-conditioning compressor
- Location of the central forced-air unit
- How the forced-air unit is vented
- Piping between the compressor and forced-air unit
- Approximate locations of diffusers
- Approximate locations of return-air grilles

✓ CHECK YOUR PROGRESS

Can you perform these tasks?

☐ Describe the type of heating system to be installed.

☐ List the locations and sizes of heating and air-conditioning diffusers and return-air grilles.

☐ List the locations of hot-water convection outlets.

☐ Describe the location of major heating and air-conditioning equipment.

☐ Trace the flow of water and antifreeze through a solar collector system.

ASSIGNMENT

Refer to the drawings for the Town House, plan A.

1. Where is the air-conditioning compressor located?

2. Where is the tubing which connects the air-conditioning compressor to the forced-air unit?

3. What size tubing connects the compressor to the forced-air unit?

4. How many diffuser outlets supply conditioned air to the rooms?

5. How many return-air grilles are there?

6. Where is the forced-air unit located?

7. What kind of fuel does the forced-air unit use for heating?

8. Where are the sizes of the diffuser outlets given?

OBJECTIVES

After completing this unit, you will be able to perform the following tasks:

• Identify the electrical symbols shown on a plan.

• Explain how the lighting circuits are to be controlled.

CURRENT, VOLTAGE, RESISTANCE, AND WATTS

To understand the wiring in a building you should know how electricity flows. Electricity is energy. To do any work (turn a motor, light a lamp, or produce heat) the electrical energy must have movement. This movement is called *current*. The amount of current is measured in *amperes*, sometimes called *amps*. A single household-type light bulb requires a current of slightly less than 1 ampere. An electric water heater might require 50 amperes.

The amount of force of pressure causing the current to flow affects the amount of current. The force behind an electric current is called *voltage*. If 115 volts causes a current flow of 5 amperes, 230 volts will cause a current flow of 10 amperes.

The ease with which the current is able to flow through the device also affects the amount of current. The **ease** or difficulty with which the current flows through the device is called the *resistance* of that device. As the resistance goes up, the current flow goes down. As the resistance goes down, the current flow goes up.

The amount of work the electricity can do in any device depends on both the amount of current (amps) and the force of the current (volts). Electrical work is measured in *watts*. The number of watts of power in a device can be found by multiplying the number of amperes by the number of volts. Stated another way the current flowing in a device can be found by dividing the number of watts by the voltage. For example, how much current flows through a 1500-watt heater at 115 volts? 1500 divided by 115 equals about 13 amperes. Figure 41-1 shows the current, wattage, and voltage of some typical electrical equipment.

Figure 41-1 Current, voltage, and power ratings of some typical electrical devices

DEVICE	AMPERES	VOLTS	WATTS
Ceiling light fixture	1.3	115	150
Vacuum Cleaner	6.1	115	700
Radio	0.4	115	4
Clock	0.4	115	4
Dishwasher	8.7	115	1,000
Toaster	13	115	1,500
Cook Top	32	230	7,450
Oven	29	230	6,600
Clothes Dryer	25	230	5,750
Washing Machine	10	115	1,150
Garbage Disposal	7.4	115	850

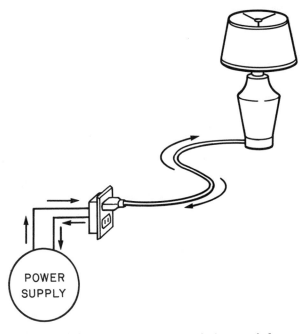

Figure 41-2 A complete circuit includes a path from the supply to the device and back again.

CIRCUITS

In order for current to flow, it must have a continuous path from the power source, through the electrical device, and back to its source. This complete path is called a *circuit*, Figure 41-2.

Many circuits include one or more switches. A switch allows the continuous path to be broken, Figure 41-3. By using two 3-way switches, the circuit can be controlled from two places, Figure 41-4. When the circuit is broken by a switch, a broken wire, or for any other reason, it is said to be *open*.

Any material that carries electric current is called a *conductor*. In Figure 41-2 each of the wires is a conductor. When two or more wire conductors are bundled together, they make a cable, Figure 41-5.

Most house wiring is done with cables containing the needed conductors plus one ground conductor. The ground conductor does not normally carry current. The *ground*, as it is usually abbreviated, connects all of the electrical devices in the house to the ground. If, because of some malfunction, the voltage reaches a part of the device that someone might touch, the ground protects them from a serious shock. The current that might

otherwise flow through the person follows the ground conductor to the earth. The earth actually carries this current back to the generating station.

Additional protection against serious shock can be provided by using a *ground-fault circuit interrupter* (GFCI or GFI.) A GFI is a device that measures the flow of current in the hot (supply) conductor and the neutral (return) conductor. If a faulty device allows some of the current to flow through a person rather than the neutral conductor, the GFI stops all current flow immediately. GFIs are so effective that the National Electric Code requires their use on circuits for outlets installed outdoors, in bathrooms, in garages, and near any other water hazards.

The electrical service entrance was discussed earlier in Unit 13. The service feeder cable ends at a distribution panel. From the distribution panel, the electrical system is split up into several branch circuits, Figure 41-6. Each branch circuit includes a circuit breaker or fuse. The circuit breaker or fuse opens the circuit if the current flow exceeds the rated capacity of the circuit. Branch circuits for special equipment such as water heaters and air conditioners serve that piece of equipment only. Branch circuits for small appliances and

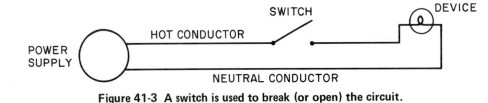

Figure 41-3 A switch is used to break (or open) the circuit.

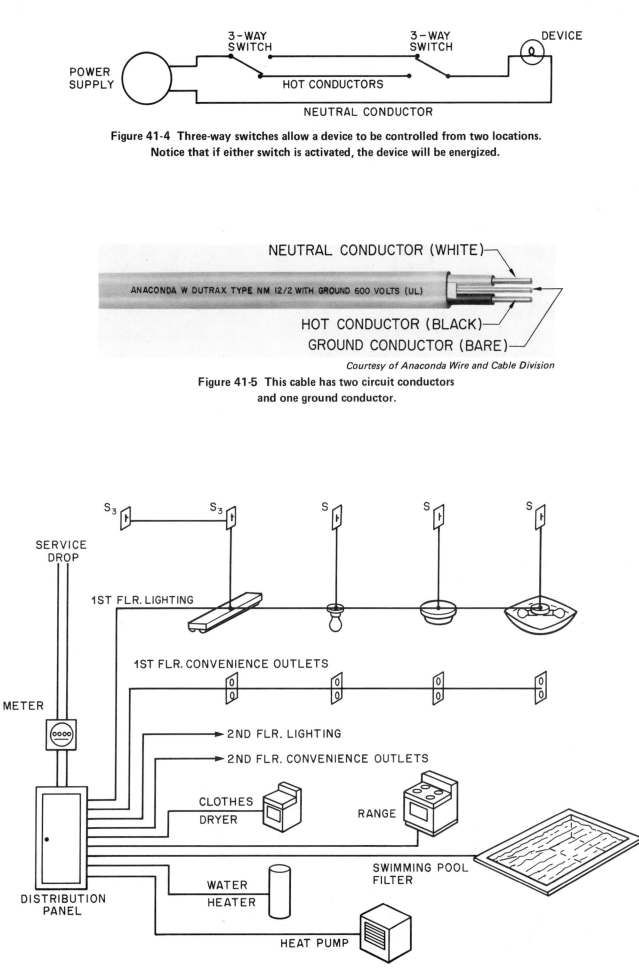

Figure 41-4 Three-way switches allow a device to be controlled from two locations. Notice that if either switch is activated, the device will be energized.

ANACONDA W DUTRAX TYPE NM 12/2 WITH GROUND 600 VOLTS (UL)

NEUTRAL CONDUCTOR (WHITE)

HOT CONDUCTOR (BLACK)

GROUND CONDUCTOR (BARE)

Courtesy of Anaconda Wire and Cable Division

Figure 41-5 This cable has two circuit conductors and one ground conductor.

Figure 41-6 The electrical service is split up into branch circuits at the distribution panel.

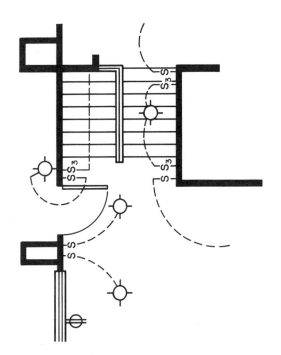

Figure 41-7 Switch legs on a plan

miscellaneous use may serve several outlets. Branch circuits for lighting are restricted to lighting only, but a single circuit may serve several lights. Lighting circuits also include switches to turn the lights on and off.

The National Fire Protection Association publishes the NATIONAL ELECTRICAL CODE which specifies the design of safe electrical systems. Electrical engineers and electricians must know this code which is accepted as the standard for all installations. The following are among the items it covers:

- Kinds and sizes of conductors
- Locations of outlets and devices
- Overcurrent protection (fuses and circuit breakers)
- Number of conductors allowed in a box
- Safe construction of devices
- Grounding
- Switches

The specifications for the structure indicate such things as the type and quality of the equipment to be used, the kind of wiring, and any other information that is not given on the drawings.

However, electricians must know the NATIONAL ELECTRICAL CODE and any state or local codes that apply because specifications sometimes refer to these codes.

ELECTRICAL SYMBOLS ON PLANS

The drawings for residential construction usually include electrical information on the floor plans. Only the symbols for outlets, light fixtures, switches, and switch wiring are included. The exact location of the device may not be dimensioned. The position of the device is determined by the electrician after observing the surrounding construction. It should also be noted that all wiring is left to the judgement of the electrician and the regulations of the electrical codes. Switch wiring for light fixtures is included only to show which switches control each light fixture. Switch wiring is shown by a broken line connecting the device and its switch, Figure 41-7.

In rooms without a permanent light fixture, one or more convenience outlets may be split wired and controlled by a switch. In split wiring one-half of the outlet is always hot; the other half can be opened by a switch, Figure 41-8.

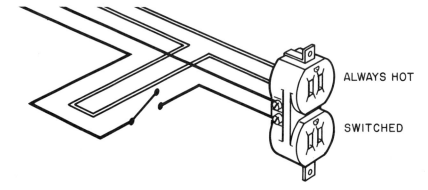

Figure 41-8 Split-wired outlet

ALWAYS HOT

SWITCHED

✓ CHECK YOUR PROGRESS

Can you perform these tasks?

☐ List all of the convenience outlets, special outlets, light fixtures, and switches shown on plans.

☐ Describe the location of the main electrical distribution panel.

ASSIGNMENT

Refer to drawings for the Town House, floor plan A, including the garage.

1. How many light fixtures are shown on the floor plan? (Include the garage.)

2. How many switches are shown on the floor plan? (Include the garage.)

3. How many duplex outlets are shown?

4. Briefly describe the location of each split-wired outlet and the switch or switches that control each.

5. List five pieces of equipment shown on the floor plan that probably require separate branch circuits.

6. How many outlets are to have ground-fault circuit interrupters included in their circuits?

7. What is the location of the switch, or switches, that control(s) the light over the stairs to the bedroom level?

8. What is the location of the switch, or switches, that control(s) the light fixture over the stairs to the garage?

9. Where is the smoke detector located?

10. Where is each of the two telephone outlets?

Glossary

Aggregate Hard materials such as sand and crushed stone used to make concrete

Ampere (AMP) Unit of measure of electric current

Anchor Bolt A bolt placed in the surface of concrete for attaching wood framing members

Apron Concrete slab at the approach to a garage door — Also the wood trim below a window stool

Architect's Scale A flat or triangular scale used to measure scale drawings

Ash Dump A small metal door in the bottom of a fireplace

Awning Window A window that is hinged near the top, so the bottom opens outward

Backfill Earth placed against a building wall after the foundation is in place

Backsplash The raised lip on the back edge of a countertop to prevent water from running down the backs of the cabinets

Balloon Framing Type of construction in which the studs are continuous from the sill to the top of the wall — Upper floor joists are supported by a let-in ribbon.

Balusters Vertical pieces which support a railing

Balustrade An assembly of balusters and a handrail

Batt Insulation Flexible, blanketlike pieces, usually of fiberglass, used for thermal or sound insulation

Batten Narrow strip of wood used to cover joints between boards of sheet materials

Batten Boards An arrangement of stakes and horizontal pieces used to attach lines for laying out a building

Beam Any major horizontal structural member

Beam Pocket A recessed area to hold the end of a beam in a concrete or masonry wall

Board Foot One hundred forty-four cubic inches of wood or the amount contained in a piece measuring 12"x12"x1"

Bottom Chord The bottom horizontal member in a truss

Box Sill The header joist nailed across the ends of floor joists at the sill

Branch Circuit The electrical circuit that carries current from the distribution panel to the various parts of the building

British Thermal Unit (BTU) The amount of heat required to raise the temperature of one pound of water one degree Fahrenheit

Building Lines The outside edge of the exterior walls of a building

Casement Window A window that is hinged at one side so the opposite side opens outward

Casing The trim around a door or window

Centerline An actual or imaginary line through the exact center of any object

Cleanout A pipe fitting with a removable plug that allows for cleaning the run of piping in which it is installed or an access door at the bottom of a chimney

Collar Beam Horizontal members that tie opposing rafters together, usually installed about half way up the rafters

Column A metal post to support an object above

Common Rafter A rafter extending from the top of the wall to the ridge

Concrete Building material consisting of fine and coarse aggregates bonded together by portland cement

Conductor Electrical wire — A cable may contain several conductors.

Contour Lines Lines on a topographic map or site plan to describe the contour of the land

Contract Any agreement in writing for one party to perform certain work and the other party to pay for the work

Convenience Outlet Electrical outlet provided for convenient use of lamps, appliances, and other electrical equipment

Cornice The construction which encloses the ends of the rafters at the top of the wall

Cornice Return The construction where the level cornice meets the sloping rake cornice

Course A single row of building units such as concrete blocks or shingles

Cove Mold Concave molding used to trim an inside corner

Damper A door installed in the throat of a fireplace to regulate the draft

Dampproofing Vapor barrier or coating on foundation walls or under concrete slabs to prevent moisture from entering the house

Datum A reference point from which elevations are measured

Detail A drawing showing special information about a particular part of the construction — Details are usually drawn to a larger scale than on other drawings and are sometimes section views.

Dormer A raised section in a roof to provide extra headroom below

Double-hung Window A window consisting of two sash which slide up and down past one another

Drip Cap A wood ledge over wall openings to prevent water from running back under the frame or trim around the opening

Drip Edge Metal trim installed at the edge of a roof to stop water from running back under the edge of the roof deck

Drywall Interior wall construction using gypsum wallboard

Elevation A drawing that shows vertical dimensions — It may also be the height of a point, usually in feet above sea level.

Fascia The part of a cornice that covers the ends of the rafters

Firestop Blocking or noncombustible material between wall studs to prevent vertical draft and flamespread

Flashing Sheet metal used to cover openings and joints in walls and roofs

Float To level concrete before it begins to cure — Floating is done with a tool called a float.

Floor Plan A drawing showing the arrangement of rooms, the locations of windows and doors, and complete dimensions — A floor plan is actually a horizontal section through the entire building.

Flue The opening inside a chimney — The flue is usually formed by a terra cotta flue liner.

Flush Door A door having flat surfaces

Footing The concrete base upon which the foundation walls are built

Frieze A horizontal board beneath the cornice and against the wall above the siding

Frost Line The maximum depth to which frost penetrates the earth

Furring Narrow strips of wood attached to a surface for the purpose of creating a plumb or level surface for attaching the wall, ceiling, or floor surface

Gable The triangular area between the roof and the top plate of the walls at the ends of a gable roof

Gable Studs The studs placed between the end rafters and the top plates of the end walls

Gauge A standard unit of measurement for the diameter of wire or the thickness of sheet metal

Girder A beam which supports floor joists

Grout A thin mixture of high-strength concrete or mortar

Gypsum Wallboard Drywall materials made of gypsum encased in paper to form boards

Header A joist fastened across the ends of regular joists in an opening, or the framing member above a window or door opening

Hearth Concrete or masonry apron in front of a fireplace

Hip Outside corner formed by intersecting roofs

Hip Rafter The rafter extending from the corner of a building to the ridge at a hip

Hose Bibb An outside faucet to which a hose can be attached

Insulated Glazing Two or more pieces of glass in a single sash with air space between them for the purpose of insulation

Isometric A kind of drawing in which horizontal lines are 30 degrees from true horizontal and vertical lines are vertical

Jack Rafter Rafter between the outside wall and a hip rafter or the ridge and a valley rafter

Jamb Side members of a door or window frame

Joists Horizontal framing members that support a floor or ceiling

Lintel Steel or concrete member that spans a clear opening — usually found over doors, windows, and fireplace openings

Masonry Cement Cement which is specially prepared for making mortar

Mil A unit of measure for the thickness of very thin sheets — One Mil equals 1.001".

Miter A 45-degree cut so that two pieces will form a 90-degree corner

Mortar Cement and aggregate mixture for bonding masonry units together

Mullion The vertical piece between two windows that are installed side by side — Window units that include a mullion are called mullion windows.

Muntin Small vertical and horizontal strips that separate the individual panes of glass in a window sash

Nailer A piece of wood used in any of several places to provide a nailing surface for other framing members

Nominal Size The size by which a material is specified — The actual size is often slightly smaller.

Nosing The portion of a stair tread that projects beyond the riser

Orthographic Projection A method of drawing that shows separate views of an object

Panel Door A door made up of panels held in place by rails and stiles

Parging A thin coat of portland-cement plaster used to smooth masonry walls

Penny Size The length of nails

Perimeter Drain (also, **Footing Drain**) An underground drain pipe around the footings to carry ground water away from the building

Pilaster A masonry or concrete pier built as an integral part of a wall

Pitch Refers to the steepness of a roof — The pitch is written as a fraction with the rise over the span.

Plate The horizontal framing members at the top and bottom of the wall studs

Platform Framing (also called **Western Framing**) A method of framing in which each level is framed separately — The subfloor is laid for each floor before the walls above it are formed.

Plenum A chamber within a forced-air heating system that is pressurized with warm air

Plumb Truly vertical or true according to a plumb bob

Portland Cement Finely powdered limestone material used to bond the aggregates together in concrete and mortar

R-value The ability of a material to resist the flow of heat

Rafter The framing members in a roof

Rail The horizontal members in a door, sash, or other panel construction

Rake The sloping cornice at the end of a gable roof

Resilient Flooring Vinyl, vinyl-asbestos, and other manmade floor coverings that are flexible yet produce a smooth surface

Ridge Board The framing member between the tops of rafters which runs the length of the ridge of a roof

Rise The vertical dimension of a roof or stair

Riser The vertical dimension of one step in a stair — The board enclosing the space between two treads is called a riser.

Rowlock Position of bricks in which the bricks are laid on edge

Run The horizontal distance covered by an inclined surface such as a rafter or stair

Sash The frame holding the glass in a window

Saturated Felt Paperlike felt which has been treated with asphalt to make it water resistant

Screed A straight board used to level concrete immediately after it is placed

Section View A drawing showing what would be seen by cutting through a building or part

Setback The distance from a street or front property line to the front of a building

Sheathing The rough exterior covering over the framing members of a building

Shim Thin pieces, usually wood, used to build up low spots between framing and finish work

Sill The framing member in contact with a masonry or concrete foundation

Sill Sealer Compressible material used under the sill to seal any gaps

Site Constructed Built on the job

Site Plan The drawing that shows the boundaries of the building, its location, site utilities

Sliding Window A window with two or more sash that slide horizontally past one another

Soffit The bottom surface of any part of a building, such as the underside of a cornice or lowered portion of a ceiling over wall cabinets

Soldier Brick position in which the bricks are stood on end

Span The horizontal dimension between vertical supports — The span of a beam is the distance between the posts that support it.

Specifications Written description of materials or construction

Square The amount of siding or roofing materials required to cover 100 square feet

Stack The main vertical pipe into which plumbing fixtures drain

Stair Carriage The supporting framework under a stair

Stile The vertical members in a sash, door, or other panel construction

Stool Trim piece that forms the finished window sill

Stop Molding that stops a door from swinging through the opening as it is closed — also used to hold the sash in place in a window frame

Stud Vertical framing members in a wall

Subfloor The first layer of rough flooring applied to the floor joists

Sweat Method of soldering used in plumbing

Termite Shield Sheet-metal shield installed at the top of a foundation to prevent termites from entering the wood superstructure

Thermal-break Window Window with a metal frame that has the interior and exterior separated by a material with a higher R-value

Thermostat An electrical switch that is activated by changes in temperature

Top Chord The top horizontal member of a truss

Trap A plumbing fitting that holds enough water to prevent sewer gas from entering the building

Tread The surface of a step in stair construction

Trimmers The double framing members at the sides of an opening

Truss A manufactured assembly used to support a load over a long span

Underlayment Any material installed over the subfloor to provide a smooth surface over which floor covering will be installed

Valley The inside corner formed by intersecting roofs

Valley Rafter The rafter extending from an inside corner in the walls to the ridge at a valley

Vapor Barrier Sheet material used to prevent water vapor from passing through a building surface

Veneer A thin covering — in masonry, a single wythe of finished masonry over a wall — in woodwork, a thin layer of wood

Vent Pipe A pipe, usually through the roof, that allows atmospheric pressure into the drainage system

Vertical Contour Interval The difference in elevation between adjacent contour lines on a topographic map or site plan

Volt The unit of measurement for electrical force

Water Closet A plumbing fixture commonly called *toilet*

Watt The unit of measurement of electrical power — One watt is the amount of power from one ampere of current with one volt of force.

Weep Hole A small hole through a masonry wall to allow water to pass

Wythe A single thickness of masonry construction

Math Reviews

MATH REVIEW 1
FRACTIONS AND MIXED NUMBERS
— MEANINGS AND DEFINITIONS

- A *fraction* is a value which shows the number of equal parts taken of a whole quantity. A fraction consists of a numerator and a denominator.

$$\frac{7 \longleftarrow \text{Numerator}}{16 \longleftarrow \text{Denominator}}$$

- *Equivalent fractions* are fractions which have the same value. The value of a fraction is **not** changed by multiplying the numerator and denominator by the same number.

    Example   Express $\frac{5}{8}$ as thirty-seconds.

    Determine what number the denominator is multiplied by to get the desired denominator. $(32 \div 8 = 4)$

    $$\frac{5}{8} = \frac{?}{32}$$

    Multiply the numerator and denominator by 4.

    $$\frac{5}{8} \times \frac{4}{4} = \frac{20}{32}$$

- The *lowest common denominator* of two or more fractions is the smallest denominator which is evenly divisible by each of the denominators of the fractions.

    Example 1   The lowest common denominator of $\frac{3}{4}, \frac{5}{8},$ and $\frac{13}{32}$ is 32, because 32 is the smallest number evenly divisible by 4, 8, and 32.

    $$32 \div 4 = 8$$
    $$32 \div 8 = 4$$
    $$32 \div 32 = 1$$

    Example 2   The lowest common denominator of $\frac{2}{3}, \frac{1}{5},$ and $\frac{7}{10}$ is 30, because 30 is the smallest number evenly divisible by 3, 5, and 10.

    $$30 \div 3 = 10$$
    $$30 \div 5 = 6$$
    $$30 \div 10 = 3$$

- *Factors* are numbers used in multiplying. For example, 3 and 5 are factors of 15.

    $$3 \times 5 = 15$$

- A fraction is in its *lowest terms* when the numerator and the denominator **do not** contain a common factor.

    Example   Express $\frac{12}{16}$ in lowest terms.

    Determine the largest common factor in the numerator and denominator. The numerator and the denominator can be evenly divided by 4.

    $$\frac{12 \div 4}{16 \div 4} = \frac{3}{4}$$

- A *mixed number* is a whole number plus a fraction.

    $$6 \frac{15}{16}$$

    Whole Number ⟶ ⟵ Fraction

    $$6 + \frac{15}{16} = 6 \frac{15}{16}$$

- *Expressing fractions as mixed numbers.* In certain fractions, the numerator is larger than the denominator. To express the fraction as a mixed number, divide the numerator by the denominator. Express the fractional part in lowest terms.

  Example Express $\frac{38}{16}$ as a mixed number.

  Divide the numerator 38 by the denominator 16.

  $$\frac{38}{16} = 2\,\frac{6}{16}$$

  Express the fractional part $\frac{6}{16}$ in lowest terms.

  $$\frac{6 \div 2}{16 \div 2} = \frac{3}{8}$$

  Combine the whole number and fraction.

  $$\frac{38}{16} = 2\,\frac{3}{8}$$

- *Expressing mixed numbers as fractions.* To express a mixed number as a fraction, multiply the whole number by the denominator of the fractional part. Add the numerator of the fractional part. The sum is the numerator of the fraction. The denominator is the same as the denominator of the original fractional part.

  Example Express $7\,\frac{3}{4}$ as a fraction.

  Multiply the whole number 7 by the denominator 4 of the fractional part ($7 \times 4 = 28$). Add the numerator 3 of the fractional part to 28. The sum 31 is the numerator of the fraction. The denominator 4 is the same as the denominator of the original fractional part.

  $$\frac{7 \times 4 + 3}{4} = \frac{31}{4}$$

  or

  $$\frac{7}{1} \times \frac{4}{4} = \frac{28}{4}$$

  $$\frac{28}{4} + \frac{3}{4} = \frac{31}{4}$$

MATH REVIEW 2
ADDING FRACTIONS

- Fractions must have a common denominator in order to be added.

- To add fractions, express the fractions as equivalent fractions having the lowest common denominator. Add the numerators and write their sum over the lowest common denominator. Express the fraction in lowest terms.

  Example Add: $\frac{3}{8} + \frac{1}{4} + \frac{3}{16} + \frac{1}{32}$

  Express the fractions as equivalent fractions with 32 as the denominator.

  Add the numerators.

  $$\frac{3}{8} = \frac{3}{8} \times \frac{4}{4} = \frac{12}{32}$$
  $$\frac{1}{4} = \frac{1}{4} \times \frac{8}{8} = \frac{8}{32}$$
  $$\frac{3}{16} = \frac{3}{16} \times \frac{2}{2} = \frac{6}{32}$$
  $$+\frac{1}{32} = \qquad \frac{1}{32}$$
  $$\overline{\qquad\qquad} \quad \frac{27}{32}$$

- After fractions are added, if the numerator is greater than the denominator, the fraction should be expressed as a mixed number.

  Example Add: $\frac{1}{2} + \frac{3}{4} + \frac{15}{16} + \frac{11}{16}$

  Express the fractions as equivalent fractions with 16 as the denominator.

  Add the numerators.

  $$\frac{1}{2} = \frac{1}{2} \times \frac{8}{8} = \frac{8}{16}$$
  $$\frac{3}{4} = \frac{3}{4} \times \frac{4}{4} = \frac{12}{16}$$
  $$\frac{15}{16} = \qquad \frac{15}{16}$$
  $$+\frac{11}{16} = \qquad \frac{11}{16}$$
  $$\overline{\qquad\qquad} \quad \frac{46}{16}$$

  Express $\frac{46}{16}$ as a mixed number in lowest terms.

  $$\frac{46}{16} = 2\,\frac{14}{16} = 2\,\frac{7}{8}$$

MATH REVIEW 3
ADDING COMBINATIONS OF FRACTIONS,
MIXED NUMBERS, AND WHOLE NUMBERS

- To add mixed numbers or combinations of fractions, mixed numbers, and whole numbers, express the fractional parts of the numbers as equivalent fractions having the lowest common denominator. Add the whole numbers. Add the fractions. Combine the whole number and the fraction and express in lowest terms.

  Example 1   Add: $3 \frac{7}{8} + 5 \frac{1}{2} + 9 \frac{3}{16}$

  Express the fractional parts as equivalent fractions with 16 as the common denominator. Add the whole numbers. Add the fractions. Combine the whole number and the fraction. Express the answer in lowest terms.

$$3 \frac{7}{8} = 3 \frac{14}{16}$$
$$5 \frac{1}{2} = 5 \frac{8}{16}$$
$$+9 \frac{3}{16} = 9 \frac{3}{16}$$
$$17 \frac{25}{16} = 17 + 1 \frac{9}{16} = 18 \frac{9}{16}$$

  Example 2   Add: $6 \frac{3}{4} + \frac{9}{16} + 7 \frac{21}{32} + 15$

  Express the fractional parts as equivalent fractions with 32 as the common denominator. Add the whole numbers. Add the fractions. Combine the whole number and the fraction. Express the answer in lowest terms.

$$6 \frac{3}{4} = 6 \frac{24}{32}$$
$$\frac{9}{16} = \frac{18}{32}$$
$$7 \frac{21}{32} = 7 \frac{21}{32}$$
$$+15 = 15$$
$$28 \frac{63}{32} = 28 + 1 \frac{31}{32} = 29 \frac{31}{32}$$

MATH REVIEW 4:
SUBTRACTING FRACTIONS FROM FRACTIONS

- Fractions must have a common denominator in order to be subtracted.

- To subtract a fraction from a fraction, express the fractions as equivalent fractions having the lowest common denominator. Subtract the numerators. Write their difference over the common denominator.

  Example   Subtract $\frac{3}{4}$ from $\frac{15}{16}$

  Express the fractions as equivalent fractions with 16 as the common denominator. Subtract the numerator 12 from the numerator 15. Write the difference 3 over the common denominator 16.

$$\frac{15}{16} = \frac{15}{16}$$
$$-\frac{3}{4} = -\frac{12}{16}$$
$$\frac{3}{16}$$

MATH REVIEW 5:
SUBTRACTING FRACTIONS AND
MIXED NUMBERS FROM WHOLE NUMBERS

- To subtract a fraction or a mixed number from a whole number, express the whole number as an equivalent mixed number. The fraction of the mixed number has the same denominator as the denominator of the fraction which is subtracted. Subtract the numerators of the fractions and write their difference over the common denominator. Subtract the whole numbers. Combine the whole number and fraction. Express the answer in lowest terms.

Example 1 Subtract $\frac{3}{8}$ from 7

Express the whole number as an equivalent mixed number with the same denominator as the denominator of the fraction which is subtracted ($7 = 6\frac{8}{8}$).

$$
\begin{aligned}
7 &= 6\frac{8}{8} \\
-\frac{3}{8} &= -\frac{3}{8} \\
\hline
&\quad\; 6\frac{5}{8}
\end{aligned}
$$

Subtract $\frac{3}{8}$ from $\frac{8}{8}$

Combine whole number and fraction.

Example 2 Subtract $5\frac{15}{32}$ from 12

Express the whole number as an equivalent mixed number with the same denominator as the denominator of fraction which is subtracted ($12 = 11\frac{32}{32}$).

$$
\begin{aligned}
12 &= 11\frac{32}{32} \\
-\;5\frac{15}{32} &= -5\frac{15}{32} \\
\hline
&\quad\; 6\frac{17}{32}
\end{aligned}
$$

Subtract fractions.

Subtract whole numbers.

Combine whole number and fraction.

MATH REVIEW 6: SUBTRACTING FRACTIONS AND MIXED NUMBERS FROM MIXED NUMBERS

- To subtract a fraction or a mixed number from a mixed number, the fractional part of each number must have the same denominator. Express fractions as equivalent fractions having a common denominator. When the fraction subtracted is larger than the fraction from which it is subtracted, one unit of the whole number is expressed as a fraction with the common denominator. Combine the whole number and fractions. Subtract fractions and subtract whole numbers.

Example 1 Subtract $\frac{7}{8}$ from $4\frac{3}{16}$

Express the fractions as equivalent fractions with the common denominator 16. Since 14 is larger than 3, express one unit of $4\frac{3}{16}$ as a fraction and combine whole number and fractions ($4\frac{3}{16} = 3 + \frac{16}{16} + \frac{3}{16} = 3\frac{19}{16}$).

$$
\begin{aligned}
4\frac{3}{16} &= 4\frac{3}{16} = 3\frac{19}{16} \\
-\;\frac{7}{8} &= \;\frac{14}{16} = -\frac{14}{16} \\
\hline
&\qquad\qquad\quad 3\frac{5}{16}
\end{aligned}
$$

Subtract.

Example 2 Subtract $13\frac{1}{4}$ from $20\frac{15}{32}$

Express the fractions as equivalent fractions with the common denominator 32.

$$
\begin{aligned}
20\frac{15}{32} &= 20\frac{15}{32} \\
-\;13\frac{1}{4} &= -13\frac{8}{32} \\
\hline
&\quad\; 7\frac{7}{32}
\end{aligned}
$$

Subtract fractions.

Subtract whole numbers.

MATH REVIEW 7:
MULTIPLYING FRACTIONS

- To multiply two or more fractions, multiply the numerators. Multiply the denominators. Write as a fraction with the product of the numerators over the product of the denominators. Express the answer in lowest terms.

  Example 1   Multiply $\frac{3}{4} \times \frac{5}{8}$

  Multiply the numerators.

  Multiply the denominators.

  Write as a fraction.

  $$\frac{3}{4} \times \frac{5}{8} = \frac{15}{32}$$

  Example 2   Multiply $\frac{1}{2} \times \frac{2}{3} \times \frac{4}{5}$

  Multiply the numerators.

  Multiply the denominators.

  Write as a fraction and express answer in lowest terms.

  $$\frac{1}{2} \times \frac{2}{3} \times \frac{4}{5} = \frac{8}{30} = \frac{4}{15}$$

MATH REVIEW 8:
MULTIPLYING ANY COMBINATION OF
FRACTIONS, MIXED NUMBERS, AND WHOLE NUMBERS

- To multiply any combination of fractions, mixed numbers, and whole numbers, write the mixed numbers as fractions. Write whole numbers over the denominator 1. Multiply numerators. Multiply denominators. Express the answer in lowest terms.

  Example 1   Multiply $3 \frac{1}{4} \times \frac{3}{8}$

  Write the mixed number $3 \frac{1}{4}$ as the fraction $\frac{13}{4}$.

  Multiply the numerators.

  Multiply the denominators.

  Express as a mixed number.

  $$3 \frac{1}{4} \times \frac{3}{8} = \frac{13}{4} \times \frac{3}{8} = \frac{39}{32} = 1 \frac{7}{32}$$

  Example 2   Multiply $2 \frac{1}{3} \times 4 \times \frac{4}{5}$

  Write the mixed number $2 \frac{1}{3}$ as the fraction $\frac{7}{3}$.

  Write the whole number 4 over 1.

  Multiply the numerators.

  Multiply the denominators.

  Express as a mixed number.

  $$2 \frac{1}{3} \times 4 \times \frac{4}{5} = \frac{7}{3} \times \frac{4}{1} \times \frac{4}{5} = \frac{112}{15}$$

  $$\frac{112}{15} = 7 \frac{7}{15}$$

MATH REVIEW 9:
DIVIDING FRACTIONS

- Division is the inverse of multiplication. Dividing by 4 is the same as multiplying by $\frac{1}{4}$. Four is the inverse of $\frac{1}{4}$ and $\frac{1}{4}$ is the inverse of 4. The inverse of $\frac{5}{16}$ is $\frac{16}{5}$.

- To divide fractions, invert the divisor, change to the inverse operation and multiply. Express the answer in lowest terms.

    Example Divide: $\frac{7}{8} \div \frac{2}{3}$

    Invert the divisor $\frac{2}{3}$

    $\frac{2}{3}$ inverted is $\frac{3}{2}$.

    Change to the inverse operation and multiply.

    Express as a mixed number.

$$\frac{7}{8} \div \frac{2}{3} = \frac{7}{8} \times \frac{3}{2} = \frac{21}{16} = 1\frac{5}{16}$$

MATH REVIEW 10:
DIVIDING ANY COMBINATION OF
FRACTIONS, MIXED NUMBERS, AND WHOLE NUMBERS

- To divide any combination of fractions, mixed numbers, and whole numbers, write the mixed numbers as fractions. Write whole numbers over the denominator 1. Invert the divisor. Change to the inverse operation and multiply. Express the answer in lowest terms.

    Example 1 Divide: $6 \div \frac{7}{10}$

    Write the whole number 6 over the denominator 1.

    Invert the divisor $\frac{7}{10}$; $\frac{7}{10}$ inverted is $\frac{10}{7}$.

    Change to the inverse operation and multiply.

    Express as a mixed number.

$$\frac{6}{1} \div \frac{7}{10} =$$

$$\frac{6}{1} \times \frac{10}{7} = \frac{60}{7} = 8\frac{4}{7}$$

    Example 2 Divide: $\frac{3}{4} \div 2\frac{1}{5}$

    Write the mixed number divisor $2\frac{1}{5}$ as the fraction $\frac{11}{5}$.

    Invert the divisor $\frac{11}{5}$; $\frac{11}{5}$ inverted is $\frac{5}{11}$.

    Change to the inverse operation and multiply.

$$\frac{3}{4} \div \frac{11}{5} =$$

$$\frac{3}{4} \times \frac{5}{11} = \frac{15}{44}$$

    Example 3 Divide: $4\frac{5}{8} \div 7$

    Write the mixed number $4\frac{5}{8}$ as the fraction $\frac{37}{8}$.

    Write the whole number divisor 7 over the denominator 1.

    Invert the divisor $\frac{7}{1}$; $\frac{7}{1}$ inverted is $\frac{1}{7}$.

    Change to the inverse operation and multiply.

$$\frac{37}{8} \div \frac{7}{1} =$$

$$\frac{37}{8} \times \frac{1}{7} = \frac{37}{56}$$

MATH REVIEW 11:
ROUNDING DECIMAL FRACTIONS

- To round a decimal fraction, locate the digit in the number that gives the desired number of decimal places. Increase that digit by 1 if the digit which directly follows is 5 or more. Do not change the value of the digit if the digit which follows is less than 5. Drop all digits which follow.

Example 1   Round 0.63861 to 3 decimal places.

Locate the digit in the third place (8). The fourth decimal-place digit, 6, is greater than 5 and increases the third decimal-place digit 8, to 9. Drop all digits which follow.

$0.63\underline{8}61 \approx 0.639$

Example 2   Round 3.0746 to 2 decimal places.

Locate the digit in the second decimal place (7). The third decimal-place digit 4 is less than 5 and does not change the value of the second decimal-place digit 7. Drop all digits which follow.

$3.0\underline{7}46 \approx 3.07$

MATH REVIEW 12:
ADDING DECIMAL FRACTIONS

- To add decimal fractions, arrange the numbers so that the decimal points are directly under each other. The decimal point of a whole number is directly to the right of the last digit. Add each column as with whole numbers. Place the decimal point in the sum directly under the other decimal points.

Example   Add: 7.65 + 208.062 + 0.009 + 36 + 5.1037

Arrange the numbers so that the decimal points are directly under each other.

Add zeros so that all numbers have the same number of places to the right of the decimal point.

Add each column of numbers.

```
    7.6500
  208.0620
    0.0090
   36.0000
+   5.1037
  _____
  256.8247
```

Place the decimal point in the sum directly under the other decimal points.

MATH REVIEW 13:
SUBTRACTING DECIMAL FRACTIONS

- To subtract decimal fractions, arrange the numbers so that the decimal points are directly under each other. Subtract each column as with whole numbers. Place the decimal point in the difference directly under the other decimal points.

    Example   Subtract: 87.4 – 42.125

    Arrange the numbers so that the decimal points are directly under each other. Add zeros so that the numbers have the same number of places to the right of the decimal point.

    $$\begin{array}{r} 87.400 \\ -\ 42.125 \\ \hline 45.275 \end{array}$$

    Subtract each column of numbers.

    Place the decimal point in the difference directly under the other decimal points.

MATH REVIEW 14:
MULTIPLYING DECIMAL FRACTIONS

- To multiply decimal fractions, multiply using the same procedure as with whole numbers. Count the number of decimal places in both the multiplier and the multiplicand. Begin counting from the last digit on the right of the product and place the decimal point the same number of places as there are in both the multiplicand and the multiplier.

    Example   Mutiply: 50.216 X 1.73

    Multiply as with whole numbers.

    Count the number of decimal places in the multiplier (2 places) and the multiplicand (3 places).

    Beginning at the right of the product, place the decimal point the same number of places as there are in both the multiplicand and the multiplier (5 places).

    $$\begin{array}{r} 50.216 \quad \text{Multiplicand (3 places)} \\ \times\ 1.73 \quad \text{Multiplier (2 places)} \\ \hline 150648 \\ 351512 \\ 50216 \\ \hline 86.87368 \quad \text{(5 places)} \end{array}$$

- When multiplying certain decimal fractions, the product has a smaller number of digits than the number of decimal places required. For these products, add as many zeros to the left of the product as are necessary to give the required number of decimal places.

    Example   Multiply: 0.27 X 0.18

    Multiply as with whole numbers.

    The product must have 4 decimal places.

    Add one zero to the left of the product.

    $$\begin{array}{r} 0.27 \quad \text{(2 places)} \\ \times\ 0.18 \quad \text{(2 places)} \\ \hline 216 \\ 27 \\ \hline 0.0486 \quad \text{(4 places)} \end{array}$$

MATH REVIEW 15:
DIVIDING DECIMAL FRACTIONS

- To divide decimal fractions, use the same procedure as with whole numbers. Move the decimal point of the divisor as many places to the right as necessary to make the divisor a whole number. Move the decimal point of the dividend the same number of places to the right. Add zeros to the dividend if necessary. Place the decimal point in the answer directly above the decimal point in the dividend Divide as with whole numbers. Zeros may be added to the dividend to give the number of decimal places required in the answer.

    Example 1   Divide: $0.6150 \div 0.75$

    Move the decimal point 2 places to the right in the divisor.

    Move the decimal point 2 places in the dividend.

    Place the decimal point in the answer directly above the decimal point in the dividend.

    Divide as with whole numbers.

```
                          0.82
Divisor ──→ 0 75. )0 61.50 ←Dividend
                          60 0
                           1 50
                           1 50
```

    Example 2   Divide: $10.7 \div 4.375$. Round the answer to 3 decimal places.

    Move the decimal point 3 places to the right in the divisor.

    Move the decimal point 3 places in the dividend, adding 2 zeros.

    Place the decimal point in the answer directly above the decimal point in the dividend.

    Add 4 zeros to the dividend. One more zero is added than the number of decimal places required in the answer.

    Divide as with whole numbers.

```
                 2.4457   ≈   2.446
 4 375. )10 700.0000
         8 750
         1 950 0
         1 750 0
           200 00
           175 00
            25 000
            21 875
             3 1250
             3 0625
               625
```

MATH REVIEW 16:
EXPRESSING COMMON FRACTIONS AS DECIMAL FRACTIONS

- A common fraction is an indicated division. A common fraction is expressed as a decimal fraction by dividing the numerator by the denominator.

    Example   Express $\frac{5}{8}$ as a decimal fraction.

    Write $\frac{5}{8}$ as an indicated division.

    $8 \overline{)5}$

    Place a decimal point after the 5 and add zeros to the right of the decimal point.

    $8 \overline{)5.000}$

    Place the decimal point for the answer directly above the decimal point in the dividend.

    $8 \overline{)5.000}^{\,.}$

    Divide.

    $8 \overline{)5.000}^{\,0.625}$

- A common fraction which will not divide evenly is expressed as a repeating decimal.

    Example   Express $\frac{1}{3}$ as a decimal.

    Write $\frac{1}{3}$ as an indicated division.

    $3\overline{)1}$

    Place a decimal point after the 1 and add zeros to the right of the decimal point.

    $3\overline{)1.0000}$

    Place the decimal point for the answer directly above the decimal point in the dividend.

    $3\overline{)1.0000}^{\,\cdot}$

    Divide.

    $\begin{array}{r} 0.3333 \\ 3\overline{)1.0000} \end{array}$

MATH REVIEW 17:
EXPRESSING DECIMAL FRACTIONS AS COMMON FRACTIONS

- To express a decimal fraction as a common fraction, write the number after the decimal point as the numerator of a common fraction. Write the denominator as 1 followed by as many zeros as there are digits to the right of the decimal point. Express the common fraction in lowest terms.

    Example 1   Express 0.9 as a common fraction.

    Write 9 as the numerator.

    Write the denominator as 1 followed by 1 zero. The denominator is 10.

    $\frac{9}{10}$

    Example 2   Express 0.125 as a common fraction.

    Write 125 as the numerator.

    Write the denominator as 1 followed by 3 zeros. The denominator is 1000.

    $\frac{125}{1000}$

    Express the fraction in lowest terms.

    $\frac{125}{1000} = \frac{1}{8}$

MATH REVIEW 18:
EXPRESSING INCHES AS FEET AND INCHES

- There are 12 inches in 1 foot.

- To express inches as feet and inches, divide the given length in inches by 12 to obtain the number of whole feet. The remainder is the number of inches in addition to the number of whole feet. The answer is the number of whole feet plus the remainder in inches.

    Example 1   Express $176\frac{7}{16}$ inches as feet and inches.

    Divide $176\frac{7}{16}$ inches by 12.

    There are 14 feet plus a remainder of $8\frac{7}{16}$ inches.

    $$\begin{array}{r} 14 \quad \text{(feet)} \\ 12\overline{)176\frac{7}{16}} \\ \underline{12} \\ 56 \\ \underline{48} \\ 8\frac{7}{16} \leftarrow \text{Remainder (inches)} \end{array}$$

    $14' - 8\frac{7}{16}''$

Example 2   Express 54.2 inches as feet and inches.

Divide 54.2 inches by 12.

There are 4 feet plus a remainder of 6.2 inches.

$$
\begin{array}{r}
4 \quad \text{(feet)} \\
12\overline{)54.2} \\
\underline{48} \\
6.2 \leftarrow \text{Remainder} \\
\text{(inches)}
\end{array}
$$

4 feet 6.2 inches

MATH REVIEW 19:
EXPRESSING FEET AND INCHES AS INCHES

- There are 12 inches in one foot.

- To express feet and inches as inches, multiply the number of feet in the given length by 12. To this product, add the number of inches in the given length.

    Example   Express 7 feet 9 $\frac{3}{4}$ inches as inches.

    Multiply 7 feet by 12. There are 84 inches in 7 feet.

    Add 9 $\frac{3}{4}$ inches to 84 inches.

$7 \times 12 = 84$

7 feet = 84 inches

84 inches + 9 $\frac{3}{4}$ inches =

93 $\frac{3}{4}$ inches

MATH REVIEW 20:
EXPRESSING INCHES AS DECIMAL FRACTIONS OF A FOOT

- An inch is $\frac{1}{12}$ of a foot. To express whole inches as a decimal part of a foot, divide the number of inches by 12.

    Example   Express 7 inches as a decimal fraction of a foot.

    Divide 7 by 12.

$7 \div 12 = 0.58$

0.58 feet

- To express a common fraction of an inch as a decimal fraction of a foot, express the common fraction as a decimal, then divide the decimal by 12.

    Example 1   Express $\frac{3}{4}$ inch as a decimal fraction of a foot.

    Express $\frac{3}{4}$ as a decimal.

    Divide the decimal by 12 .

$3 \div 4 = 0.75$

$0.75 \div 12 = 0.06$

0.06 feet

    Example 2   Express 4 $\frac{3}{4}$ inches as a decimal fraction of a foot.

    Express 4 $\frac{3}{4}$ as a decimal.

    Divide the decimal inches by 12.

$4 + \frac{3}{4} = 4 + 0.75 = 4.75$

$4.75 \div 12 = 0.39$

0.39 feet

MATH REVIEW 21:
EXPRESSING DECIMAL FRACTIONS OF A FOOT AS INCHES

- To express a decimal part of a foot as decimal inches multiply by 12.

    Example   Express 0.62 feet as inches.

    Multiply 0.62 by 12.

    $0.62 \times 12 = 7.44$
    7.44 inches

- To express the decimal fraction of an inch as a common fraction, see Math Review 17.

MATH REVIEW 22:
AREA MEASURE

- A surface is measured by determining the number of surface units contained in it. A surface is two dimensional. It has length and width, but no thickness. Both length and width must be expressed in the same unit of measure. Area is expressed in square units. For example, 5 feet $\times$ 8 feet equals 40 square feet.

- *Equivalent Units of Area Measure:*

    1 square foot (sq ft) =
       12 inches $\times$ 12 inches = 144 square inches (sq in)

    1 square yard (sq yd) =
       3 feet $\times$ 3 feet = 9 square feet (sq ft)

- To express a given unit of area as a larger unit of area, divide the given area by the number of square units contained in one of the larger units.

    Example 1   Express 648 square inches as square feet.

    Since 144 sq in = 1 sq ft, divide 648 by 144.

    $648 \div 144 = 4.5$
    648 square inches = 4.5 square feet

    Example 2   Express 28.8 square feet as square yards.

    Since 9 sq ft = 1 sq yd, divide 28.8 by 9.

    $28.8 \div 9 = 3.2$
    28.8 square feet = 3.2 square yards

- To express a given unit of area as a smaller unit of area, multiply the given area by the number of square units contained in one of the larger units.

    Example 1   Express 7.5 square feet as square inches.

    Since 144 sq in = 1 sq ft, multiply 7.5 by 144.

    $7.5 \times 144 = 1080$
    7.5 square feet = 1080 square inches

    Example 2   Express 23 square yards as square feet.

    Since 9 sq ft = 1 sq yd, multiply 23 by 9.

    $23 \times 9 = 207$
    23 square yards = 207 square feet

• *Computing Areas of Common Geometric Figures:*

1. **Rectangle**   A rectangle is a four-sided plane figure with 4 right (90°) angles.

    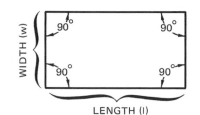

    The area of a rectangle is equal to the product of its length and its width.

    Area = length X width (A = l X w)

    Example   Find the area of a rectangle 24 feet long and 13 feet wide.

    A = l X w

    A = 24 ft X 13 ft

    A = 312 square feet

2. **Triangle**   A triangle is a plane figure with 3 sides and 3 angles.

    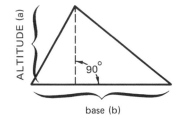

    The area of a triangle is equal to one-half the product of its base and altitude.

    $A = \frac{1}{2}$ base X altitude (A = $\frac{1}{2}$ b X a)

    Example   Find the area of a triangle with a base of 16 feet and an altitude of 12 feet.

    $A = \frac{1}{2}$ b X a

    $A = \frac{1}{2}$ X 16 ft X 12 ft

    A = 96 square feet

3. **Circle**   The area of a circle is equal to π times the square of its radius.

    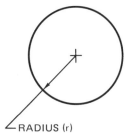

    Area = π X radius^2 (A = π X r^2)

    Note: π (pronounced "pi") is approximately equal to 3.14. Radius squared (r^2) means r X r.

    Example   Find the area of a circle with a 15-inch radius.

    A = π X r^2

    A = 3.14 X (15 in)^2

    A = 3.14 X   225 sq in

    A = 706.5 square inches

MATH REVIEW 23:
VOLUME MEASURE

- A solid is measured by determining the number of cubic units contained in it. A solid is three dimensional; it has length, width, and thickness or height. Length, width, and thickness must be expressed in the same unit of measure. Volume is expressed in cubic units. For example, 3 feet X 5 feet X 10 feet = 150 cubic feet.

- *Equivalent Units of Volume Measure:*

  1 cubic foot (cu ft) =
  12 in X 12 in X 12 in = 1728 cubic inches (cu in)

  1 cubic yard (cu yd) =
  3 ft X 3 ft X 3 ft = 27 cubic feet (cu ft)

- To express a given unit of volume as a larger unit of volume, divide the given volume by the number of cubic units contained in one of the larger units.

  Example 1   Express 6048 cubic inches as cubic feet.

  Since 1728 cu in = 1 cu ft, divide 6048 by 1728.

  $6048 \div 1728 = 3.5$
  6048 cubic inches = 3.5 cubic feet

  Example 2   Express 167.4 cubic feet as cubic yards.

  Since 27 cu ft = 1 cu yd, divide 167.4 by 27.

  $167.4 \div 27 = 6.2$
  167.4 cubic feet = 6.2 cubic yards

- To express a given unit of volume as a smaller unit of volume, multiply the given volume by the number of cubic units contained in one of the larger units.

  Example 1   Express 1.6 cubic feet as cubic inches.

  Since 1728 cu in = 1 cu ft, multiply 1.6 by 1728.

  $1.6 \times 1728 = 2764.8$
  1.6 cubic feet = 2764.8 cubic inches

  Example 2   Express 8.1 cubic yards as cubic feet.

  Since 27 cu ft = 1 cu yd, multiply 8.1 by 27.

  $8.1 \times 27 = 218.7$
  8.1 cubic yards = 218.7 cubic feet

- *Computing Volumes of Common Solids*

- A prism is a solid which has two identical faces called bases and parallel lateral edges. In a right prism, the lateral edges are perpendicular (at 90°) to the bases. The altitude or height (h) of a prism is the perpendicular distance between its two bases. Prisms are named according to the shapes of their bases.

- The volume of any prism is equal to the product of the area of its base and altitude or height.

  Volume = area of base X altitude ($V = A_B \times h$)

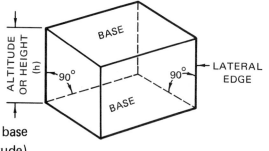

- *Right Rectangular Prism.*

    A right rectangular prism has rectangular bases.

    Volume = area of base × altitude

    $$V = A_B \times h$$

    Example Find the volume of a rectangular prism with a base length of 20 feet, a base width of 14 feet and a height (altitude) of 8 feet.

    $$V = A_B \times h$$

    Compute the area of the base (A_B):

    Area of base = length × width

    $$A_B = 20\ ft \times 14\ ft$$

    $$A_B = 280\ sq\ ft$$

    Compute the volume of the prism:

    $$V = A_B \times h$$

    $$V = 280\ sq\ ft \times 8\ ft$$

    $$V = 2240\ cu\ ft$$

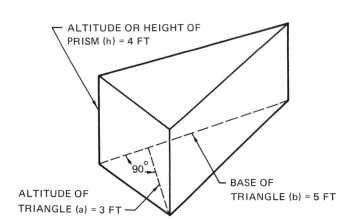

- *Right Triangular Prism.*

    A right triangular prism has triangular bases.

    Volume = area of base × altitude

    $$V = A_B \times h$$

    Example Find the volume of a triangular prism in which the base of the triangle is 5 feet, the altitude of the triangle is 3 feet, and the altitude (height) of the prism is 4 feet. Refer to the accompanying figure.

    Volume = area of base × altitude

    $$V = A_B \times h$$

    Compute the area of the base:

    Area of base = $\frac{1}{2}$ base of triangle × altitude of triangle

    $$A_B = \frac{1}{2} b \times a$$

    $$A_B = \frac{1}{2} \times 5\ ft \times 3\ ft$$

    $$A_B = 7.5\ sq\ ft$$

    Compute the volume of the prism:

    $$V = A_B \times h$$

    $$V = 7.5\ sq\ ft \times 4\ ft$$

    $$V = 30\ cubic\ feet$$

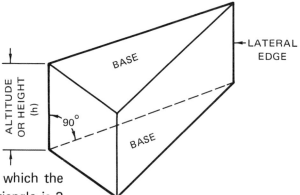

• *Right Circular Cylinder*

A right circular cylinder has circular bases.

Volume = area of base × altitude

$$V = A_B \times h$$

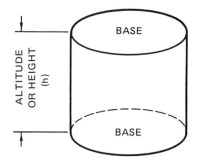

Example Find the volume of a circular cylinder 1 foot in diameter and 10 feet high.

Note: Radius = $\frac{1}{2}$ Diameter; Radius = $\frac{1}{2}$ × 1 ft = 0.5 ft.

$$V = A_B \times h$$

Compute the area of the base:

Area of base = π × radius squared

$$A_B = 3.14 \times (0.5 \text{ ft})^2$$

$$A_B = 3.14 \times 0.5 \text{ ft} \times 0.5 \text{ ft}$$

$$A_B = 3.14 \times 0.25 \text{ sq ft}$$

$$A_B = 0.785 \text{ sq ft}$$

Compute the volume of the cylinder:

$$V = A_B \times h$$

$$V = 0.785 \text{ sq ft} \times 10 \text{ ft}$$

$$V = 7.85 \text{ cubic feet}$$

MATH REVIEW 24:
FINDING AN UNKNOWN SIDE OF A RIGHT TRIANGLE, GIVEN TWO SIDES

• If one of the angles of a triangle is a right (90°) angle, the figure is called a right triangle. The side opposite the right angle is called the hypotenuse. In the figure shown, c is opposite the right angle; c is the hypotenuse.

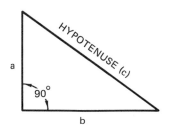

• In a right triangle, the square of the hypotenuse is equal to the sum of the squares of the other two sides:

$$c^2 = a^2 + b^2$$

If any two sides of a right triangle are known, the length of the third side can be determined by one of the following formulas:

$$c = \sqrt{a^2 + b^2}$$

$$a = \sqrt{c^2 - b^2}$$

$$b = \sqrt{c^2 - a^2}$$

Example 1   In the right triangle shown, a = 6 ft, b = 8 ft, find c.

$$c = \sqrt{a^2 + b^2}$$

$$c = \sqrt{6^2 + 8^2}$$

$$c = \sqrt{36 + 64}$$

$$c = \sqrt{100}$$

$$c = 10 \text{ feet}$$

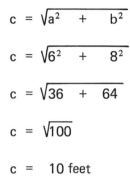
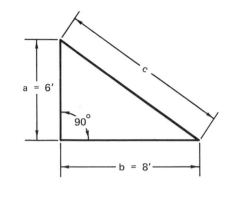

Example 2   In the right triangle shown, c = 30 ft, b = 20 ft, find a.

$$a = \sqrt{c^2 - b^2}$$

$$a = \sqrt{30^2 - 20^2}$$

$$a = \sqrt{900 - 400}$$

$$a = \sqrt{500}$$

$$a = 22.36 \text{ feet (to 2 decimal places)}$$

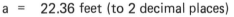
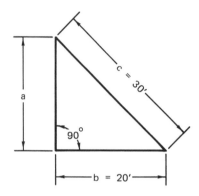

Example 3   In the right triangle shown, c = 18 ft, a = 6 ft, find b.

$$b = \sqrt{c^2 - a^2}$$

$$b = \sqrt{18^2 - 6^2}$$

$$b = \sqrt{324 - 36}$$

$$b = \sqrt{288}$$

$$b = 16.97 \text{ feet (to 2 decimal places)}$$

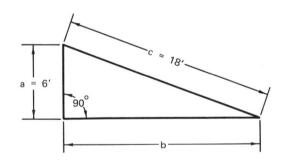

Material Symbols in Sections

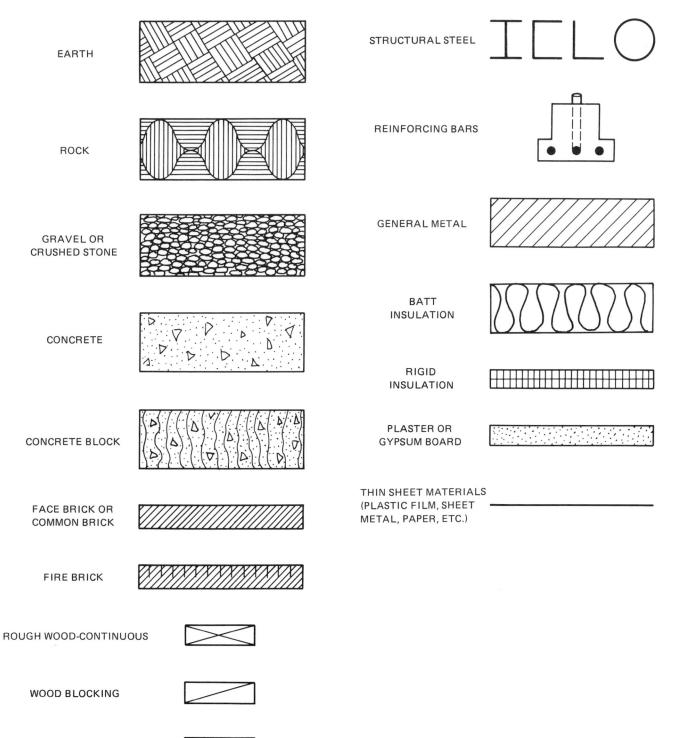

EARTH

ROCK

GRAVEL OR CRUSHED STONE

CONCRETE

CONCRETE BLOCK

FACE BRICK OR COMMON BRICK

FIRE BRICK

ROUGH WOOD-CONTINUOUS

WOOD BLOCKING

FINISH WOOD

STRUCTURAL STEEL

REINFORCING BARS

GENERAL METAL

BATT INSULATION

RIGID INSULATION

PLASTER OR GYPSUM BOARD

THIN SHEET MATERIALS (PLASTIC FILM, SHEET METAL, PAPER, ETC.)

Plumbing Symbols

PIPING

DRAIN OR WASTE ABOVE GROUND	
DRAIN OR WASTE BELOW GROUND	
VENT	
COLD WATER	
HOT WATER	
HOT WATER HEAT SUPPLY	HW
HOT WATER HEAT RETURN	HWR
GAS	G
PIPE TURNING DOWN OR AWAY	
PIPE TURNING UP OR TOWARD	
BREAK—PIPE CONTINUES	

FITTINGS	SOLDERED	SCREWED
T		
WYE		
ELBOW — 90°		
ELBOW — 45°		
CAP		
UNION CLEANOUT		
STOP VALVE		

Electrical Symbols

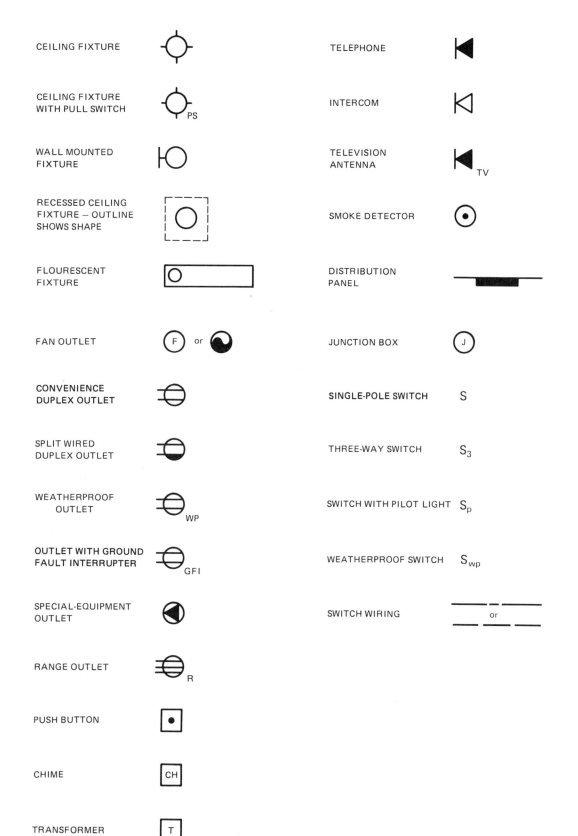

CEILING FIXTURE	TELEPHONE
CEILING FIXTURE WITH PULL SWITCH	INTERCOM
WALL MOUNTED FIXTURE	TELEVISION ANTENNA
RECESSED CEILING FIXTURE – OUTLINE SHOWS SHAPE	SMOKE DETECTOR
FLOURESCENT FIXTURE	DISTRIBUTION PANEL
FAN OUTLET or	JUNCTION BOX
CONVENIENCE DUPLEX OUTLET	SINGLE-POLE SWITCH S
SPLIT WIRED DUPLEX OUTLET	THREE-WAY SWITCH S_3
WEATHERPROOF OUTLET	SWITCH WITH PILOT LIGHT S_p
OUTLET WITH GROUND FAULT INTERRUPTER	WEATHERPROOF SWITCH S_{wp}
SPECIAL-EQUIPMENT OUTLET	SWITCH WIRING or
RANGE OUTLET	
PUSH BUTTON	
CHIME	
TRANSFORMER	

ABBREVIATIONS

A.B. — anchor bolt
A.C. — air conditioning
AL. or ALUM. — aluminum
BA — bathroom
BLDG. — building
BLK. — block
BLKG. — blocking
BM. — beam
BOTT. — bottom
B.PL. — base plate
BR — bedroom
BRM. — broom closet
BSMT. — basement
CAB. — cabinet
₵ — centerline
CLNG. or CLG. — ceiling
C.M.U. — concrete masonry unit (concrete block)
CNTR. — center or counter
COL. — column
COMP. — composition
CONC. — concrete
CONST. — construction
CONT. — continuous
CORRUG. — corrugated
CRNRS. — corners
CU — copper
d — penny (nail size)
DBL. — double
DET. — detail
DIA or ∅ — diameter
DIM. — dimension
DN. — down
DP. — deep or depth
DR. — door
D.W. — dishwasher
ELEC. — electric
ELEV. — elevation
EQ. — equal
EXP. — exposed or expansion

EXT. — exterior
F.G. — fuel gas
FIN. — finish
FL. or FLR. — floor
FOUND. or FDN. — foundation
F.P. — fireplace
FT. — foot or feet
FTG. — footing
GAR. — garage
G.F.I. — ground fault interrupter
G.I. — galvanized iron
GL. — glass
GRD. — grade
GYP.BD. — gypsum board
H.C. — hollow core door
H.C.W. — hollow core wood
HDR. — header
H.M. — hollow metal
HORIZ. — horizontal
HT. or HGT. — height
H.W. — hot water
H.W.M. — high water mark
IN. — inch or inches
INSUL. — insulation
INT. — interior
JSTS. — joists
JT. — joint
LAV. — lavatory
L.H. — left hand
LIN. — linen closet
LT. — light
MANUF. — manufacturer
MAS. — masonry
MATL. — material
MAX. — maximum
MIN. — minimum
MTL. — metal
NAT. — natural
N/F — now or formerly

N.I.C. — not in contract
o/ — overhead or over
O.C. — on centers
O.H. DOOR — overhead door
PERF. — perforated
℞ — plate
PLYWD. — plywood
P.T. — pressure-treated lumber
R — risers
REF. — refrigerator
REINF. — reinforcement
REQ. — requirement
R.H. — right hand
RM — room
R.O.B. — run of bank (gravel)
R.O.W. — right of way
SCRND. — screened
SHT. — sheet
SHTG. — sheathing
SHWR. — shower
SIM. — similar
SL. — sliding
S&P — shelf and pole
SQ. or ⊘ — square
STD. — standard
STL. — steel
STY. — story
T&G — tongue and groove
THK. — thick
T'HOLD. — threshold
TYP. — typical
V.B. — vapor barrier
w/ — with
WARD. — wardrobe
W.C. — water closet
WD. — wood
WDW. — window
W.H. — water heater
W.I. — wrought iron

Index

Trees:
    as sunshade, 63
    to be saved, symbol for, 64
    wells for breathing of, 186, 187
    wooded areas, 64

Unions. *See* Plumbing
Utilities, piping for, 76, 77

Valley rafters, 148–150
    discussion, 148, 150
    framing of, 149
    and spans and pitches of roofs, 148, 149
Vapor barriers, 93
Visualization, of walls and
                partitions, 124–126
Volume, measurement of, 297–299

Walls, covering for, 204–205
Watertight molding, 205
Weepholes, 188
Welded wire fabric, in concert, 94
Wind, bracing for, 101
Windows:
    detail drawings of, 162, 163
    installation, details of, 164
    measurements for, 32
    metal, 161, 163, 164
    sample catalog page about, 130
    schedules of for project, 129
    symbols for, 29, 30, 31
    thermal-break, 161
    wood, components, 160–161, 163, 164
Wracking, 100, 101

Zoning laws, 4